# Cities and Nature



**Series Editors**

Peter Newman, Sustainability Policy Institute, Curtin University, Perth, WA, Australia

Cheryl Desha, School of Engineering and Built Environment, Griffith University, Nathan, QLD, Australia

Alessandro Sanches-Pereira, Instituto 17, São Paulo, São Paulo, Brazil

**Cities and Nature** fosters high-quality multi-disciplinary research addressing the interface between cities and the natural environment. It provides a valuable source of relevant knowledge for researchers, planners and policy-makers. The series welcomes empirically based, cutting-edge and theoretical research in urban geography, urban planning, environmental planning, urban ecology, regional science and economics. It publishes peer-reviewed edited and authored volumes on topics dealing with the urban and the environment nexus, including: spatial dynamics of urban built areas, urban and peri-urban agriculture, urban greening and green infrastructure, environmental planning, urban forests, urban ecology, regional dynamics and landscape fragmentation.

Indexed in Scopus!

Runming Yao
Editor

# Resilient Urban Environments

## Planning for Livable Cities

Springer

*Editor*
Runming Yao
Chongqing University
Chongqing, China

University of Reading
Reading, UK

Professor Runming Yao contributes the chapter Outdoor Thermal Comfort and Heat Exposure Risks with Xizhen Huang, Tiantian Xu, and Yuening Zhu, and contributes the chapter Modelling Urban Microclimates with Wenbo Wang.

ISSN 2520-8306 ISSN 2520-8314 (electronic)
Cities and Nature
ISBN 978-3-031-55484-1 ISBN 978-3-031-55482-7 (eBook)
https://doi.org/10.1007/978-3-031-55482-7



This Springer imprint is published by the registered company Springer Nature Switzerland AG
The registered company address is: Gewerbestrasse 11, 6330 Cham, Switzerland

Paper in this product is recyclable.

# Foreword by Prof. C. Alan Short

It is a great pleasure to be invited to write the foreword for this epic scaled, thoroughly interdisciplinary assault on the problem of adapting and remaking cities across all climate types to become benign and health-giving. This is an immense challenge which requires the intense and meaningful collaboration of very many disciplines and professions, some surprising and apparently peripheral but as the following twenty chapters demonstrate, all are indispensable. Here are contributions from urban designers, architects, social scientists, social anthropologists, academics in environmental and more broadly in public health, environmental engineers, air quality scientists, energy physicists, computational modelers, acousticians, hydrologists, earth scientists, specialists in respiratory disease and air quality, all at the top of their professions across some of the world's most prestigious institutions. One must commend Professor Yao for her indefatigable wrangling of such a broad list of authors in the pursuit of excellent contributions. The book is burgeoning with content.

The fact is that 'cities', the dense 'downtowns' so beloved of the *flaneurs* and 'architectural philosophers' reacting against the post-war planning mantra of land use zoning and atomized city centers robbed of spontaneous life, can turn nasty. Increasing density through policy, opportunism or the self-replicating phenomenon of informal settlement through a regionally changing climate amplified by the self-generation of localized heat islands will harm the very old, the very young, vulnerable and disadvantaged. As I write this the President of France is promoting UNESCO recognition of the zinc-covered mansard roofs of Paris beneath which so many died in the 2003 heatwave, the decidedly non-interdisciplinary collision of historical geography with the physics of climate change, exactly what Prof. Yao sets out to avoid with this carefully edited volume.

I must echo Prof. Yao's thanks to China's Ministry of Science and Technology which has supported the National Centre for Sustainable Development in Buildings and Built Environment in its pursuit of knowledge and understanding of these extraordinarily complex problems within a truly global interdisciplinary context.

May 2023

Prof. C. Alan Short
MA, Dip.Arch., Ph.D., Litt.D. (Cantab)
President of Clare Hall Cambridge
The Professor of Architecture Emeritus
University of Cambridge
Cambridge, UK

# Foreword by Prof. Baizhan Li

I am much honored to be invited to write a foreword for this book *Sustainable and Resilient Urban Environments—Planning for Livable Cities.* I, as Prof. Runming Yao's colleague at the National Centre for International Research on Low-carbon and Green Buildings, am very proud of such delicate and meaningful work. I witnessed Prof Yao's tireless work on the proposal, structuring and forming the author team and her enormous communications and discussions with her team authors over the past couple of years. The book was completed and to be on the shelf soon after Prof. Yao and her team's hard work. I would like to take this opportunity to congratulate her and all chapter authors.

This is truly a production of international and multidisciplinary teamwork. The authors include urban planners, environmental scientists, social scientists, health scientists, building designers and environmental engineers from academia and practice in the UK, Europe, India, Japan and China.

The book sets out a clear aim and focused themes that help readers to be aware of the impact of climate change on health and well-being and understand modeling technologies. The book focuses on the heat, air, noise, water and flood-related urban environments. The evidence shows that advanced modeling technologies play a pivotal role in providing evidence-based support to environmental interventions. Urban green infrastructure contributes to supporting biodiversity, urban ecosystem, carbon sequestration, Urban Heat Island mitigation and flood protection so as to enhance urban resilience.

Climate change is caused by the excessive use of fossil fuels. The deployment of renewable energy particularly in cities is challenging but it is a very trendy topic. Low-carbon city is one of the important elements of a resilient city. Policies play an essential role in climate change mitigation strategies. Case studies for achieving United Nations Sustainable Development Goals particularly demonstrate the successful stories in transforming resilient cities and regions.

I am confident this is a very timely and useful reference book for readers who are willing to commit to a resilient city development career. The book is particularly suitable for people including researchers, students and practitioners, who engage in the field of urban resilience study.

December 2023

Prof. Baizhan Li
Director of National Centre for International Research on Low-Carbon and Green Buildings
Ministry of Science and Technology
Sustainable Development of Building and Environment (SuDBE)
Chongqing University
Chongqing, China

# Preface

The urban environment is a human-made setting to accommodate human activities. Global urbanization accelerated the adverse climate impact. Cities' urban microclimates accordingly are changing. Extreme weather events such as heatwaves and flooding occur frequently, and air pollution, heat stress and noise significantly affect people's living quality and become a threat to vulnerable groups of people. It is important to take steps to ensure a city is capable of delivering a cleaner and greener city through urban environmental interventions to make cities more sustainable and resilient.

The book *Resilient Urban Environments* aims to provide evidence of the impact of climate change and urbanization on cities' urban environments, thus on human health and well-being. It introduces principles and methods for the improvement of the resilience of a city to extreme weather and long-term climatic changes through case studies. The book focuses on the following three themes:

(1) The urban environment and its impact on human health and well-being.
(2) The analysis of adaptation and mitigation measures through modeling technologies.
(3) Cases studies of the implementation of the concept of sustainability and resilience in cities.

In summary, the book *Resilient Urban Environments* is the collective production of the authors from inter-disciplines of urban planners, social and health scientists and building and urban engineers, from academia and professional practitioners. We hope readers will benefit from gaining knowledge in creating and assessing sustainable and resilient urban environments in the context of improving living environments, thus benefiting human health and well-being. We hope the case studies in mitigation and adaptation of climate change impact will help students, academics and practitioners to be useful in their learning and teaching and practicing in the real world. I would like to thank all chapter authors for their contributions. I would like to thank the financial support from the Chinese Ministry of Science and Technology of the People's Republic of China for the International Collaboration Programme for the project titled 'Low carbon resilient cities and built environment'. I would like to thank the administrative support from the National Centre for International Research—Sustainable Development in Buildings and Built Environment (SuDBE) based at Chongqing University, China.

Chongqing, China
Reading, UK
June 2023

Runming Yao

# Contents

# Editor and Contributors

## About the Editor

**Prof. Runming Yao** [Ph.D., M.Sc., B.Eng., C.Eng. FCIBSE, FRSA, FCIOB, FHEA, MASHRAE] is currently Global Expert Professor, Lead Academic of Design and Management of Sustainable Built Environments at Chongqing University and Professor of the Sustainable Built Environment at the University of Reading, at the School of the Built Environment, University of Reading. She is Fellow of the Chartered Institution of Building Services Engineers (FCIBSE), Member of the American Society of Heating, Refrigerating and Air Conditioning Engineers (ASHRAE), Fellow of the Chartered Institute of Building (FCIOB) and Fellow of Higher Academy (FHEA). She has been invited as Visiting Fellow of Clare Hall at the University of Cambridge. She has broad research interests in the sustainable built environment with a focus on energy efficiency and the environmental quality in buildings and urban environments. She has successfully led and completed numerous research projects funded by the UK, EU and China. The most recent one is the largest Chinese national R&D project on the solutions for heating and cooling of buildings in the Yangtze River region (125 million Yuan). She published numerous refereed journal articles in the areas of thermal comfort and well-being, building performance assessment, urban environment, energy modeling and energy systems and management. She has been frequently invited to present her research at conferences, seminars and lectures internationally. She has been International Reviewer for numerous research funding bodies and served as International Assessment Panelist for major research projects internationally. She is Co-editor-in-Chief of the *Journal of Building Engineering (JoBE).*

## Contributors

**K. V. Abhijith** Global Centre for Clean Air Research (GCARE), School of Sustainability, Civil and Environmental Engineering, Faculty of Engineering and Physical Sciences, University of Surrey, Guildford, UK

**Francesco Aletta** UCL Institute for Environmental Design and Engineering, University College London (UCL), Central House, London, UK

**Derek Clements-Croome** University of Reading, Reading, UK

**Liangjun Da** Institute of Science and Engineering of Ecology in Arid and Semi-arid Areas, Xi'an University of Architecture and Technology, Xi'an, China

**Sisay Debele** Global Centre for Clean Air Research (GCARE), School of Sustainability, Civil and Environmental Engineering, Faculty of Engineering and Physical Sciences, University of Surrey, Guildford, UK

**Timothy J. Dixon** Sustainable Futures in the Built Environment, School of the Built Environment, University of Reading, Reading, UK

**Fangxin Fang** Department of Earth Science and Engineering, Imperial College London, London, UK

**Dawei Han** School of Civil, Aerospace and Design Engineering, University of Bristol, Bristol, UK

**Jian Hang** School of Atmospheric Sciences, Sun Yat-Sen University, Southern Marine Science and Engineering Guangdong Laboratory (Zhuhai), Zhuhai, China

**Peter Head** CBE FREng FRSA 2022, London, UK

**Xizhen Huang** School of Civil Engineering, Chongqing University, Chongqing, China

**Siwei Hu** Key Laboratory of the Three Gorges Reservoir Region's Eco-Environment, Ministry of Education, Chongqing University, Chongqing, China

**Yuandong Hu** College of Landscape Architecture, Northeast Forestry University, Harbin, China

**Cheng Jin** Key Laboratory of the Three Gorges Reservoir Region's Eco-Environment, Ministry of Education, Chongqing University, Chongqing, China

**Soteris A. Kalogirou** Department of Mechanical Engineering and Materials Sciences and Engineering, Cyprus University of Technology, Limassol, Cyprus; Cyprus Academy of Sciences, Letters, and Arts, Nicosia, Cyprus

**Haidong Kan** Key Lab of Public Health Safety of the Ministry of Education, NHC Key Lab of Health Technology Assessment, IRDR ICoE on Risk Interconnectivity and Governance on Weather/Climate Extremes Impact and Public Health, Department of Environmental Health, School of Public Health, Fudan University, Shanghai, China

**Jian Kang** UCL Institute for Environmental Design and Engineering, University College London (UCL), Central House, London, UK

**Soheila Khalili** Global Centre for Clean Air Research (GCARE), School of Sustainability, Civil and Environmental Engineering, Faculty of Engineering and Physical Sciences, University of Surrey, Guildford, UK

**Maria Kolokotroni** Brunel University London, London, UK

**Manish Kumar** Sustainability Cluster, University of Petroleum and Energy Studies, Dehradun, Uttarakhand, India;
Escuela de Ingeniería y Ciencias, Tecnologico de Monterrey, Monterrey Nuevo Leon, Mexico

**Prashant Kumar** Global Centre for Clean Air Research (GCARE), School of Sustainability, Civil and Environmental Engineering, Faculty of Engineering and Physical Sciences, University of Surrey, Guildford, UK

**Rakesh Kumar** Department of Biosystems Engineering, Auburn University, Auburn, AL, United States

**Jian Lei** Key Lab of Public Health Safety of the Ministry of Education, NHC Key Lab of Health Technology Assessment, IRDR ICoE on Risk Interconnectivity and Governance on Weather/Climate Extremes Impact and Public Health, Department of Environmental Health, School of Public Health, Fudan University, Shanghai, China

**Zhiwen Luo** Welsh School of Architecture, Cardiff University, Cardiff, UK

**Shuo-Jun Mei** School of Atmospheric Sciences, Sun Yat-Sen University, Southern Marine Science and Engineering Guangdong Laboratory (Zhuhai), Zhuhai, China

**Jeetendra Sahani** Global Centre for Clean Air Research (GCARE), School of Sustainability, Civil and Environmental Engineering, Faculty of Engineering and Physical Sciences, University of Surrey, Guildford, UK

**Stefán Thor Smith** School of the Built Environment, University of Reading, Reading, UK

**Ziona Strelitz** ZZA Responsive User Environments, London, UK

**Arvind Tiwari** Global Centre for Clean Air Research (GCARE), School of Sustainability, Civil and Environmental Engineering, Faculty of Engineering and Physical Sciences, University of Surrey, Guildford, UK

**Simone Torresin** Department of Civil Environmental and Mechanical Engineering, University of Trento, Trento, Italy;
UCL Institute for Environmental Design and Engineering, University College London (UCL), Central House, London, UK

**Wenbo Wang** Chongqing University, Chongqing, China;
University of Reading, Reading, UK

**Tiantian Xu** School of Civil Engineering, Chongqing University, Chongqing, China

**Wei Yang** Wei Yang & Partners, London, UK

**Yongchuan Yang** Key Laboratory of the Three Gorges Reservoir Region's Eco-Environment, Ministry of Education, Chongqing University, Chongqing, China

**Runming Yao** School of Civil Engineering, Chongqing University, Chongqing, China;
School of the Built Environment, University of Reading, Reading, UK

**Hiroshi Yoshino** Tohoku University, Sendai, Japan

**Jian Zang** School of Civil Engineering, Chongqing University, Chongqing, China

**Lu Zhou** Key Lab of Public Health Safety of the Ministry of Education, NHC Key Lab of Health Technology Assessment, IRDR ICoE on Risk Interconnectivity and Governance on Weather/Climate Extremes Impact and Public Health, Department of Environmental Health, School of Public Health, Fudan University, Shanghai, China

**Yuening Zhu** School of Civil Engineering, Chongqing University, Chongqing, China

# Introduction

**Runming Yao**

**Abstract** This introductory chapter sets the scene for the book, providing an overview of sustainability and resilience in the urban environment. It illustrates the impact of climate change on the living environment, thus affecting people's health and well-being. Interventions to mitigate the impact through instruments of policy and technical solutions are provided. Computer modelling and simulations provide tools for scenario analysis to assist risk forecasting and decision-making of mitigation strategies. Case studies provide examples of sustainable and resilient urban design and policy implementation which are hoped to be useful to the readers.

**Keywords** Sustainability · Resilience · Health and well-being · Heat stress · Modelling · Air quality · Noise · Biodiversity · Green infrastructure · Flood · Water · Policy

## 1 Background

Nowadays, more than half of the population is living in cities, and it is estimated that the number will reach 60% by 2030 and further increase to 68% by 2050 (UN 2015). Citizens' daily activities in a city are relying on living environment conditions and resources like pleasant temperatures, clean air, clean water, flood resilience, acceptable acoustic environment, sufficient energy supplies, affordable living supplies and so on (IPCC 2018). Those elements are fundamental for providing a healthy environment for people to live in and maintaining city sustainability and resilience.

---

R. Yao (✉)
School of Civil Engineering, Chongqing University, Chongqing, China
e-mail: r.yao@cqu.edu.cn; r.yao@reading.ac.uk

School of the Built Environment, University of Reading, Reading, UK

 
R. Yao (ed.), *Resilient Urban Environments*, Cities and Nature,
https://doi.org/10.1007/978-3-031-55482-7_1

### Temperature

Climate change increases the likelihood of heatwave events, posing great threats to human health as well as resulting in increased mortality and human thermal discomfort. Extreme temperature events have been classified as one of the most dangerous global disasters (WMO 2021). Between 1998 and 2017 more than 166,000 people died due to extreme temperatures, and especially in 2003 over 70,000 people were killed in heatwaves in Europe (CRED and UNISDR 2018). In the UK, heatwaves, in June and August 2020, caused more than 2000 deaths (PHE 2020). Heatwaves would make the Urban Heat Island (UHI) intensity substantial, bringing heat-related health issues for city dwellers worldwide (Macintyre et al. 2018; Iping et al. 2019; He et al. 2022).

Heat has direct and indirect impacts on health issues (Fig. 1). Excessive heat can directly cause dehydration and even the development of emergency hospitalizations and require immediate management, such as heat stroke, heat exhaustion and heat syncope (WMO and WHO 2015). Indirectly, it can alter human behaviours, increase the transmission of diseases, disrupt critical infrastructure and so on. Thus establishment of the protection in addressing the above issues is very necessary, as does increasing the resilience of the city as well as protecting citizen health.

### Air

Air pollution is the presence of toxic chemicals or compounds (including those of biological origin) in the air, at levels that are one of the greatest environmental risks to health (Seinfeld and Pandis 1998). Epidemiological studies revealed that air pollution is a risk factor for diseases of public health importance such as cardiovascular diseases for instance stroke and ischaemic heart disease, cancers and respiratory diseases

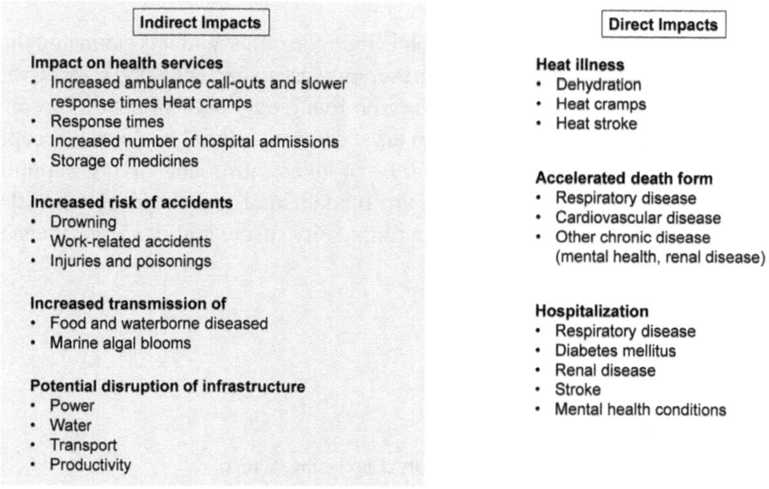


**Fig. 1** Impacts of extreme heat on health (WHO 2023c)

(Cohen et al. 2015). Ambient (outdoor) air pollution is estimated to have caused 4.2 million premature deaths worldwide in 2019 (WHO 2023a).

World Health Organization (WHO) data show that almost all of the global population (99%) breathes highly polluted air that exceeds WHO limits (WHO 2023b). WHO aggregates global particulate matter data and collates them into reports. Figure 1 shows the annual mean concentration and the coverage of the ground measurements of PM2.5 for countries around the world from 2010 to 2016. It shows that countries with low and middle income suffer from the highest exposures.

Referring to sources of air pollutants, the burning of fossil fuels, motor vehicles and power stations are the major source, which is one of the biggest causes of air pollution in developing countries. As a typical large developing country, China's growing energy consumption, reliance on coal and rapidly increasing vehicle population cause severe air pollution. Air pollution concentration levels tend to increase appreciably or even rise sharply if no effective measures are taken when economic development reaches an intermediate stage (WHO 2013). Hence, there is a growing need to create better living spaces and ensure human well-being. Sustainable urban planning and climate control measures are required to maintain and improve public health.

As shown in Fig. 2, the location of the PM2.5 concentration in more than 4000 monitoring stations was listed on the map. Particulate matter (PM), carbon monoxide (CO), ozone ($O_3$), nitrogen dioxide ($NO_2$) and sulphur dioxide ($SO_2$) are the main pollutants of major public health concern (WHO 2023c). Harmful pollutants exposure is proven to be associated with respiratory illness, hospital admission and premature death (Huangfu and Atkinson 2020). The health impacts of ambient PM2.5 have become a great concern worldwide. Epidemiological investigations have shown that PM2.5 exposure contributes to cardiopulmonary morbidity and mortality, the incidence and development of diabetes mellitus and adverse birth outcomes (Feng et al. 2016). Considering the significant health risks posed by pollutants, interventions at regional, national and international levels are needed to reduce levels of air pollution.

## Water

Water in urban is necessary in terms of human daily activities needs and sanitation needs. The water quality and quantity played a vital role in city resilience. Hygiene behaviour by using clean water can significantly reduce respiratory infections (Howard et al. 2020). Water quantity is limited, freshwater accounts for only 2.5% of the global water (Tang et al. 2022), and the reliability of the water is heavily connected with human well-being. United Nations Sustainable Development Goals emphasize the sanitation and quantity needs for water in the city, to satisfy SDG 6, the Guidelines for Drinking Water Quality (GDWQ) provide the recommendations of the World Health Organization for managing the risk from hazards that may compromise the safety of drinking water and assist water and health regulators and policymakers to maintain public health by supporting the development of national regulations and standards.

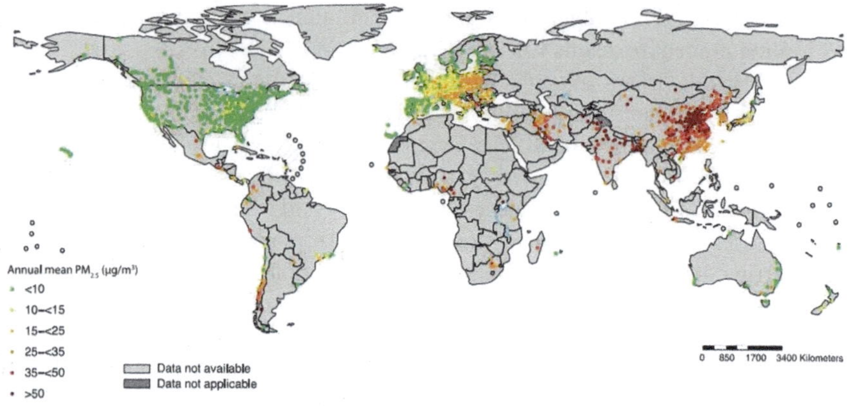


**Fig. 2** Location of the monitoring stations and PM2.5 concentration in more than 4000 human settlements, 2010–2016 (WHO 2018a, b)

### Flood

Over the last 20 years, flood events have occurred in 49 countries in the WHO European Region. These have caused more than 2000 deaths, other health effects, property losses, damage to health facilities, displacement and enormous economic costs estimated at €70 billion (WHO 2017). Flooding is widely regarded as extreme water stress or land loss disaster in urban or rural areas on the mainland. Understanding flooding and increasing the robustness capacity against flood risk are important for human well-being. Flooding effects included accident and injury risk; diarrhoeal/respiratory diseases and skin/eye infections; chemical poisoning; stress; and negative health effect linked to overcrowding (WHO 2017). The second effect can occur after floods, for example, damage to health care infrastructures, living supplies, water shortage, communication crash, destruction of property and vital community facilities, damage to crops, disruption of food supplies and disruption of livelihoods and income. A flood warning measures system is, therefore, very necessary for city resilience, strengthening the monitoring of flood hazard points, establishing and improving monitoring systems, observing models, forecasting models, and early warning systems, group measurement and prevention work are now considered as potential solutions.

### Acoustic

Noise pollution in urban environments caused by traffic is nowadays recognized as one of the top public health threats across all ages (UNEP 2022). Prolonged exposure to high levels of noise impairs human health and well-being. Traffic and other urban noises affect not only human well-being but also disturb and endanger the survival of species crucial to the urban environment (Francis and Barber 2013).

Estimates suggest that in Europe 22 million and 6.5 million people suffer from chronic noise annoyance and sleep disturbance, respectively (EEA 2020). The

**Table 1** Prevalence of hearing loss (of moderate or higher grade) across WHO regions (WHO 2021)

| Area | Quantity/million | Proportion (%) |
|---|---|---|
| Americas | 62.7 | 6.2 |
| African | 39.9 | 3.6 |
| European | 57.3 | 6.2 |
| Eastern Mediterranean | 22.1 | 3.1 |
| Southeast Asia | 109.4 | 5.5 |
| Western Pacific | 136.5 | 7.1 |

**Table 2** Illustrative combined direct, indirect and intangible costs of hearing loss (in billion dollars) (David et al. 2021)

| Area | Health | Education | Productivity | Intangible | All costs |
|---|---|---|---|---|---|
| Americas | 88.6 | 3.8 | 44.1 | 125.3 | 262 |
| African | 7.3 | 3.6 | 3.5 | 12.4 | 27.1 |
| European | 74.5 | 3.2 | 21.1 | 125.6 | 224.5 |
| Eastern Mediterranean | 9.0 | 1.6 | 5.2 | 13.9 | 29.8 |
| Southeast Asia | 32.2 | 7.0 | 29.7 | 38.7 | 107.7 |
| Western Pacific | 101.8 | 7.4 | 78.5 | 140.5 | 328.3 |
| World | 313.6 | 26.8 | 182.4 | 456.5 | 979.6 |

elderly, pregnant woman and shift workers are among those at risk of noise-induced sleep disturbance (Halperin 2014). Noise-induced awakenings can trigger a range of physiological and psychological stress responses because sleep is necessary for hormonal regulation and cardiovascular functioning (Münzel et al. 2014). There is increasing evidence that traffic noise exposure is a risk factor for the development of cardiovascular and metabolic disorders such as elevated blood pressure, arterial hypertension, coronary heart disease and diabetes (Münzel et al. 2018).

Table 1 shows the prevalence of hearing loss across WHO regions, the highest proportion is Western Pacific (accounting for 7.1%, 136.5 million), and the lowest is in Southeast Asia (3.1%). Also, European and the Americas should concern due to the relatively high index.

Table 2 indicated hearing loss has a considerable economic impact on society as a whole. WHO data reveal that the overall global cost of unaddressed hearing loss is greater than $980 billion annually.

## 2 Energy and Environment

With the rapid urbanization and urban sprawl, global demand and consumption of energy by cities are constantly increasing. Cities consume approximately 67% of the world's energy, and their carbon emissions have surpassed 70% of the global total (IEA 2021). It is estimated that by 2040, the building sector will account for 35% of the global electricity demand growth. The rapid increase in urban energy consumption and carbon emissions is of concern, and there is an urgent need to take measures to promote energy conservation and emission reduction in cities.

Fossil fuel combustion is the main contributor to carbon emissions to meet urban energy demand. The combustion of carbon-based substances causes a large amount of carbon dioxide emissions into the air and bring potential negative health effect on the human. Mitigation of carbon emissions needs to implement a zero-energy building policy. Clean and renewable energy is the most important issue for a city's sustainability and resilience.

Human activities and urbanization have a negative impact on biodiversity comparing its original nature elements. Landscape plant species with their notable influence on human well-being are important components of the urban ecosystem. Urban greening is expected to avoid biotic homogenization. This will help the urban designer thinking as well as the decision-making on protecting human well-being and environmental biodiversity. Urban biodiversity study can determine the practical relevance of tools for different types of audiences, including designers, planners and ecology working in different stages of urban development and design processes.

## 3 Sustainability and Resilience

Urban sustainability consists of appreciating and balancing the three pillars of sustainable development (economic, social and environmental), while also being inclusive and equitable. While city resilience indicates the city's capacity to adapt to changing conditions and withstand and rapidly recover from disruption due to emergencies. A sustainable and resilient city is the development path for future cities facing increasingly unknown urbanization challenges. Responding to climate change, policies are recognized as mandatory guiding decisions that can safeguard the fair interests of the majority of people and thus play a vital role in society. United Nations Sustainable Development Goal (UN SDG) is promoted for providing a sustainable index and goals for humans, requiring people in all stages together to build a sustainable world for humans and the next generation. The target of 'Good health and well-being' is one of the United Nations' 17 Sustainable Development Goals.

The holistic system thinking approach is considered in the improvement of urban living environments. Cities require a systems approach for planning, design and management. The living environment as a whole in a city covers all aspects like a community unit together as a pyramid, and it needs intelligent structures and infrastructures, social provision, amenities and basic property rights for its citizens. It is essential to set a vision for all the stakeholders, so they feel comfortable and have easy access to clean, sufficient and daily needs. The sustainable and resilient city should have functions of providing sufficient energy, clean air and water, flood resilience and be accessible to everyone. System thinking is a key improvement in an urban living environment.

## 4 Summary

The liveable cities include factors affecting liveability; ecological and biophilic cities; economic values; health and well-being; and opportunities for people. There is a basic human need for a social dynamic, and cities provide this for people. Cities are like organisms pulsing with life bounded by economic, ecological, health, and well-being requirements. Physical and social health is addressed in the context of resilient cities in Chapter "Outdoor Thermal Comfort and Heat Exposure Risks". This chapter defines the features of sustainable liveable cities and reviews how cities can function in a climate change-respectful way but also be joyful places for people to be.

People in cities use outdoor space for a wide range of activities such as exercise, rest and socializing. Climate change, Urban Heat Islands and extreme heatwave events are threats to pedestrians' health and well-being. Good urban planning and design can provide a high-quality urban living environment by meeting the need for outdoor thermal comfort and thermal resilience. Chapter "Liveable, Joyful and Soulful Cities" presents the human adaptation mechanism for achieving outdoor thermal comfort, the factors that influence it, strategies for thermal resilience, the management of heat risk and thermal comfort assessment methods to use.

Heatwaves and high temperatures are associated with increased mortality risks, and the burden may keep rising in the future. Chapter "Mortality Risks and Burden of Heatwaves and High Temperature" provides evidence of climate change and the mortality risks of heatwaves and high temperatures. This study examined the population's vulnerability to heatwaves and high temperatures, including demographic and climatic modifications, highlighting the necessity for adaptation measures against heat stress to contribute to the reduction of the heat-related mortality burden. The method of estimating the city-specific risks has been presented.

The computer modelling method is one of the effective measures that can help assess urban environments and decision-making of urban intervention strategies for the improvement of people's health and well-being. It plays an important role in the resilience of cities. It has been widely conducted in many aspects including assessment of outdoor thermal comfort, Urban Heat Islands, urban air quality, urban flooding, urban ventilation design, urban acoustic and flooding. It can assist policy-makers, urban designers and city managers to work collaboratively to deliver sustainable urban planning and regeneration from the earliest stage. Chapter "Modelling Urban Microclimates" introduces a robust and fast-running tool that can be used to simulate urban solar radiation, surface temperatures and air temperature. It can be used in a variety of scenarios to assist in the assessment of urban planning/ building design projects related to the outdoor thermal environment. The case study presented briefly demonstrates some of the functions that can be used in the planning and design phase. Chapter "Predictive Modelling in Urban Environments" presents the challenges of modelling complex physical phenomena in terms of the prediction of characteristics and evolution with adequate accuracy and reliability. Data assimilation techniques incorporate information from experiments and observations to reduce uncertainties in numerical prediction that has been presented. The resolutions of modelling techniques on an urban scale have been demonstrated with case studies of simulations of urban air pollution, urban green–blue infrastructure, land surface thermal dynamics and urban flooding forecasting.

Inhabitants of high-rise and high-density cities are at greater risk of severe environmental hazards, such as air pollution. Typically, high-density cities have high pollutant emission rates, attributable to high traffic volume per unit area. Urban ventilation is one of the practical measures through proper urban and building design that can mitigate urban heat and air pollution. Chapter "Urban Design to Reduce Air Pollution Exposure in High-Rise and High-Density Cities" presents six design principles that can improve urban ventilation in high-density cities and therefore reduce air pollution exposure. By incorporating climate-sensitive design strategies, high-density cities can potentially reduce both indoor and outdoor exposure to air pollution, leading to improved sustainability and resilience. Chapter "Urban Ventilation" presents the impact of the urban context in terms of air and thermal pollution on ventilation strategies. Three case studies of the performance evaluation of buildings in urban areas are presented.

The acoustic quality of living and working environments can affect the health, well-being and quality of life of building occupants. Chapter "Balancing Outdoor and Indoor Soundscapes in a Warming Climate" introduces the soundscape methodologies that provide a comprehensive characteristic of acoustic perception by people in a specific context, thus guiding the design of built environments to shape resilient cities, able to address the main global challenges posed by global warming. The consideration of soundscape aspects from the earliest stages of urban planning and building design would favour conditions that allow for the adoption of passive ventilation (and cooling) strategies for occupants' health and well-being.

Rapid urbanization has greatly affected global biodiversity. Human preferences for the plant species of urban greening have greatly facilitated the spread of non-native species, resulting in the homogenization of landscape plant communities across spatial scales. Chapter "Biotic Homogenization and Rewilding of Urban Greening" explores the climatic suitability and sustainability of landscape plants for urban biodiversity. The application of biodiversity-friendliness spontaneous urban plants (SUPs) has been proven an effective way to urban diversity in 59 major cities in China.

Nature-based measures like green infrastructures have been regarded as a multifunctional approach to urban resilience and sustainability for capturing particulate matters and mitigating the impact of Urban Heat Islands at the city scale, acting as nature-based solutions to abate the impact of natural hydro-meteorological hazards such as flooding and heatwaves, to indirectly foster biodiversity. Chapter "Urban Green Infrastructure" summarizes the design principles and the role of green infrastructure (GI) in the abatement of air pollution, urban heat and natural hazards. The

GI method has been proven to contribute to improved health and well-being in terms of sustainable and resilient cities.

Floods threaten cities in various forms whose magnitude and frequency are likely exacerbated by climate change and increasing urbanization. Chapter "Urban Flood Resilience" explains their causes, consequences, observation methods, hazard modelling techniques and mitigation infrastructures relevant to urban flood resilience. It introduces hybrid infrastructure (grey, green and blue) systems for urban resilience to flooding. It is addressed that flooding risk has to be managed through resilient and sustainable planning, especially in fast-developing areas.

The global success of mitigating climate change is dependent on transitions to net zero carbon energy systems. The increasing urbanization of global populations inevitably increases the energy demand for working, living and improving living environments, transportation, entertainment and so on. Chapter "Energy Demand and Cities: Understanding the Complexity of Reduction Potential" presents the relationship of different causal pathways (top-down and bottom-up) to demand reduction in cities. It presents a matrix of the five demand reduction sectors and seven demand reduction trends as of the Low Energy Demand Scenarios (CREDS) for the potentials of energy demand reduction in the context of urban sustainability.

Mitigation of carbon emissions needs to implement a zero-energy building policy. Clean and renewable energy is the most important issue for a city's sustainability and resilience. Chapter "Clean and Renewable Energy" introduces the applications of renewable energy including solar water heating, space heating and cooling systems, photovoltaics and hybrid photovoltaic and solar thermal (PV/T) systems which provide both thermal and electrical energy and the geothermal energy systems which employ ground heat exchangers in combination with heat pumps. Renewable energy systems are used not only in buildings directly but nowadays are also used in district heating and cooling networks in the urban/community context for a sustainable future.

The evolution of policies by setting frameworks plays a pivotal role in the delivery of sustainable and resilient cities. Chapter "Global Policies and Practices for Transforming Resilient City-Regions" presents examples of implementations of policies and sustainable development guides as well as tools in cities towards a more resilient, sustainable and liveable future. It includes exemplary cases from London, Shanghai, Accra, Singapore, Amsterdam and Melbourne. These include urban forests, urban agriculture, water storage and reuse, flood management using nature-based solutions, walking, cycling and electric public transport for mobility and using waste as a resource.

The case study of the implementation of energy policy for zero-carbon cities (communities) in Japan has been introduced in Chapter "Zero-Carbon City and Community in Japan—Policies, Proposals, and Examples". The proposals made by the Architectural Institute of Japan and the Science Council of Japan regarding zero-carbon cities and communities are explained and discussed. An overview of the decarbonization efforts in 26 regions selected as decarbonization leading areas and specific examples of four regions is provided. It further demonstrates the policy's role in delivering sustainable cities.

The scarcity of freshwater across the world has become one of the major crises exacerbated by a growing population, changing consumption patterns and the rise in irrigated agriculture. Chapter "Urban Maintenance and Robustness" presents the developing country, Indian case studies of household end-use water demand and the innovative decentralized water system, which combined rainwater harvesting (RWH) with wastewater. Guidance of five points for urban water resilience has been proposed to increase city resilience in the context of end-user and management through scenario analysis.

A long-term integrated approach is crucial in planning and implementing green and liveable cities. On the one hand, this requires cross-departmental and multidisciplinary collaboration, as well as wider citizen engagement; on the other hand, it requires a long-term strategy that considers the whole lifecycle of creating sustainable and resilient places starting from visioning, planning, implementation, enforcement, management and maintenance, as well as monitoring and feedback. Chapter "A Long-Term Integrated Approach to Plan and Implement Green and Liveable Cities: Case Studies in the UK" introduces two UK case studies to discuss how the long-term integrated approach can be utilized: the first case study of Cornwall Council is focused on its inclusive and integrated policymaking process; the second case study of Milton Keynes is focused on the long-term strategy of planning and implementing its city-wide green infrastructure.

Chapter "Wellbeing and Social Health in the Built Environment" looks beyond well-recognized factors like thermal quality, addressing lifestyle and balance and social influences on 'well-being in use', including location, spatial scale, biophilia, 'active by design', community and social memory. This has been demonstrated by drawing on the author's rich repertoire of research and practical experience in people's use of buildings and urban spaces. It implicitly proposes a breadth of professional expertise to shape healthy environments.

Chapter "Reading, UK: A Case Study in Urban Futures" addresses the importance of city governance in shaping the future of a city. It introduces the overall concept of urban future thinking. The Reading case study in the UK demonstrates that the cities with the most robust and effective 'hard' and 'soft' climate change and governance structures will be the cities that continue to attract investment. The key points are addressed that can be learned by other cities and towns in achieving sustainable and resilient cities, towns and communities. They include integrated approach, role of stakeholders, power relations, R&D opportunities, COVID-19, resilience and net zero. Ultimately people, businesses, local government, academia and other key city stakeholders must all be part of helping shape that future.

## References

Cohen J et al (2015) Estimates and 25-year trends of the global burden of disease attributable to ambient air pollution: an analysis of data from the Global Burden of Diseases Study 2015. Lancet 389(10082):1907–1918. https://doi.org/10.1016/S0140-6736(17)30505-6

CRED, UNISDR (2018) Economic losses, poverty & disasters: 1998–2017. Centre for Research on the Epidemiology of Disasters (CRED), United Nations Office for Disaster Risk Reduction (UNISDR). https://www.undrr.org/publication/economic-losses-poverty-disasters-1998-2017

David M, Park AL, Shelly C (2021) Estimating the global costs of hearing loss. Int J Audiol 60(3):162–170. https://doi.org/10.1080/14992027.2021.1883197

EEA (European Environment Agency) (2020) Environmental noise in Europe—2020. Publications Office of the European Union, Luxembourg. https://doi.org/10.2800/686249

Feng S, Gao D, Liao F, Zhou F, Wang X (2016) The health effects of ambient PM2.5 and potential mechanisms. Ecotoxicol Environ Saf 128:67–74. https://doi.org/10.1016/j.ecoenv.2016.01.030

Francis C, Barber J (2013) A framework for understanding noise impacts on wildlife: an urgent conservation priority. Front Ecol Environ 11(6):305–313. https://doi.org/10.1890/120183

Halperin D (2014) Environmental noise and sleep disturbances: a threat to health? Sleep Sci 7(4):209–212. https://doi.org/10.1016/j.slsci.2014.11.003

He B, Wang J, Zhu J, Qi J (2022) Beating the urban heat: situation, background, impacts and the way forward in China. Renew Sustain Energy Rev 161:112350. https://doi.org/10.1016/j.rser.2022.112350

Howard G, Bartram J, Williams A, Overbo A, Fuente D, Geere J (2020) Domestic water quantity, service level and health, 2nd edn. World Health Organization, Copenhagen. ISBN 978-92-4-001524-1

Huangfu P, Atkinson R (2020) Long-term exposure to $NO_2$ and $O_3$ and all-cause and respiratory mortality: a systematic review and meta-analysis. Environ Int 144:105998. https://doi.org/10.1016/j.envint.2020.105998

IEA (International Energy Agency) (2021) Empowering cities for a net zero future: unlocking resilient, smart, sustainable urban energy systems. OECD Publishing. https://iea.blob.core.windows.net/assets/4d5c939d-9c37-490b-bb53-2c0d23f2cf3d/G20EmpoweringCitiesforaNetZeroFuture.pdf. Accessed 18 May 2023

IPCC (Intergovernmental Panel on Climate Change) (2018) Special report on global warming of 1.5°C. https://www.ipcc.ch/sr15/. Accessed 3 May 2023

Iping A, Kidston-Lattari J, Simpson-Young A, Duncan E, McManus P (2019) (Re)presenting urban heat islands in Australian cities: a study of media reporting and implications for urban heat and climate change debates. Urban Clim 27:420–429. https://doi.org/10.1016/j.uclim.2018.12.014

Macintyre HL, Heaviside C, Taylor J, Picetti R, Symonds P, Cai XM, Vardoulakis S (2018) Assessing urban population vulnerability and environmental risks across an urban area during heatwaves—implications for health protection. Sci Total Environ 610:678–690. https://doi.org/10.1016/j.scitotenv.2017.08.062

Münzel T, Gori T, Babisch W, Basner M (2014) Cardiovascular effects of environmental noise exposure. Eur Heart J 35(13):829–836. https://doi.org/10.1093/eurheartj/ehu030

Münzel R, Schmidt F, Steven S, Herzog J, Daiber A, Sørensen M (2018) Environmental noise and the cardiovascular system. J Am Coll Cardiol 71(6):688–697. https://doi.org/10.1016/j.jacc.2017.12.015

PHE (2020) Heatwave mortality monitoring report: 2020. Public Health England (PHE). https://www.gov.uk/government/publications/phe-heatwave-mortality-monitoring/heatwave-mortality-monitoring-report-2020. Accessed 22 Nov 2022

Seinfeld J, Pandis S (1998) Atmospheric chemistry and physics: from air pollution to climate change. Wiley

Tang W, Pei Y, Zheng H, Zhao Y, Shu L, Zhang H (2022) Twenty years of China's water pollution control: experiences and challenges. Chemosphere 295:133875. https://doi.org/10.1016/j.chemosphere.2022.133875

UN (United Nations) (2015) World population prospects: the 2015 revision, key findings and advance tables. Working paper no. ESA/P/WP.241. https://population.un.org/wpp/publications/files/key_findings_wpp_2015.pdf. Assessed 18 May 2023

UNEP (United Nations Environment Programme) (2022) Frontiers 2022: noise, blazes and mismatches—emerging issues of environmental concern. Nairobi

WHO (2013) Health effects of particulate matter: policy implications for countries in eastern Europe, Caucasus and central Asia. World Health Organization, WHO Regional Office for Europe, Copenhagen

WHO (2017) Flooding: managing health risks in the WHO European region. World Health Organization, Copenhagen. ISBN 978289052795

WHO (2018a) Heat and health. https://www.who.int/news-room/fact-sheets/detail/climate-change-heat-and-health. Accessed 4 May 2023

WHO (2018b) WHO ambient (outdoor) air quality database. Summary results, update 2018. https://cdn.who.int/media/docs/default-source/air-quality-database/aqd-2018/aap_database_summary_results_2018_final2.pdf?sfvrsn=7b92eafc_3. Accessed 15 May 2023

WHO (2021) World report on hearing. World Health Organization, Geneva. ISBN 978-92-4-002048-1

WHO (2023a) Ambient (outdoor) air pollution. World Health Organization. https://www.who.int/news-room/fact-sheets/detail/ambient-(outdoor)-air-quality-and-health. Accessed 9 May 2023

WHO (2023b) Air pollution. World Health Organization. https://www.who.int/health-topics/air-pollution#tab=tab_1. Accessed 9 May 2023

WHO (2023c) Heat and health. World Health Organization. https://www.who.int/news-room/fact-sheets/detail/climate-change-heat-and-health, Accessed 15 May 2023

WMO (2021) WMO atlas of mortality and economic losses from weather, climate and water extremes (1970–2019). World Meteorological Organization (WMO). https://library.wmo.int/index.php?lvl=notice_display&id=21930#.ZFPHTnbMJPa

WMO, WHO (2015) Heatwaves and health: guidance on warning-system development. World Meteorological Organization (WMO), World Health Organization (WHO). https://www.who.int/publications/m/item/heatwaves-and-health--guidance-on-warning-system-development

# Outdoor Thermal Comfort and Heat Exposure Risks

**Runming Yao, Xizhen Huang, Tiantian Xu, and Yuening Zhu**

**Abstract** People in cities use outdoor space for a wide range of activities such as exercise, rest and socializing. Climate change, Urban Heat Islands and extreme heat-wave events are threats to pedestrians' health and well-being. Good urban planning and design can provide a high-quality urban living environment by meeting the need for outdoor thermal comfort and thermal resilience. People's thermal sensations and perceptions are greatly affected by the local microclimate and their thermal experience while, in turn, outdoor thermal comfort affects the decisions on whether to use a space. This chapter presents the human adaptation mechanism for achieving outdoor thermal comfort, the factors that influence it, strategies for thermal resilience, the management of heat risk and thermal comfort assessment methods to use. Finally, a case study for evaluating thermal perception in public squares is introduced.

**Keywords** Climate change · Urban Heat Islands · Outdoor thermal comfort · Urban planning and design · Thermal stress · Physiological adaptation

## 1 Introduction

Outdoor thermal comfort is important in the context of sustainability and resilience. A pleasant microclimate contributes to the quality of outdoor spaces and attracts outdoor activities, such as walking, cycling, rest, exercise and socializing. In this section, the main factors that impact on the comfort of outdoor environment will be presented.

R. Yao (✉) · X. Huang · T. Xu · Y. Zhu
School of Civil Engineering, Chongqing University, Chongqing, China
e-mail: r.yao@cqu.edu.cn; r.yao@reading.ac.uk

R. Yao
School of the Built Environment, University of Reading, Reading, UK

 
R. Yao (ed.), *Resilient Urban Environments*, Cities and Nature,
https://doi.org/10.1007/978-3-031-55482-7_2

## 1.1 Climate Warming and UHI Caused by Urbanization

Since the nineteenth century, the process of urban growth has caused a positive heat balance and a more documented Urban Heat Island (UHI) effect, which reflects higher air and surface temperatures in urban areas than in the surrounding suburban ones (Mills et al. 2010; Oke 1982; Santamouris 2001).

It is predicted that, in future, warmer climates and more frequent excessive heat events will pose great heat threats to human comfort and health (Grimmond et al. 2010). The urban overheating phenomenon is a potential risk faced by outdoor space users, which greatly impacts urban liveability and sustainability.

## 1.2 Mortality Due to Heatwave Events

In recent years urban heatwaves and Urban Heat Islands have become a critical problem for human health in many cities. In China, heatwaves cause severe health impacts on females and the elderly, and the illiterates are particularly vulnerable (Yang et al. 2019). Also, in 2019, European summer heatwaves saw Germany, the Netherlands and the UK experiencing record-breaking temperatures exceeding normal ones by 4.7 °C (Ma et al. 2020). In 2020, UK heatwaves returned in June and August causing more than 2000 deaths (PHE 2020; Thompson et al. 2022). In extreme heat environment, the body's thermoregulatory system becomes significantly stressed which may lead to conditions such as heat syncope, where the brain lacks sufficient blood and oxygen, thermal exhaustion, cardiovascular stress and, in the worst cases, heat stroke which can subsequently lead to respiratory distress syndrome, kidney and liver failure, blood clotting and, ultimately, death. Needless to say, physiological discomfort is accompanied by significant psychological discomfort in the form of emotional and work stress, anxiety and mental fatigue which are all exacerbated by sleep deprivation during hot nights (Yao and Steemers 2013). Extremes of temperature can cause a multitude of detrimental effects on human physiology and psychology.

## 1.3 Heat-Resilient Cities

The alarming danger of climate change has promoted further research on creating an evidence base for response plans, which include threshold temperatures to trigger heat health alerts and heat health messages about different adaptation methods (Lowe et al. 2011). So, revealing the adaptation mechanism of the human body for thermal sensation is critical for monitoring thermal health and combatting city overheating risks. Therefore, under the challenges of climate change, an understanding of the dynamic characteristics of physiological and thermal responses to varying outdoor heat stress conditions can improve the prediction of outdoor dynamic thermal perceptions in advance and enhance the heat resilience of cities.

## 2 Factors Influencing Outdoor Thermal Comfort

### *2.1 Thermal Comfort Definition*

Although the human sense of "comfort" encompasses many environmental factors, thermal comfort is one of the most important sensations for human health and happiness (Gagge et al. 1967). Thermal comfort is a state for which people actively strive for when they feel discomfort (Parsons 2002). Often, it is generally agreed that thermal comfort is a "condition of mind that expresses satisfaction with the thermal environment and is assessed by subjective evaluation", according to a definition by the American Society for Heating, Refrigerating and Air-Conditioning Engineers (ASHRAE 2021). Therefore, the main disciplines of thermal comfort are engineered perceptions.

### *2.2 Pedestrians' Adaptation to Thermal Comfort*

A comfortable outdoor thermal environment is conducive to outdoor activities, bringing social, health and environmental benefits (Huang et al. 2016). The influence of thermal comfort on outdoor activities is a complex issue comprising both climatic and behavioural aspects. Thus, it is necessary to understand the factors influencing outdoor thermal comfort.

Human thermal adaptation procedures can be attributed to different processes for perceived outdoor thermal comfort conditions. Stemming from studying the indoor environment of buildings, these adaptative processes are behavioural, physiological and psychological (Brager and de Dear 1998). The achievement of a state of thermal comfort is complex, as it includes cultural, climatic, social, psychological and behavioural factors as part of a person's adaptive process (Yao et al. 2009).

Besides, thermal comfort has three general definitions in terms of a heat balance approach based on the human body: a psychological comfort approach, a thermo-physiological comfort approach based on the rates of nervous signals and outdoor psychological aspects (Höppe 2002). Because the physiological approach and physical models for designing outdoor spaces are inadequate for characterizing thermal comfort conditions, it is necessary to understand the dynamic human parameters involved in psychological adaptation, such as available choice, thermal history and expectations (Nikolopoulou et al. 2001). So, the mechanism for heat stress relief involves not only a subjective paradigm to assess this perception by thermal sensation votes, but also an objective paradigm to investigate the thermal environment in different physical conditions.

## 2.3 Factors Influencing the Human Adaptation Process

Based on the previous assertion, outdoor thermal comfort can be influenced by physical, psychological, physiological, behavioural and personal factors (Fig. 1). These factors mutually interact, constituting the factors to apply in hardware and software of an urban system for enhancing a city's resilience capacity to urban and metabolic heat.

**Fig. 1** Factors influencing human adaptation to secure outdoor thermal comfort

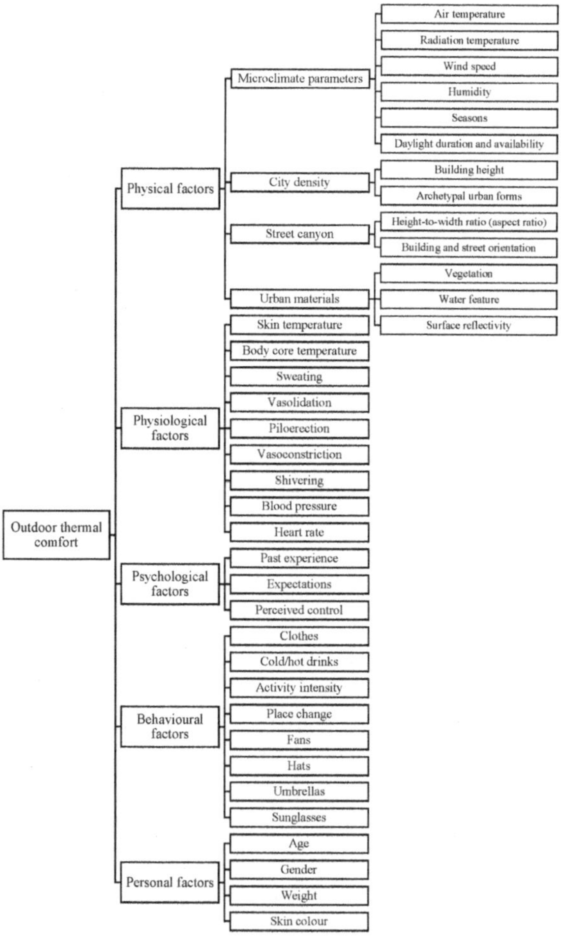

### Physical Factors

Physical factors that affect thermal comfort in an outdoor environment include the meteorological parameters and the urban spaces for adaptation. Designing comfortable microclimates is the foundation for pleasant outdoor spaces that encourage people's outdoor activity. The physical parameters describing microclimates include air temperature, humidity, radiant temperature and air velocity. In urban spaces, microclimates can be influenced by spatial categories related to the enclosure, and a correlation between these physical parameters and outdoor perceived comfort could be found for evaluating their impacts (Ahmed 2003). These urban textures and their impact have consequences for thermal conditions since they include the availability of shade, direct solar radiation and diffuse daylight, breezes arising from the effects of wind and airborne pollutant movements. Furthermore, these urban forms are constituents of street canyons, along with the building height physically shape the city density.

### Physiological Factors

Physiological factors represent the human body's thermoregulatory system. Microclimate conditions would impose thermal stress on the human body (Oke et al. 2017). In adapting to ambient temperatures higher than skin and body core, the body causes skin vasodilation or sweating to increase heat loss. When the skin becomes cold, piloerection or "hairs standing on end" occurs to reduce heat loss, and vasoconstriction and shivering can also occur. In turn, these changes also impact on blood pressure and heart rate.

Moreover, physiological adaptation can be changed through acclimatization after long-term exposure to a microclimate as shown in studies of people with a long-term thermal history background (Lam et al. 2021b).

### Psychological Factors

Psychological factors represent the psychological adaptation parameters that would result in a sensation of thermal comfort. Psychological factors of outdoor thermal comfort involve the naturalness of the context, past experience, perceived control, length of exposure to the environment, environmental stimulation and thermal expectations (Nikolopoulou and Steemers 2003). These comprehensive parameters of psychological adaptation would jointly result in a sensation of thermal comfort which can be measured on a seven-point thermal sensation scale by asking participants to rate their sensation from "hot" to "cold" (ASHRAE 2021).

### Behavioural and Personal Factors

Behavioural adaptation by pedestrians can easily be observed in daily life, such as changing clothes, the consumption of cold or hot drinks, variations in the intensity of activity, moving places and cultural responses (e.g. having a siesta). Some individual measures include using fans and parasols and/or wearing hats and sunglasses.

As for individual differences, human subjects can have various thermal perceptions due to different ages, genders, weights and skin colours (Lai et al. 2020).

# 3 Strategies for Thermal Resilience and Heat Risk Management

The adaptive process implies a bottom-up perspective for designing an urban microclimate and providing facilities to manage a city's overheating risk. In an urban system, the hardware includes the city's infrastructure as well as public spaces functioning as adaptation places, and the analysis software can be viewed as human subjects and their interaction with, and adaptation to, urban climates. They need to be taken together for assessing building thermal resilience. The mechanism of human-urban interaction and adaptation for managing heat exposure risks is revealed in Fig. 2. In this system, three agents: the city's infrastructure, adaptation places and human subjects can be identified. A city's infrastructure could not only regulate microclimates and offer facilities but also provides pedestrians with adaptation places. Human subjects constantly interact and adapt to heat exposure risks.

## *3.1 A City's Infrastructure*

### Urban Forms

Urban forms physically constitute the spatial patterns of a city that can have an overarching influence on outdoor thermal comfort. Over time, the engineering community has developed various methods to help mitigate UHI effects. Two principal strategies are commonly adopted: the first is to increase vegetation coverage, mainly in the form of parks, recreational spaces and gardens, in order to maximize the control of

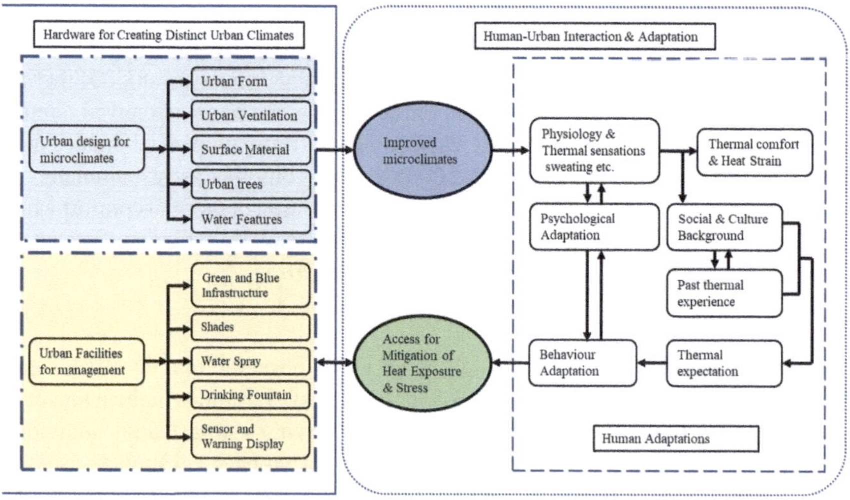


**Fig. 2** Human-urban interaction and adaptation for managing heat exposure

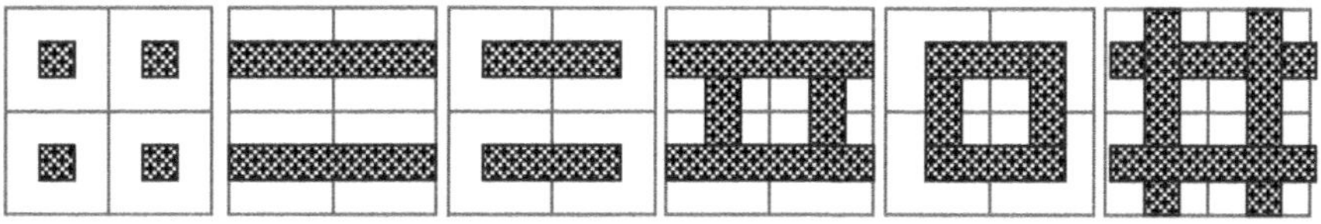

**Fig. 3** Archetypal urban forms; from left to right: pavilions, slabs, terraces, terrace courts, pavilion courts and courts *Source* after Ratti et al. (2003)

temperature rises through increasing evaporation surfaces; foliage can also capture incoming solar radiation which is essential for photosynthesis. The second is to increase surface reflectivity by increasing the albedo in order to reduce radiation absorption from urban surfaces.

The urban form is complex. Martin and March (1972) examined six simplified or archetypal urban forms in order to limit the complexities found in real urban textures: pavilions, slabs, terraces, terrace courts, pavilion courts and courts (Fig. 3). These forms are still used to examine and compare the impact of geometry alone and are common in generic studies and have been extensively adopted (Ratti et al. 2003).

The skeleton of urban forms contains the urban canyon or street ratio and the building's height and orientation. The height-to-width ratio ($H/W$), also known as the *aspect ratio*, and building orientation are found to have a significant influence on the variations of reflectance that occur on urban surfaces (Nunez and Oke 1977). Further studies have focused on the role of aspect ratio and street orientation in assessing the availability of solar radiation on building surfaces and pedestrian paving (Arnfield 1990). Additional complexities are encountered due to the dynamic nature of solar radiation and pedestrian motion which can limit the number of different situations observable.

## Urban Materials: Physical Infrastructure

Urban materials differentiate the surface reflectivity of pavements and buildings. Reflective building surfaces can be achieved simply by painting them a light colour, typically white in many countries with hot climates, such as Greece. Another method is to form a reflective membrane using elastomeric coatings; these have been reported to increase the albedo of commercial roofs from 10 to 70% (Rosenfeld et al. 1995). However, both techniques are mainly applied to roofs and pavements. Indeed, shielding roofs from radiation is especially important in commercial and residential buildings where significant energy demand for cooling can be offset by reducing the direct heat gains of a building. Similarly, constructing "cool" pavements is achieved by lightening the colour of asphalt by the inclusion of white chips. Alternatively, a light concrete covering, using Portland cement, can be used; this is called *whitetopping*.

**Green and Blue Infrastructure**

Green and blue infrastructure (GBI) can be also viewed as a natural material for influencing outdoor thermal comfort in an urban environment. They can be represented by vegetation and water features. Green infrastructure (GI) systems can be categorized into urban trees, urban green spaces and building greenery. They represent the greening actions for mitigating urban heat stress and improving perceived outdoor thermal comfort, and they are supported by scientific evidence published over the past two decades which has seen the three vegetation types representing hypothetical GI scenarios (green roof, grasslands, trees) tested in experiments as effective engineered facilities for mitigating the UHI effect (Tiwari et al. 2021). For the maximum cooling effect of using green systems, combining trees and grasslands has been recommended by scholars (Lobaccaro and Acero 2015; Lee et al. 2016).

Designers also frequently encounter conflicting design principles due to seasonal variations and aesthetics. For example, vegetation patterns (green spaces) can have both positive and negative impacts depending on seasons. Lin et al. (2008) compare three cases and conclude that vegetation is desirable in summer for localized shading, but creates windbreaks that reduce wind speeds which is undesirable. On the contrary, during winter, trees affect solar radiation distribution and reduce solar gains; however, their wind-breaking effects can then be beneficial.

## 3.2 *Adaptation Places*

The adaptation places provide multiple functions for using urban outdoor spaces. They can be open spaces, shaded spaces, public spaces and residential spaces. On the other hand, semi-outdoor spaces, often as an architectural feature mediating between the indoors and the outdoors, can provide effective shading and rain protection (Gamero-Salinas et al. 2021). In addition, especially in outdoor environments, vegetation and water are the natural materials that can characterize the adaptation places, such as parks, greenspaces and waterfront spaces. For example, in heatwaves, people tend to relax in tree-shaded areas and engage in activities involving water as a means of thermal comfort adaptation (Fig. 4).

## 3.3 *Human Subjects*

The risk of urban overheating threatens human life. Warning systems and intellectual tools can be developed, given the current advanced and low-cost sensors. For instance, Internet of Things (IoT) and Big Data can be integrated to characterize urban heat exposure (Nazarian et al. 2022). Since the human body is itself a system for thermal regulation as stated in Sect. 6.2.2, thermal adaptation can be adjusted through its in-built physiological regulation. Also, psychological adaptation

(a) (b) (c)

**Fig. 4** Relaxing in different urban infrastructures for outdoor thermal comfort, such as **a**, **b** recreation under tree shading on a day of Chongqing heatwaves and **c** water activities in Reading during a heatwave. *Source* Authors' own photographs

and behavioural adaptation can be adopted. A method involving this procedure for predicting outdoor heat stress will be introduced in Sect. 6.4.4 where a regression-based model is developed to capture human physiological responses for evaluating dynamic thermal sensations.

In addition to the monitoring tools, joint efforts from sectors and stakeholders such as governance actors and citizen engagement are needed for managing overheating risks in urban systems. As stated in Sect. 6.3.1, implementing GI can be one strategy for enhancing a city's resilience in terms of counteracting the UHI effect and improving outdoor thermal comfort. A hypothetical scenario of optimizing GI for heat exposure management could evoke engagement and discussion among stakeholders. Meerow and Newell (2019) applied the "Five Ws" as questions about urban resilience. The "Five Ws of urban heat resilience" for GI planning stand for "Who" and/or "What" will be made heat resilient, "When" will heat resilience happen and "Where", and "Why" is heat resilience needed, which can help participants focus GI implementation on empirical contexts. Moreover, the focus group technique for scenario planning and visioning workshops can help to shape the policy for managing overheating risks (Iwaniec et al. 2020).

## 4 Thermal Comfort Assessment Techniques

Outdoor thermal comfort assessment provides climate-responsive design strategies for urban open spaces. Proper assessment methods and tools for the evaluation of outdoor thermal comfort will support landscape architects and urban designers to design places with comfortable microclimates that encourage outdoor activities.

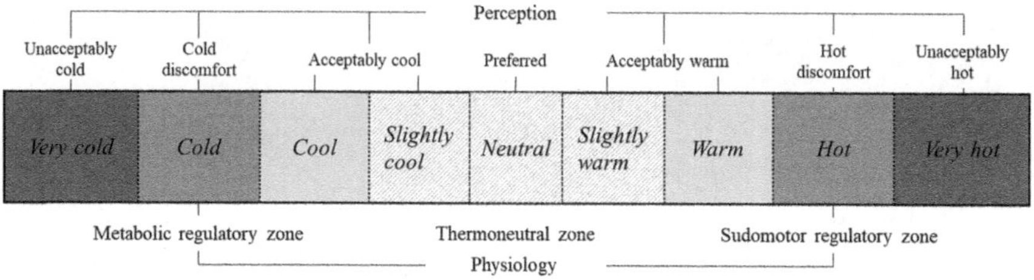


**Fig. 5** Thermal comfort perceptions described through a nine-point thermal sensation scale that pairs physiological zones

## 4.1 Thermal Comfort Perception Scales

For identifying participants' perception of thermal comfort, several scales have been proposed to address thermal sensation and comfort sensation (Fig. 5). The thermal sensation scales, usually with seven or nine points, are continuous with respondents identifying the point matching their thermal perception (Zhang 2003). Numerous studies have applied the ASHRAE seven-point scale with numerical values ranging from cold (− 3), cool (− 2), slightly cool (− 1), neutral (0), slightly warm (1) and warm (2) to hot (3). The nine-point scale adds the perceptions of "very cold" and "very hot" to allow for extreme conditions (Xu et al. 2022).

## 4.2 Thermal Indexes Based on Equivalent Temperature

Models are based on outdoor thermal indices, such as Physiologically Equivalent Temperature (PET), Standard Effective Temperature (SET*), Outdoor Standard Effective Temperature (OUT_SET*) and the Universal Thermal Climate Index (UTCI) (Potchter et al. 2018), which are developed from the human body heat balance. These thermal indices consider the heat exchange between the human body and the surrounding thermal environment and are sensitive to changes in thermal environmental parameters. For outdoor thermal comfort evaluation, UTCI and PET are the most widely used indexes.

PET is based on the Munich Energy-Balance Model for Individuals (MEMI) that defines the balance equation of the human body (Höppe 1999). The UTCI index is derived by the International Society of Biometeorology based on the advanced 340-node "Fiala" thermoregulation model in combination with an adaptive clothing model (Huang et al. 2017). The UTCI has been recognized and validated in various outdoor thermal environments, including extreme weather conditions (Fiala et al. 2012). UTCI is defined as the air temperature of the reference environment, which produces the same thermal stress in an actual environment. This index is presented by the combination of $T_a$, RH, $v_{10}$ and mean radiation temperature ($T_{mrt}$) and is selected as the thermal index to represent the comprehensive effect of the thermal environment

**Table 1** Original UTCI range

| UTCI (°C) range | Stress category |
|---|---|
| Above + 46 | Extreme heat stress |
| + 38 to + 46 | Very strong heat stress |
| + 32 to + 38 | Strong heat stress |
| + 26 to + 32 | Moderate heat stress |
| + 9 to + 26 | No thermal stress |
| 0 to + 9 | Slight cold stress |
| − 13 to 0 | Moderate cold stress |
| − 27 to − 13 | Strong cold stress |
| − 40 to − 27 | Very strong cold stress |
| Below − 40 | Extreme cold stress |

UTCI categorized in terms of thermal stress. *Source* UTCI (2014)

conditions. The original UTCI scale was related to thermal stress instead of thermal sensation, while the original PET scales were involved with thermal sensation, which may be one of the reasons why UTCI is superior to PET in describing outdoor thermal stress classification (Chen et al. 2020). The original UTCI range is shown in Table 1.

Table 2 shows the outdoor thermal comfort evaluation based on UTCI ranges in different studies. It can be seen that the UTCI range varies in different climatic regions. For example, the UTCI temperature range for a certain thermal stress category obtained by surveys in Asia and the original UTCI range for Europe are different. According to the principle whereby China is divided into climate zones, it can also be found that the UTCI range for the hot summer and cold winter region is different from that for the hot summer and warm winter region, as well as being different from that for cold regions. The difference can be explained by the local culture, traditional customs and the adaptability of local residents to their local climate (Lam et al. 2021a). Therefore, the outdoor thermal environment for urban planning and design should be evaluated based on the thermal adaptation characteristics of local people.

## 4.3 Models of Regression

Depending on its type, an empirical regression model directly regresses the thermal sensation in terms of environmental and other variables. Various regression models have been generated by different studies conducted in different regions. These regression models usually consider the environmental parameters (mean radiant temperature, air temperature, relative humidity, global temperature, wind velocity, etc.) as independent variables. The typical studies include Givoni et al. (2003) in Japan, regression models by Nikolopoulou et al. in five different European countries (Nikolopoulou and Lykoudis 2006), a study by Spagnolo et al. in Australia (Spagnolo

**Table 2** UTCI ranges in different regions

| Stress category | Europe (Bröde et al. 2012) | Xi'an (Xu et al. 2019) | Tianjin (Lai et al. 2014) | Guangzhou (Fang et al. 2019) | Harbin (Jin et al. 2019) |
|---|---|---|---|---|---|
| Extreme cold stress | < − 40 | < 2.0 | < − 21 | / | < − 30.2 |
| Very strong cold stress | − 40 ~ − 27 | 2.0 ~ 4.2 | − 21 ~ − 16 | / | − 30.2 ~ − 25.6 |
| Strong cold stress | − 27 ~ − 13 | 4.2 ~ 7.8 | − 16 ~ − 11 | / | − 25.6 ~ − 18.3 |
| Moderate cold stress | − 13 ~ 0 | 7.8 ~ 11.3 | − 11 ~ − 6 | / | − 18.3 ~ − 7.2 |
| Slight cold stress | 0 ~ 9 | 11.3 ~ 18 | − 6 ~ 12 | / | − 7.2 ~ − 3.8 |
| No thermal stress | 9 ~ 26 | 18 ~ 29.1 | 12 ~ 25 | / | − 3.8 ~ 23.0 |
| Moderate heat stress | 26 ~ 32 | 29.1 ~ 36 | 25 ~ 33 | 32.3 ~ 36.5 | 23.0 ~ 29.1 |
| Strong heat stress | 32 ~ 38 | 36 ~ 40.6 | 33 ~ 39 | 36.5 ~ 40.7 | 29.1 ~ 40.9 |
| Very strong heat stress | 38 ~ 46 | 40.6 ~ 42.1 | 39 ~ 47 | 40.7 ~ 44.9 | 40.9 ~ 49.4 |
| Extreme heat stress | > 46 | > 42.1 | > 47 | > 44.9 | > 49.4 |

and de Dear 2003) and many studies in different areas of China (Zhao et al. 2016; Liu et al. 2016).

Many studies have established outdoor thermal sensation models in various climate zones based on the microclimate parameters, see Table 3 (Feng et al. 2021).

A recent study (Xu et al. 2022) explored the physiological factors in the prediction of overall thermal sensation. It aimed to develop a dynamic thermal sensation model based on the relationship between the processes for dynamic physiological adjustment and dynamic thermal perception response. The lower-arm skin temperature is identified as most sensitive and used as a representative physiological factor in predicting outdoor thermal sensation.

The dynamic thermal sensation can be expressed as the sum of the static and dynamic terms (Zhang et al. 2010). Based on the static and dynamic terms of the thermal sensation model above, the dynamic outdoor thermal sensation, namely Overall Thermal Sensation Vote (Xu et al. 2022), can be described by:

$$\mathrm{OTSV} = 4\left(1 - \frac{2}{1 + \exp\left(0.511\Delta' T_{\text{lower arm}} + 0.0227\Delta' T_{\text{ty}}\right)}\right) + 1.724\frac{\Delta T_{\text{lower arm}}}{\Delta t} + 4.306\frac{\Delta T_{\text{ty}}}{\Delta t} \tag{1}$$

**Table 3** Regional thermal sensation regression models

| Location | Regression TSV models | Research time |
|---|---|---|
| Yokohama, Japan (Givoni et al. 2003) | $TSV = 0.1115T_a + 0.00198SR - 0.3185V_a + 1.2$ | All year |
| Hong Kong, China (Cheng et al. 2012) | $TSV = 0.1185T_a + 0.0025SR - 0.2256V_a - 0.0079RH + 3.2464$ | Summer |
| Changsha, China (Liu et al. 2016) | $TSV = 0.146T_a + 0.007Ah - 0.308V_a + 0.097T_{ga} - 2.819$ | All year |
| Guangzhou, China (Zhao et al. 2016) | $TSV = 0.245T_a + 0.059T_{mrt} - 0.457V_a + 0.013RH - 8.527$ | Summer |
| Tehran, Iran (Hadianpour et al. 2018) | $TSV = 0.101T_a + 0.028T_{mrt} - 0.233V_a + 0.002RH - 2.681$ | All year |
| Mendoza, Argentina (Ruiz and Correa 2015) | $TSV = 0.0621T_a - 0.3257V_a + 0.0079RH - 0.9796$ | Summer and winter |
| Birmingham, UK (Metje et al. 2008) | $TSV = 0.0665T_a + 0.0011SR - 0.2256V_a - 0.0079RH + 3.2464$ | All year |
| Combined model for Europe (Nikolopoulou and Lykoudis 2006) | $TSV = 0.049T_a + 0.001SR - 0.051V_a + 0.014RH - 2.079$ | All year |

Note that $T_a$ is the air temperature, SR is the solar radiation, $V_a$ is the wind velocity, RH is the humidity, and $T_{mrt}$ is the mean radiation temperature

where $\Delta' T_{\text{lower arm}}$ is the difference between the actual and neutral lower-arm skin temperature and $\Delta' T_{ty}$ is the difference between the actual and neutral core temperature. $\Delta T_{\text{lower arm}}/\Delta t$ and $\Delta T_{ty}/\Delta t$ are the change rate of the lower-arm skin temperature and the change rate of the core temperature, respectively. When the body's core temperature remains neutral, skin temperature is a good indication of the thermal sensation under steady-state conditions.

These formulae reveal the relationship between dynamic thermal sensation and the physiological adjustment process. When a heatwave is present, the lower-arm skin temperature is high and the change rate of the lower-arm skin temperature will also change greatly, resulting in great changes in people's thermal perception. Being in such an uncomfortable environment for a long time can have an adverse effect on people's health. The wearable sensor technology can measure pedestrians' skin temperature and provide overheating risk alerts for people engaged in activities outdoors, especially during heatwaves. The outdoor thermal sensation prediction can provide guidance for heat stress resilience strategies and sustainable urban development.

## 5 A Case Study in Public Squares

Understanding the relationship between the microclimate parameters and peoples' thermal perceptions as well as thermal evaluation benchmarks for assessing thermal risk in outdoor environments, especially for hot summers, are particularly significant. This chapter presents a case study of an environmental campaign and pedestrian outdoor thermal comfort survey in summer 2020 and 2021 in Chongqing, a city in the southwest of China. In order to get a sufficient sample size, the survey sites for this study are pedestrian streets in Sanxia square (S-square) in Shapingba district, Jiefangbei square (J-square) in Yuzhong district and Guanyinqiao square (G-square) in Jiangbei district respectively, as shown in Fig. 6.

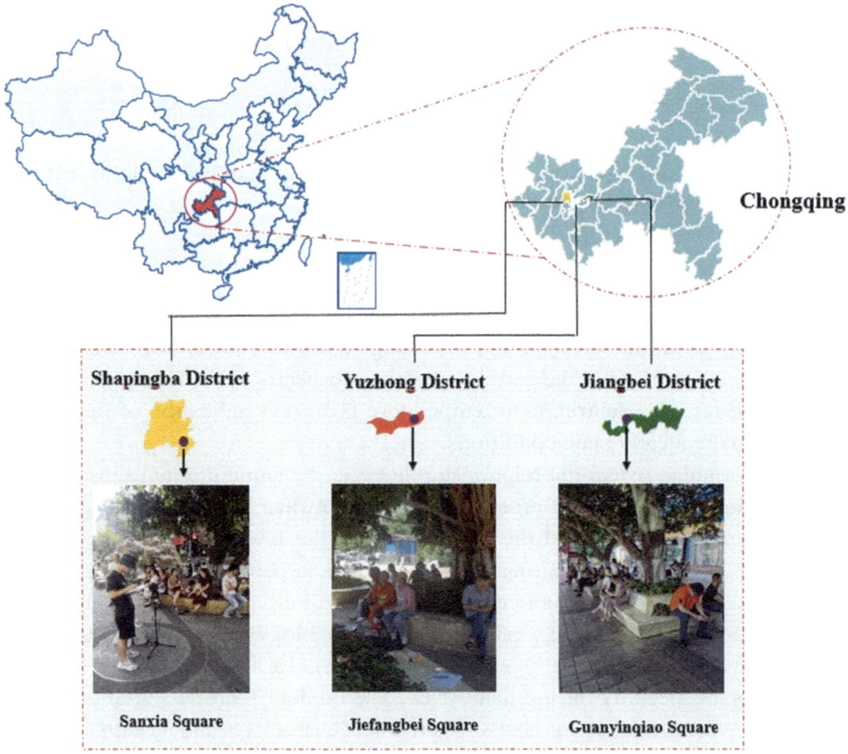


**Fig. 6** Schematic diagram of surveys in public squares in Chongqing

## 5.1 *Spatial Characteristics of Street Canyon and Microclimate Parameters*

The measuring points were located at different spatial locations of the street canyon in G-square; the Sky View Factors (SVF) for these locations are shown in Fig. 7. The SVF reflects the degree of openness of three-dimensional space to the sky, and its value is between 0 and 1. The degree of openness to the sky in G-square at measuring points 1, 2 and 3 is 0.848, 0.343 and 0.675, respectively. Point 1 has the highest and Point 2 the lowest SVF values.

The onsite measurements of microclimate parameters at pedestrian height including air temperature ($T_a$), globe temperature ($T_g$) (representing the actual perceived temperature of people under the combined effect of radiation and convection), relative humidity (RH) and ground surface temperature ($T_{sur}$) in G-square are shown in Fig. 8. It can be seen that the variation trends of the microclimates at the three points are similar. The lowest air temperature $T_a$ and globe temperature $T_g$ are before sunrise in the morning, and the highest $T_a$, $T_g$ and $T_{sur}$ are from 14:00 to 16:00 h. The trend of humidity RH is opposite to that of the air temperature. However, the microclimates at the three points show significant differences. On the whole, for the daytime, Point 1 with the highest SVF has the highest air, globe and surface temperatures at the same time while the humidity is the lowest, while Point 2, with lowest SVF, has the lowest air, globe and surface temperatures at the same time while the humidity is the highest.

Table 4 shows the maximum, minimum and mean values of $T_a$, $T_g$, RH and $T_{sur}$ for different points. The maximum differences measured at Point 1 and Point 2 are

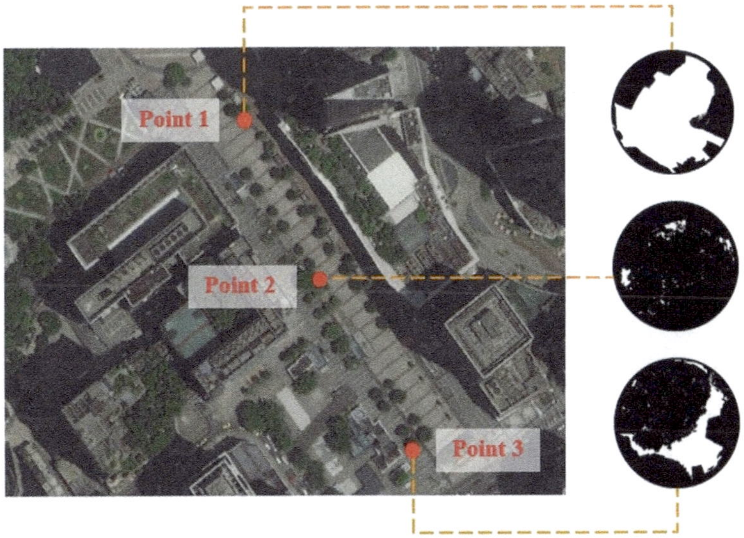


**Fig. 7** Sky View Factor at the 3 monitoring points in G-square

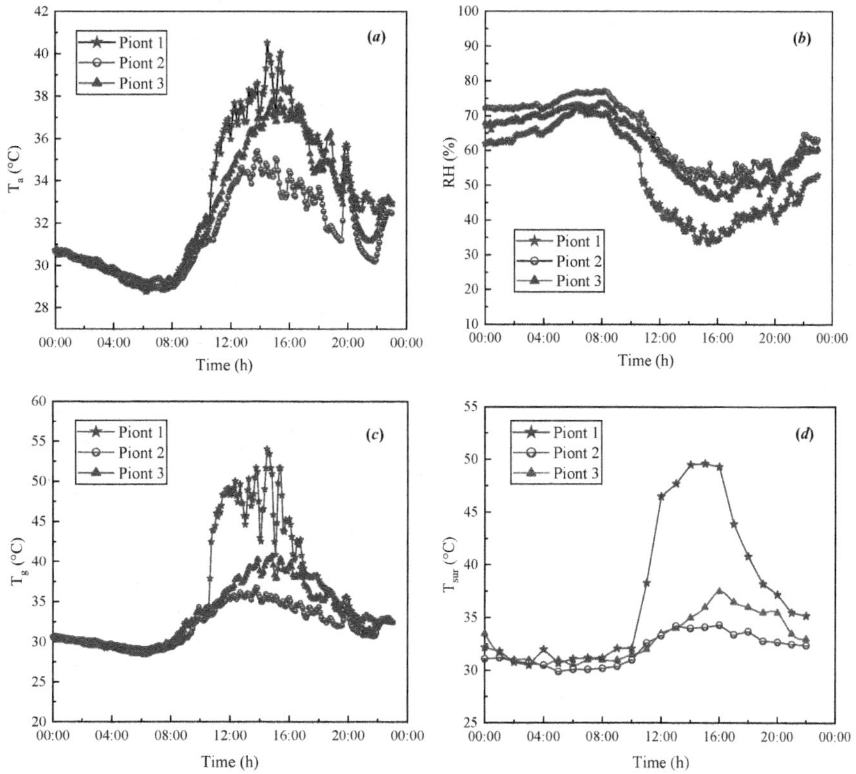


**Fig. 8** Microclimate parameter variations for **a** $T_a$, **b** RH, **c** $T_g$, **d** $T_{sur}$ for the full day

significant, showing an air temperature of 5.2 °C; a globe temperature of 17.6 °C, an air humidity of about 4.5% and a surface temperature of 15.3 °C.

**Table 4** Max., min. and mean values of $T_a$, $T_g$, RH and $T_{sur}$ for different measuring points

| Point | $T_a$ (°C) | | | $T_g$ (°C) | | | RH (%) | | | $T_{sur}$ (°C) | | |
|---|---|---|---|---|---|---|---|---|---|---|---|---|
| | Max. | Min. | Mean | Max. | Min. | Mean | Max. | Min. | Mean | Max. | Min. | Mean |
| 1 | 40.64 | 28.75 | 33.08 | 54.40 | 28.40 | 35.88 | 73.14 | 33.09 | 53.33 | 49.60 | 30.50 | 37.14 |
| 2 | 35.40 | 28.82 | 31.48 | 36.80 | 28.80 | 32.10 | 77.65 | 47.62 | 64.62 | 34.30 | 29.90 | 32.00 |
| 3 | 37.93 | 29.12 | 32.59 | 40.80 | 29.00 | 33.35 | 74.09 | 44.00 | 61.16 | 37.50 | 30.50 | 33.10 |

## 5.2 Correlation Between Microclimate Parameters and Outdoor Thermal Perception

Figure 9 displays the distribution characteristics of the microclimate parameters in the three squares, including air temperature ($T_a$), relative humidity (RH), mean radiation temperature ($T_{mrt}$) and wind velocity ($v$) under different outdoor thermal sensation votes for all questionnaires. The distribution characteristics include the maximum, minimum and mean values (the black dots in Fig. 9) as well as the relative proportion of different microclimate parameter values for a certain outdoor thermal sensation scale. The black lines show the variations for the mean value of $T_a$, RH, $T_{mrt}$ and $v$.

In terms of maximum and minimum values, during the survey, $T_a$, RH, $T_{mrt}$ and $v$ are within the ranges between 26 °C and 39 °C, 25% and 85%, 25 °C and 42 °C and between 0 m/s and 4 m/s, respectively. The microclimate parameters' distribution range corresponding to the thermal sensation voting (TSV) value of − 1 is slightly

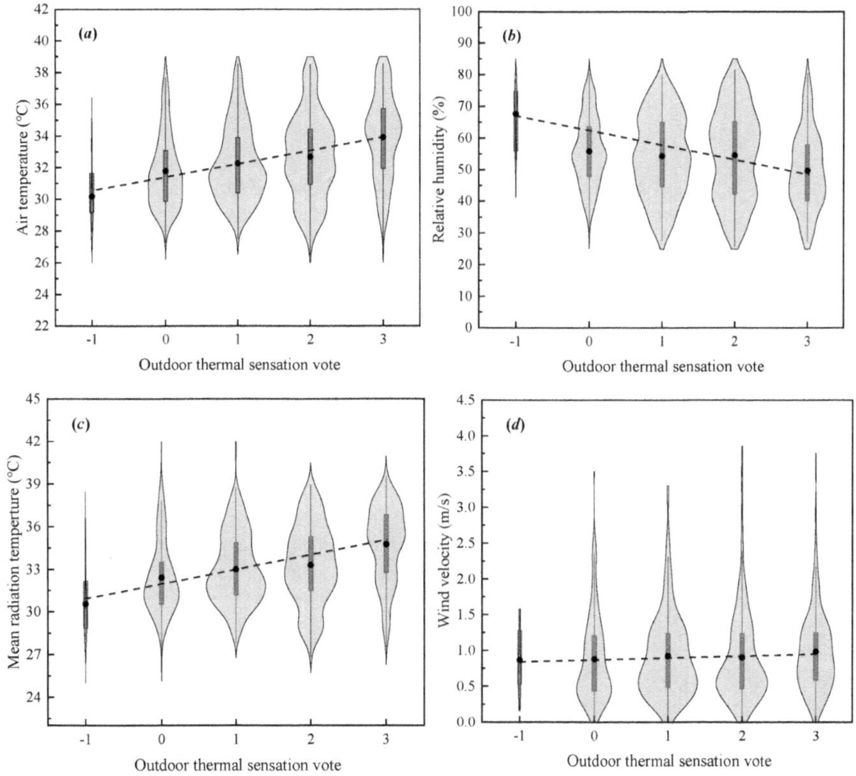


**Fig. 9** Distribution characteristics of microclimate parameters **a** air temperature, **b** relative humidity, **c** mean radiation temperature and **d** wind velocity under different outdoor thermal sensation vote *Note* The different shapes of the dots represent the different TSV values from − 1 (slightly cool) to 3 (very hot)

different from those corresponding to other TSV values. When the TSV values are 0, 1, 2 and 3, there are no obvious differences in the distribution range of microclimate parameters. However, the mean values of microclimate parameters corresponding to different TSV values are different. The TSV value is positively correlated with $T_a$ and $T_{mrt}$, which increases with the increase of $T_a$ and $T_{mrt}$. In terms of RH, RH is negatively correlated with TSV. The TSV value decreases with the increase of $RH_{rt}$. In terms of $v$, due to regional characteristics (the wind velocity in Chongqing is relatively stable and low), the mean $v$ values for each TSV value are all about 1.0 m/s without significant change.

The Spearman correlation coefficient was calculated using IBM SPSS v.25.0 to further explore the correlation between the microclimate parameters and TSV. Table 5 shows the Spearman correlation coefficients between TSV and $T_a$, RH, $v$ and $T_{mrt}$. It can be seen that $T_a$, RH, $v$ and $T_{mrt}$ are all significantly correlated with TSV, in which the correlation between $v$ and TSV is the lowest among all parameters. In terms of Spearman correlation coefficients, the correlation between $T_{mrt}$ and TSV is highest, followed by $T_a$ and RH.

Based on the Spearman correlation analysis, the Spearman correlation coefficient between PET, SET* or UTCI and mean TSV (MTSV) can be obtained.

$$\text{MTSV} = 0.11\text{PET} - 1.68\left(R^2 = 0.808\right) \tag{2}$$

$$\text{MTSV} = 0.14\text{SET} * -2.67\left(R^2 = 0.811\right) \tag{3}$$

$$\text{MTSV} = 0.16\text{UTCI} - 4.14\left(R^2 = 0.944\right) \tag{4}$$

The thermal indexes of SET*, PET and UTCI can be used to evaluate outdoor thermal sensation in summer. Table 6 shows the values of $R^2$ and Spearman correlation coefficients for the relationship between TSV and PET, SET* and UTCI. The results show that among these thermal indexes, the correlation between UTCI and TSV is the strongest.

**Table 5** Spearman correlation coefficients between microclimate parameters and TSV

| | TSV | $T_a$ | RH | $v$ | $T_{mrt}$ |
|---|---|---|---|---|---|
| TSV | 1.000 | | | | |
| $T_a$ | 0.284** | 1.000 | | | |
| RH | – 0.167** | – 0.785** | 1.000 | | |
| $v$ | 0.061** | 0.248** | – 0.163** | 1.000 | |
| $T_{mrt}$ | 0.295** | 0.937** | – 0.693** | 0.247** | 1.000 |

Note that ** represents significant at $p < 0.01$

**Table 6** Values of $R^2$ and Spearman correlation coefficients for the relationship between TSV and PET, SET* and UTCI

| Thermal comfort indicator | PET | SET* | UTCI |
|---|---|---|---|
| $R^2$ | 0.808 | 0.811 | 0.944 |
| Spearman correlation coefficients | 0.849 | 0.890 | 0.951 |

## *5.3 Correlation Analysis Between Thermal Perceptions*

Thermal perception can be described by the thermal sensation vote, thermal comfort vote and thermal acceptability vote. The thermal sensation vote can be different to the thermal comfort vote (Dzyuban et al. 2022). The relationship between the mean thermal comfort vote (MTCV) and TSV can be determined by regression analysis. The results show that the $R^2$ of the fitting curve between MTCV and TSV is greatest when performing linear fitting. The linear relationship between MTCV and TSV is given as Eq. 5.

$$\mathrm{MTCV} = 0.4055\mathrm{TSV} + 1.3773\left(R^2 = 0.968\right) \tag{5}$$

The respondents' evaluations of the thermal environment acceptability vote are either acceptable (1) or unacceptable (0). The thermal unacceptable vote percentage (TUP) is regarded as the ratio of the number of votes who cannot accept the current thermal environment to the total number of surveys under a category, which can measure whether people accept the thermal environment which they are currently exposed to (ASHRAE 2021). The TUP value of each TSV interval was calculated. The relationship between TUP and TSV is determined based on the regression method, as given in Eq. 6.

$$\mathrm{TUP} = \left(0.052\mathrm{TSV}^2 - 0.023\mathrm{TSV} + 0.112\right) \times 100\left(R^2 = 0.957\right) \tag{6}$$

There is a strong quadratic relationship between TUP and TSV with an $R^2$ value of 0.957. Figure 10 is the fitting diagram for TUP and TSV, in which point "a" represents the lowest TUP value and the corresponding TSV of point "a" is about 0.22. When TSV $\geq$ 0.22, the TUP value increases with increasing TSV.

According to ASHRAE standard 55, a comfortable thermal environment usually allows for 20% of people feeling dissatisfied, which means that the thermal comfort temperature range can be considered as the acceptable temperature range for 80% of the respondents. In this case study, the 20% thermal acceptable percentage in ASHRAE standard 55 is referenced as the limit. Point "b" in Fig. 10 represents the 20% thermal unacceptable rate. Thus, when the TSV value is between about − 1.0 and 1.5, the corresponding TUP value is below 20%. When the TSV value is equal to 3, the TUP value in the outdoor thermal environment will exceed 50%.

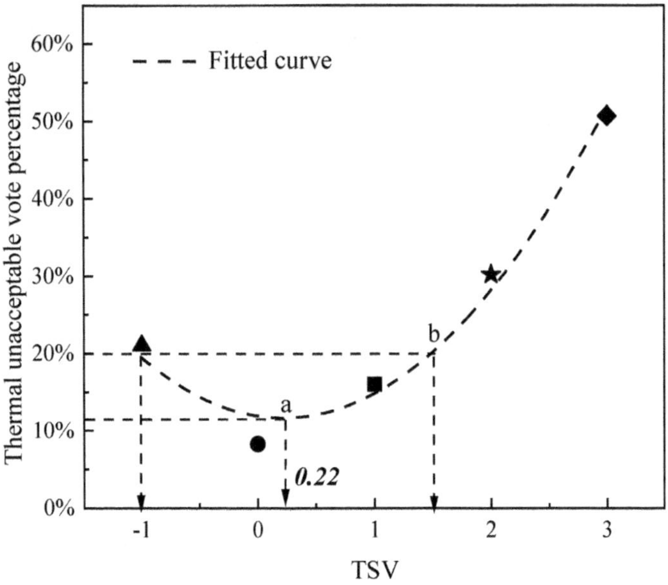


**Fig. 10** Fitting diagram of the relationship between TSV and TUP. *Note* The different shapes of the dots represent different TSV scales, in which 0, 1, 2 and 3 represent neutral, warm, hot and very hot, respectively

## 5.4 *The Evaluation Benchmark for Outdoor Thermal Stress*

Thermal benchmarks are valuable references for designing and optimizing an outdoor thermal environment to meet users' requirements and expectations. They can be commonly determined by understanding people's thermal perception, such as the acceptable temperature range, comfortable temperature range, neutral temperature range, the neutral temperature and the preferred temperature.

The evaluation for outdoor thermal stress often takes the percentage of peoples thermal unacceptability vote (TUP) as the index for identifying different categories of outdoor thermal stress. According to the relationship between TUP and UTCI in this study and the original UTCI scale in previous studies, the outdoor thermal stress is divided into five different levels. Based on the principle of dividing the TUP at 20% intervals, five sections are produced, i.e. TUP ranges of < 20%, 20 ~ 40%, 40 ~ 60%, 60 ~ 80% and > 80%. The UTCI groups corresponding to the above TUP ranges are defined as "no thermal stress", "moderate heat stress", "strong heat stress", "very strong heat stress" and "extreme heat stress" categories.

In this study, taking 0.2 °C as the temperature interval, the TUP values for different UTCI values were determined. The relationship between TUP and UTCI can be obtained using a regression method. There is a quadratic relationship between TUP and UTCI, which can be shown as in Fig. 11. The fitted curve can be expressed as Eq. 7:

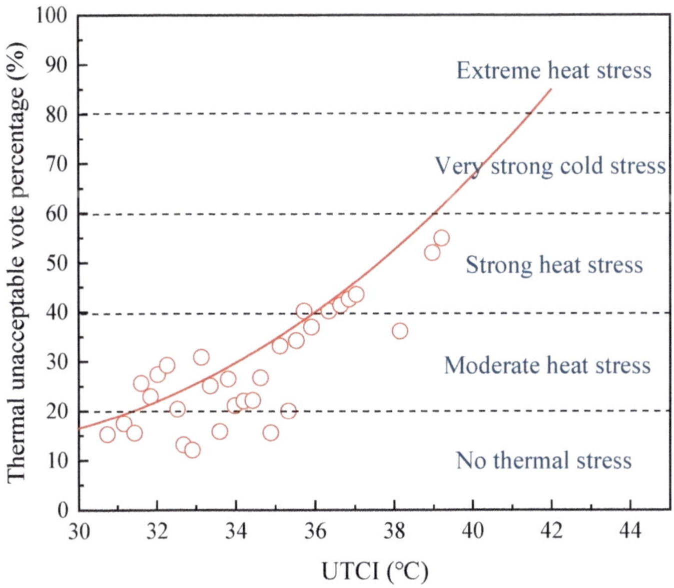


**Fig. 11** Fitting diagram of UTCI and the thermal unacceptable vote percentage

$$TUP = (0.003UTCI^2 - 0.159UTCI + 2.235) \times 100\left(R^2 = 0.716\right) \quad (7)$$

According to Eq. 7, the temperature range of 20% thermal unacceptability for the current thermal environment for the respondents in Chongqing in summer ranges from 17.5 to 31.4 °C. The UTCI ranges in the Chongqing region for the "no thermal stress", "moderate heat stress", "strong heat stress", "very strong heat stress" and "extreme heat stress" categories are in the range from 17.5 °C to 31.4 °C, from 31.4 °C to 36.0 °C, from 36.0 °C to 39.1 °C, from 39.1 °C to 41.6 °C and > 41.6 °C, respectively.

## 6 Summary

In an urbanizing era with increasing heatwave events, promoting a heat-resilient city becomes an urgent agenda for all city dwellers. Cities have the potential for acting as refuge places from heat exposure risks through advanced monitoring technologies and good design. Thus, outdoor thermal comfort can be a critical indicator for the quality of outdoor spaces. This chapter introduced the perception of outdoor thermal comfort and the approaches to its assessment as well as the mitigation strategies for enhancing the resilience of cities to heat exposure risks.

The human body achieves a state of thermal comfort through an adaptative process. The factors influencing outdoor thermal comfort can be categorized as physical, psychological, physiological, behavioural and personal. These factors mutually interact with each other, constituting the hardware and software of an urban analysis system for enhancing a city's resilience capacity to urban and metabolic heat.

As for mitigation strategies for thermal resilience and heat risk management, a trinity consisting of a city's infrastructure, adaptation places and human subjects needs to be involved. A city's infrastructure contains urban forms and materials, as well as green and blue infrastructure. Adaptation places provide multi-functional public spaces, such as squares and parks, for accommodating various outdoor activities. As for human subjects, warning systems and intellectual tools can be developed for monitoring dynamic thermal sensations under heat risk exposure, taking advantage of the regulation system of human body itself to thermal adaptation. Besides, joint efforts from relevant sectors and stakeholders, such as governance actors and citizens, should be involved and consulted for managing overheating risks in urban systems.

Thermal comfort assessment techniques combine the measured data from the subjective evaluation of thermal comfort perception and the objective measurements of microclimate parameters. The former collects pedestrians' thermal perception and often uses a seven- or nine-point assessment scale. The latter is measured with portable equipment during field investigations and then usually used in calculating thermal indexes. Additionally, the models for the thermal index can be built based on equivalent temperature, meteorological variables and physiological parameters. Regression-based analysis can be applied for predicting human thermal sensations by analysing the correlation between the dependent variables of thermal perception and the independent variables of objective parameters. To summarize, thermal indexes can be extremely useful for objectively evaluating the subjective thermal sensation and, when applied at an urban scale, they can provide a city monitoring system for managing heat exposure risks.

The outdoor thermal comfort assessment process has been illustrated through a case study conducted in public squares in two summers in Chongqing. The benchmarks for outdoor thermal stress in the Chongqing region are determined through the Universal Thermal Climate Index (UTCI) ranges with values between 17.5 and 31.4 °C indicating no thermal stress in Chongqing city. The methodology of the case study can be implemented for any other cities in the world.

## References

Ahmed K (2003) Comfort in urban spaces: defining the boundaries of outdoor thermal comfort for the tropical urban environments. Energy Build 35:103–110. https://doi.org/10.1016/S0378-7788(02)00085-3

Arnfield A (1990) Street design and urban canyon solar access. Energy Build 14:117–131. https://doi.org/10.1016/0378-7788(90)90031-D

ASHRAE (2021) ANSI/ASHRAE standard 55-2020. Thermal environmental conditions for human occupancy. ASHRAE, Atlanta

Brager G, de Dear R (1998) Thermal adaptation in the built environment: a literature review. Energy Build 27:83–96. https://doi.org/10.1016/S0378-7788(97)00053-4

Bröde P, Fiala D, Błażejczyk K, Holmér I, Jendritzky G, Kampmann B, Tinz B, Havenith G (2012) Deriving the operational procedure for the Universal Thermal Climate Index (UTCI). Int J Biometeorol 56:481–494. https://doi.org/10.1007/s00484-011-0454-1

Chen X, Gao L, Xue P, Du J, Liu J (2020) Investigation of outdoor thermal sensation and comfort evaluation methods in severe cold area. Sci Total Environ 749:141520. https://doi.org/10.1016/j.scitotenv.2020.141520

Cheng V, Ng E, Chan C, Givoni B (2012) Outdoor thermal comfort study in a sub-tropical climate: a longitudinal study based in Hong Kong. Int J Biometeorol 56:43–56. https://doi.org/10.1007/s00484-010-0396-z

Dzyuban Y, Ching G, Yik S, Tan A, Banerjee S, Crank P, Chow W (2022) Outdoor thermal comfort research in transient conditions: a narrative literature review. Landsc Urban Plan 226. https://doi.org/10.1016/j.landurbplan.2022.104496

Fang Z, Xu X, Zhou X, Deng S, Wu H, Liu J, Lin Z (2019) Investigation into the thermal comfort of university students conducting outdoor training. Build Environ 149:26–38. https://doi.org/10.1016/j.buildenv.2018.12.003

Feng X, Zheng Z, Yang Y, Fang Z (2021) Quantitative seasonal outdoor thermal sensitivity in Guangzhou, China. Urban Clim 39:100938. https://doi.org/10.1016/j.uclim.2021.100938

Fiala D, Havenith G, Bröde P, Kampmann B, Jendritzky G (2012) UTCI-Fiala multi-node model of human heat transfer and temperature regulation. Int J Biometeorol 56:429–441. https://doi.org/10.1007/s00484-011-0424-7

Gagge A, Stolwijk J, Hardy J (1967) Comfort and thermal sensations and associated physiological responses at various ambient temperatures. Environ Res 1:1–20. https://doi.org/10.1016/0013-9351(67)90002-3

Gamero-Salinas J, Kishnani N, Monge-Barrio A, López-Fidalgo J, Sánchez-Ostiz A (2021) Evaluation of thermal comfort and building form attributes in different semi-outdoor environments in a high-density tropical setting. Build Environ 205:108255. https://doi.org/10.1016/j.buildenv.2021.108255

Givoni B, Noguchi M, Saaroni H, Pochter O, Yaacov Y, Feller N, Becker S (2003) Outdoor comfort research issues. Energy Build 35:77–86. https://doi.org/10.1016/S0378-7788(02)00082-8

Grimmond C, Roth M, Oke T, Au Y, Best M, Betts R, Carmichael G, Cleugh H, Dabberdt W, Emmanuel R, Freitas E, Fortuniak K, Hanna S, Klein P, Kalkstein L, Liu C, Nickson A, Pearlmutter D, Sailor D, Voogt J (2010) Climate and more sustainable cities: climate information for improved planning and management of cities (producers/capabilities perspective). Procedia Environ Sci 1:247–274. https://doi.org/10.1016/j.proenv.2010.09.016

Hadianpour M, Mahdavinejad M, Bemanian M, Nasrollahi F (2018) Seasonal differences of subjective thermal sensation and neutral temperature in an outdoor shaded space in Tehran, Iran. Sustain Cities Soc 39:751–764. https://doi.org/10.1016/j.scs.2018.03.003

Höppe P (1999) The physiological equivalent temperature—a universal index for the biometeorological assessment of the thermal environment. Int J Biometeorol 43:71–75. https://doi.org/10.1007/s004840050118

Höppe P (2002) Different aspects of assessing indoor and outdoor thermal comfort. Energy Build 34:661–665. https://doi.org/10.1016/S0378-7788(02)00017-8

Huang J, Zhou C, Zhuo Y, Xu L, Jiang Y (2016) Outdoor thermal environments and activities in open space: an experiment study in humid subtropical climates. Build Environ 103:238–249. https://doi.org/10.1016/j.buildenv.2016.03.029

Huang T, Li J, Xie Y, Niu J, Mak C (2017) Simultaneous environmental parameter monitoring and human subject survey regarding outdoor thermal comfort and its modelling. Building Environ 125. https://doi.org/10.1016/j.buildenv.2017.09.015

Iwaniec D, Cook E, Davidson M, Berbés-Blázquez M, Georgescu M, Krayenhoff E, Middel A, Sampson D, Grimm N (2020) The co-production of sustainable future scenarios. Landsc Urban Plan 197:103744. https://doi.org/10.1016/j.landurbplan.2020.103744

Jin H, Liu S, Kang J (2019) Thermal comfort range and influence factor of urban pedestrian streets in severe cold regions. Energy Build 198:197–206. https://doi.org/10.1016/j.enbuild.2019.05.054

Lai D, Guo D, Hou Y, Lin C, Chen Q (2014) Studies of outdoor thermal comfort in northern China. Build Environ 77:110–118. https://doi.org/10.1016/j.buildenv.2014.03.026

Lai D, Lian Z, Liu W, Guo C, Liu W, Liu K, Chen Q (2020) A comprehensive review of thermal comfort studies in urban open spaces. Sci Total Environ 742:140092. https://doi.org/10.1016/j.scitotenv.2020.140092

Lam C, Cui S, Liu J, Kong X, Ou C, Hang J (2021a) Influence of acclimatization and short-term thermal history on outdoor thermal comfort in subtropical South China. Energy Build 231. https://doi.org/10.1016/j.enbuild.2020.110541

Lam C, Gao Y, Yang H, Chen T, Zhang Y, Ou C, Hang J (2021b) Interactive effect between long-term and short-term thermal history on outdoor thermal comfort: comparison between Guangzhou, Zhuhai and Melbourne. Sci Total Environ 760. https://doi.org/10.1016/j.scitotenv.2020.144141

Lee H, Mayer H, Chen L (2016) Contribution of trees and grasslands to the mitigation of human heat stress in a residential district of Freiburg, Southwest Germany. Landsc Urban Plan 148:37–50. https://doi.org/10.1016/j.landurbplan.2015.12.004

Lin B, Li X, Zhu Y, Qin Y (2008) Numerical simulation studies of the different vegetation patterns' effects on outdoor pedestrian thermal comfort. J Wind Eng Ind Aerodyn 96:1707–1718. https://doi.org/10.1016/j.jweia.2008.02.006

Liu W, Zhang Y, Deng Q (2016) The effects of urban microclimate on outdoor thermal sensation and neutral temperature in hot-summer and cold-winter climate. Energy Build 128:190–197. https://doi.org/10.1016/j.enbuild.2016.06.086

Lobaccaro G, Acero J (2015) Comparative analysis of green actions to improve outdoor thermal comfort inside typical urban street canyons. Urban Clim 14:251–267. https://doi.org/10.1016/j.uclim.2015.10.002

Lowe D, Ebi K, Forsberg B (2011) Heatwave early warning systems and adaptation advice to reduce human health consequences of heatwaves. Int J Environ Res Public Health 8:4623–4648. https://doi.org/10.3390/ijerph8124623

Ma F, Yuan X, Jiao Y, Ji P (2020) Unprecedented Europe heat in June–July 2019: risk in the historical and future context. Geophys Res Lett 47:e2020GL087809. https://doi.org/10.1029/2020GL087809

Martin L, March L (1972) Urban space and structures. Cambridge University Press, UK

Meerow S, Newell J (2019) Urban resilience for whom, what, when, where, and why? Urban Geogr 40:309–329. https://doi.org/10.1080/02723638.2016.1206395

Metje N, Sterling M, Baker C (2008) Pedestrian comfort using clothing values and body temperatures. J Wind Eng Ind Aerodyn 96:412–435. https://doi.org/10.1016/j.jweia.2008.01.003

Mills G, Cleugh H, Emmanuel R, Endlicher W, Erell E, McGranahan G, Ng E, Nickson A, Rosenthal J, Steemer K (2010) Climate information for improved planning and management of mega cities (needs perspective). Procedia Environ Sci 1:228–246. https://doi.org/10.1016/j.proenv.2010.09.015

Nazarian N, Krayenhoff E, Bechtel B, Hondula D, Paolini R, Vanos J, Cheung T, Chow W, de Dear R, Jay O, Lee J, Martilli A, Middel A, Norford L, Sadeghi M, Schiavon S, Santamouris M (2022) Integrated assessment of urban overheating impacts on human life. Earth's Future 10:e2022EF002682. https://doi.org/10.1029/2022EF002682

Nikolopoulou M, Lykoudis S (2006) Thermal comfort in outdoor urban spaces: analysis across different European countries. Build Environ 41:1455–1470. https://doi.org/10.1016/j.buildenv.2005.05.031

Nikolopoulou M, Steemers K (2003) Thermal comfort and psychological adaptation as a guide for designing urban spaces. Energy Build 35:95–101. https://doi.org/10.1016/S0378-7788(02)00084-1

Nikolopoulou M, Baker N, Steemers K (2001) Thermal comfort in outdoor urban spaces: understanding the human parameter. Sol Energy 70:227–235. https://doi.org/10.1016/S0038-092X(00)00093-1

Nunez M, Oke TR (1977) The energy balance of an urban canyon. J Appl Meteorol Climatol 16:11–19. https://doi.org/10.1175/1520-0450(1977)016%3c0011:TEBOAU%3e2.0.CO;2

Oke TR (1982) The energetic basis of the urban heat island. Q J R Meteorol Soc 108:1–24. https://doi.org/10.1002/qj.49710845502

Oke T, Mills G, Christen A, Voogt J (2017) Urban climates, 1st edn. Cambridge University Press, Cambridge

Parsons K (2002) Human thermal environments: the effects of hot, moderate, and cold environments on human health, comfort, and performance, 3rd edn. CRC Press, Boca Raton

PHE (2020) Heatwave mortality monitoring report: 2020. Public Health England (PHE). https://www.gov.uk/government/publications/phe-heatwave-mortality-monitoring/heatwave-mortality-monitoring-report-2020. Accessed 22 Nov 2022

Potchter O, Cohen P, Lin T, Matzarakis A (2018) Outdoor human thermal perception in various climates: a comprehensive review of approaches, methods and quantification. Sci Total Environ 631–632:390–406. https://doi.org/10.1016/j.scitotenv.2018.02.276

Ratti C, Raydan D, Steemers K (2003) Building form and environmental performance: archetypes, analysis and an arid climate. Energy Build 35:49–59. https://doi.org/10.1016/S0378-7788(02)00079-8

Rosenfeld A, Akbari H, Bretz S, Fishman B, Kurn D, Sailor D, Taha H (1995) Mitigation of urban heat islands: materials, utility programs, updates. Energy Build 22:255–265. https://doi.org/10.1016/0378-7788(95)00927-P

Ruiz M, Correa E (2015) Adaptive model for outdoor thermal comfort assessment in an Oasis city of arid climate. Build Environ 85:40–51. https://doi.org/10.1016/j.buildenv.2014.11.018

Santamouris M (2001) Heat-island effect. In: Santamouris M (ed) Energy and climate in the urban built environment, 1st edn. Routledge, London, pp 48–68

Spagnolo J, de Dear R (2003) A field study of thermal comfort in outdoor and semi-outdoor environments in subtropical Sydney Australia. Build Environ 38:721–738. https://doi.org/10.1016/S0360-1323(02)00209-3

Thompson R, Landeg O, Kar-Purkayastha I, Hajat S, Kovats S, O'Connell E (2022) Heatwave mortality in summer 2020 in England: an observational study. Int J Environ Res Public Health 19. https://doi.org/10.3390/ijerph19106123

Tiwari A, Kumar P, Kalaiarasan G, Ottosen T (2021) The impacts of existing and hypothetical green infrastructure scenarios on urban heat island formation. Environ Pollut 274:115898. https://doi.org/10.1016/j.envpol.2020.115898

UTCI (2014) Universal Thermal Climate Index documents. COST Action 730. http://www.utci.org/utci_doku.php. Accessed 14 Oct 2021

Xu M, Hong B, Jiang R, An L, Zhang T (2019) Outdoor thermal comfort of shaded spaces in an urban park in the cold region of China. Build Environ 155:408–420. https://doi.org/10.1016/j.buildenv.2019.03.049

Xu T, Yao R, Du C, Huang X (2022) A method of predicting the dynamic thermal sensation under varying outdoor heat stress conditions in summer. Build Environ 223:109454. https://doi.org/10.1016/j.buildenv.2022.109454

Yang J, Yin P, Sun J, Wang B, Zhou M, Li M, Tong S, Meng B, Guo Y, Liu Q (2019) Heatwave and mortality in 31 major Chinese cities: definition, vulnerability and implications. Sci Total Environ 649:695–702. https://doi.org/10.1016/j.scitotenv.2018.08.332

Yao R, Steemers K (2013) Urban microclimates and simulation. In: Yao R (ed) Design and management of sustainable built environments. Springer London, London, pp 77–97

Yao R, Li B, Liu J (2009) A theoretical adaptive model of thermal comfort—Adaptive Predicted Mean Vote (aPMV). Build Environ 44:2089–2096. https://doi.org/10.1016/j.buildenv.2009.02.014

Zhang H (2003) Human thermal sensation and comfort in transient and non-uniform thermal environments. Ph.D., University of California, Berkeley

Zhang H, Arens E, Huizenga C, Han T (2010) Thermal sensation and comfort models for non-uniform and transient environments: part I: local sensation of individual body parts. Build Environ 45:380–388. https://doi.org/10.1016/j.buildenv.2009.06.018

Zhao L, Zhou X, Li L, He S, Chen R (2016) Study on outdoor thermal comfort on a campus in a subtropical urban area in summer. Sustain Cities Soc 22:164–170. https://doi.org/10.1016/j.scs.2016.02.009

# Liveable, Joyful and Soulful Cities

**Derek Clements-Croome**

> *A City should be built to give its inhabitants security and happiness.*
> *Aristotle*
> *A healthy city is one that continually creates and improves its physical and social environments and expands the community resources that enable people to mutually support each other in performing all the functions of life and developing to their maximum potential.*
> *United Nations* (UN 2018)

**Abstract** Cities are growing in size and number to accommodate the growing urban populations in spite of there being more homeworking since COVID-19. There is a basic human need for a social dynamic, and cities provide this whether people are living or working. Cities are like organisms pulsing with life bounded by economic, ecological, health and wellbeing requirements. This Chapter reviews how cities can function in a climate change respectful way but also be joyful places for people to be.

**Keywords** Liveable city · Sustainable urbanisation · Carbon emission · Wellbeing · Adaptable city

## 1 Introduction

Cities have been centres for people to come together to live and work for thousands of years. Some of the oldest cities like Jericho in Palestine and Catalhoyok in Anatolia date back some 10,000 years. Centralisation of business, marketing, food, education and culture has led to a steady growth as cities provide the opportunities for development in all these matters. By 2050 it is expected that 70–75% of the world population will live in cities with the biggest growth in China, India and parts of Africa.

D. Clements-Croome (✉)
University of Reading, Reading, UK
e-mail: d.j.clements-croome@reading.ac.uk

 
R. Yao (ed.), *Resilient Urban Environments*, Cities and Nature,
https://doi.org/10.1007/978-3-031-55482-7_3

Of course, the recent COVID-19 pandemic has brought about a rethink on how flexible working between home and offices may be achieved but for many there is no choice. It has also been confirmed via the pandemic lockdown experiences that communication in person generates a social dynamic not reproducible online. People need to meet people but not continually on screens. Online does offer flexibility to interact with people afar and can save commuting hassles and time. Another realisation is that Nature with its green space in parks, gardens and allotments however small is so important to our mental health.

Cities if not well planned, designed and maintained can be sources of pollution, high carbon emissions, high crime rates and stress. Urban sprawl tends to generate bad health and consequently increase the demands on the national health systems besides causing heavy traffic congestion, so mobility and access are slow but even more important give rise to considerable human stress and frustration.

On the other hand, cities can be modified in sustainable ways to yield positive advantages. The cooling effects of greenery in biophilic cities can offset the heat which builds up in the day in hot weather so decreasing air-conditioning needs and hence reducing $CO_2$ emissions. Cities also bring people together for leisure or business so creating a social dynamic vivacity or 'buzz' which is good for mental and social wellbeing.

Cities need reliable integrated infrastructures to make the city functional via the utility services as well as high-quality sustainable but liveable buildings within them for people to live and work. However, the transient or mobility spaces between places for work, leisure, education or living must not be forgotten. With the evolution of AI and specific digital technologies it is more likely that planning, construction, management and operation processes will be improved as they become more streamlined and more effectively inter-connected between the various phases of the processes.

There are new cities emerging but most cities exist already, and some of these are very old but need to undergo periodic restoration to adapt them to modern needs whilst respecting the many fine attributes of vernacular architecture. There are about 35–37 megacities (over 10 million people) in 2022 but this is expected to grow to about 43 by 2030 according to the UN Department of Economic and Social Affairs, and others predict 50 metacities by 2050 (by Hoornweg and Pope 2014).

## 2 Challenges for Cities

Food, energy and water demands are increasing and expected to be 20–40% higher by 2050. Cities occupy about 3% of the earth but consume about 60–80% of the energy and are responsible for about 75–80% of all carbon emissions as indicated by KPMG International (2013) and World Economic Forum (WEF) (2022).

Climate change is bringing more unpredictable weather conditions. Temperatures rise and this means overheating is an increasing problem. Extreme events like droughts, storms, floods, forest fires and earthquakes are becoming more common. Water is predicted to rise by 30–60 cm by 2100. Water conservation, amphibian homes, attaining zero waste, green walls and roofs and zero net carbon are some of the solutions offered but it really needs countries to unite in their strategies for resilient cities if worldwide measures are to be effective. All these climate-driven events fuel human stress and life itself.

Our views on what constitutes economic value must change as discussed in the book on Value(s): Building a Better World for All by Mark Carney the former Governor of the Bank of England in 2021. He stresses the need for a social value approach. In the UK we have guidance on achieving social value using a social value portal framework called TOMS—Themes, Outcomes and Measures (info@socialvalueportal.com) which describes practical ways of delivering excellence in measuring and reporting social value.

Environment, Social and Governance (ESG) must be evaluated to benefit communities and attain a high level of health and wellbeing nationally. Air pollution is an environmental factor of climate change but directly affects health and links with the need to reduce the use of fossil fuels to achieve net zero carbon and zero waste. Social aspects emphasise the need for inclusivity. Exclusivity increases social tensions. Governance is about ensuring that effective regulations are in place but also the realisation that human behaviour plays a significant role in ESG, and this requires not only good leadership by national and local authorities but also individual responsibility. Within this, processes of management and user-centric design as well as innovation are important, but this needs an open mind by all of us to consider different approaches and ideas to deal with the challenges of climate change and its effects on health and wellbeing.

Cities need to be intelligent by embracing efficient and effective connectivity to ensure human, and sustainability values are realised. They also need to be smart so this means that Sensing, Measurement, Asset Value, Reliability and Tracking are all important. Cohesive data sets can help maintain high standards by exposing trends and weak points in the systems that affect how the city functions.

## 3 Aims, Objectives and Guidance

The United Nations has issued several Sustainable Development Goals (SDGs), and SDG 11 aims to make cities and human settlements inclusive, safe, resilient and sustainable. It has 10 targets measured with 15 indicators. Its targets aim to ensure access for all to adequate, safe and affordable housing and basic services and to upgrade slums, provide access to safe, affordable, accessible and sustainable transport systems for all, improving road safety, and enhance inclusive and sustainable urbanisation and capacity for participatory, integrated and sustainable human settlement planning and management in all countries, by 2030. This goal also calls for strengthening efforts to protect and safeguard the world's cultural and natural heritage, increasing resilience and significantly reducing the number of deaths caused and the number of people affected by disasters by 2030.

**Sustainable Goal 11: To Make Cities and Communities Sustainable**

- support links to urban peri-urban and rural areas;
- increase adoption of city integrated policies;
- support low carbon resilient buildings;
- ensure access safe housing services;
- provide access to sustainable transport;
- enhance inclusive and sustainable urbanisation;
- protect and safeguard cultural natural heritage;
- reduce deaths and losses of infrastructure due to disasters;
- reduce per capita city environmental impact;
- provide accessible green public spaces.

Underlying all these targets is the overall aim to increase the quality of life for all and central to this is the health and wellbeing of people. SDG 3 is about Good Health and Wellbeing for people of all ages. There are strong links between SDGs 3 and 11.

Targets are ideals to be aimed for, and the intentions of them are good but achieving them in practice is much more difficult. Standard ISO 37106: 2018 and 2021(en) describe smart city operating models to help leaders to develop sustainable communities. There has to be a desire by all stakeholders that cities are created and maintained for people.

## 4 Defining Sustainable Liveable Cities

An intelligent liveable city is one that is:

- Accessible and socially inclusive.
- Attracts business.
- Attractive to residents and visitors.
- Has good-quality affordable housing.
- Is fair and just for its citizens.
- Provides learning opportunities for all ages.
- Is rich in cultural activities.

To this I would add that we need:

- A Feeling City which provides a diversity of stimuli that give joy, delight and affect people's spirits and soul.
- A healthy and wellbeing city.

Rogers (1998), Paskaleva (2011) and Deakin and AlWaer (2014) express in their various ways that the Sustainable Liveable City should have these basic attributes:

**A Just City**, where justice, food, shelter, education, health and hope are fairly distributed and where all people participate in government;

**A Beautiful City**, where art, architecture and landscape spark the imagination and uplift the spirits.
**A Creative City**, where open-mindedness and innovation allow lil the full potential of its human resources to blossom and permit a fast response to change.
**An Enterprising City**: successful places are good for business because they sustain local economies by encouraging economic capital; successful places have thriving communities based on social and values.
**An Innovative City**, which enables citizens to build the social capital and capacity required to become co-creators and co-producers of new and innovative services with the means to ensure that they are delivered in more effective and inclusive ways, taking full advantage of new technologies and processes. However the durable values offered by vernacular architecture must not be forgotten.
**An Inclusive City**, where social sustainability through participation of the public in local decision-making is key to increasing democracy and governance.
**An Adaptable City**: a successful city is resilient and robust; it adapts well to change and setbacks and sustains itself in various guises over generations.
**An Ecological City** is one which minimizes its ecological impact, where green space and built form are balanced and where buildings and infrastructures are not polluting, safe and resource-efficient (Beatley 2016).
**A Connected City** is a City of Easy Contact and Mobility, which protects the countryside, focuses and integrates communities within neighbourhoods and maximises proximity. Poor transportation infrastructures can cause a lot of stress and also exclusivity.
**A Cultural City** provides a diversity of activities in the arts, sports, cuisines and crafts to engage communities and individuals beyond their work.
**A Sensory City** that provides a diversity of stimuli that give delight, joy and so affect people's spirits and souls in positive ways.
**Liveability** is the sum of the aspects that add up to the quality of life of a place, including its economy, amenity, environmental sustainability, health and well-being, equity, education and learning and leadership. The way individuals measure liveability varies, for some it lies in the amount of local green space, and others the diversity of jobs, range of entertainment options, extent of the public transport or quality of schools. Broadly speaking cities should be planned to achieve the following goals:

**Economic**—such as economic revitalization and development besides offering job opportunities.
**Land use**—such as compact, mixed use development, caring for green space.
**Transportation**—such as walkability, accessibility, and transportation choices.
**Equity**—such as affordable housing, educational opportunities, inclusive health systems and mixed-income communities.
**Community development**—such as sense of place, safety, and public health.

**Quality of Life**: is similar in intention to liveability. The term quality of life is a very general one that can mean different things to different people (Forkenbrock and Weisbrod 2001) and covers a variety of domains (Hagerty et al. 2001). Broadly, quality of life refers to the general wellbeing of individuals and societies.

Maslow (1943) mapped a pyramid of needs which describes the essential factors which define wellbeing for an individual ranging over physiological and psychological factors (see Fig. 1). If these factors are satisfied then broadly speaking a person will be content, happy and more likely to be highly motivated. The city environment can affect physiological, safety and self-actualisation needs. Even a sense of belonging to a place as well as the social cohesion that city may bring about is important to the mental wellbeing of individuals.

To achieve the pyramid of needs at a community level a city needs to provide things like intelligent structures and infrastructures, social provision, amenities and basic property rights for its citizens. Transport, housing, schools, hospitals and safety are fundamental. Cities need to be just, fair, clean and accessible to all ages and cultures. Citizens too have responsibilities to look after their urban inheritance. The challenge is to bring all this together into a harmonious whole.

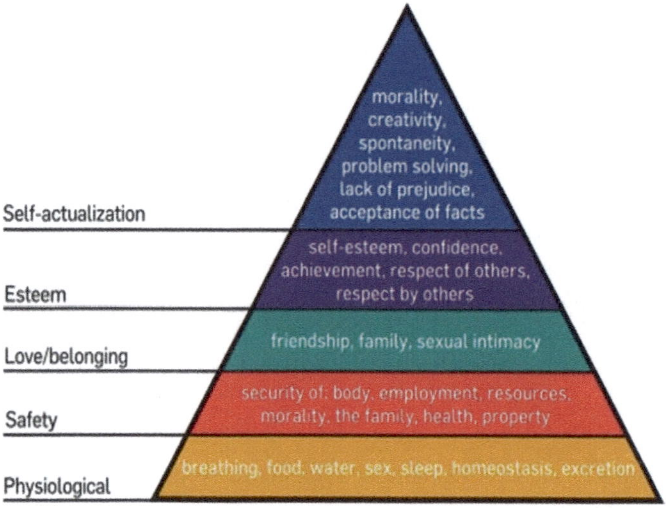


**Fig. 1** Maslow pyramid of needs (adapted by Clements-Croome 2020)

# 5 Making Cities Healthy and Wellbeing Places

People move between their home and places of work or leisure inside buildings. In focusing on the workplace one has to remember that the transit journey can affect ones wellbeing besides whether one has slept and eaten well. Our spirits and motivation can be affected by many things before we reach our workplace. All this means we must design and manage the whole system of interaction including the home, the city and the workplace in an effective way. Health and wellbeing in buildings is studied in Clements-Croome (2020) and Crisp (2021). Our focus here is towns and cities. A vibrant city can uplift our spirits but they can also be stressed by poor transport links, lack of job opportunities, too few centres of learning and culture. One thing that can soften the irritations of a city is Nature.

## *5.1 The Value of Green Space*

Many studies have demonstrated the restorative powers of green space but before this research urbanists like Ebenezer Howard had developed the garden city in the UK with examples like Letchworth and Welwyn Garden Cities. In these places housing, schools, theatres, commerce and industry were all there but spaced out with trees and parks making healthy and beautiful intercessions between them.

Green space invites walking and social conviviality. Loneliness tends to be lower, and there is some evidence that crime is reduced too. The World Economic Forum (WEF) has issued an Insight Report in 2022 titled BiodiversCities by 2030: Transforming Cities' Relationship with Nature in collaboration with Arup and AlphaBeta (WEF 2022). Forests and woodlands reduce $CO_2$ in the atmosphere as well as providing havens for plants and wild life and hence enhance biodiversity.

Nature shows us optimisation can be achieved and how social insects appear to work effortlessly in integrated teams with swarm intelligence but for humans these things are spoken about but they are not easy to achieve in practice. Most of us live in societies where money defines economic growth, and this is in conflict with the needs for sustainable development (Schumacher 1973) and ensuring health and wellbeing perspectives are taken in to account. In Nature the basic needs are the values of natural growth and sustenance, but for humans our values are viewed in very different ways by not just different cultures but also by individuals within these cultures. Too often values are sacrificed for short-term financial returns. Learning from nature and aspiring to design in a way that supports it offers us an opportunity to rethink some of our strategies in architecture but we need to approach the opportunities biomimetics and biophilia offer us with open minds. In the words of Sim Van der Ryn and Stuart Cowan

> It is time to stop designing in the image of the machine and start designing in a way that honours the complexity of life itself … we must mirror nature's deep interconnections in our own epistemology of design. (Van der Ryn and Cowan 1996 in Ellin, 2006)

Effectively integrating Nature into design can reduce the use of materials, energy and also pollutants by cleansing the air of ozone, sulphur dioxide and other volatile organic compounds. Eco-cities encourage walking and social interaction, produce shade, provide food supplies as well as recreational opportunities for all ages and significantly raise property values (Ellin 2006). Open spaces with parks, trees and water features are calming and have a direct effect on our mental and spiritual health (Chua et al. 2004; Clements-Croome 2020). Climate change means cities are becoming warmer but greenery can help to offset this Urban Heat Island effect as the leaves of plants and trees are several degrees below the surface temperatures of brick and concrete surfaces in hot weather.

WEF (2022) reckon that Nature-based Solutions (NbS) are on average 50% more costly but add 28% more added value than other approaches but spending on NbS is only about 0–3% of overall urban infrastructure in 2021. There needs to be a shift in investment towards NbS. What else can improve health and wellbeing; lower pollution; offer biogas fuel and give us new enlightened insights into systems design as much as NbS do?

The EU Horizon 2020 Programme funded the Connecting Nature Framework Guidebook 2021 to assist us in developing nature based solutions which will bring environmental, social and economic benefits and help to build resilience (ISBN 978-1-916145 https://connectingnature.eu/innovations/connecting-nature-framework).

## 5.2 *Mobility*

People want to travel to places but passing through the interconnecting spaces between them can be frustrating, time-consuming and stressful. Better transport is an almost universal demand by people in new and old cities. Walkable cities with lanes dedicated to cycling are more friendly, less polluting and healthier but this needs to be partnered by an effective public transport system that is affordable for all. We also need to remember the less mobile people like the elderly, the disabled and the infirm.

These means are easier to achieve in smaller cities. Historic London is known as a collection of villages, and you can walk and enjoy these; they are connected to each other by the underground which opened in about 1850 and bus networks. Before the pandemic there were up to 12 million passengers a day in London carried by the underground train and bus systems. The tram system feeding the radially planned city of Vienna is much envied too. However, maintenance, safety and upgrading of transport systems are a major challenge for historic cities which have become commercially successful. In the countryside transport is often lacking and cars become more necessary.

The outstanding example of public transport often quoted is the TransMilenio bus rapid transit system (BRT), in Bogata since 2000, which has reduced pollution, car volumes and accidents drastically although car ownership is quite low compared with the US. Bogota also introduced cycle routes, pedestrian-friendly measures and car-free days.

With the drive to reduce pollution from fossil fuels electric transport is being developed and beyond that there is the use of hydrogen for buses and trains being tested. One example is the hydrogen train project led by Professor John Irvine being tested at the School of Chemistry at St Andrews University in Scotland which will provide clean green transport. Hydrogen homes are also being tested in Levenmouth Scotland (H100 Fife Project) as part of net zero carbon programmes and the need to reduce pollution.

There is a role for driverless cars as well as transport on demand systems. The Institution of Civil Engineers in London has published a book in 2021 Cities for Driverless Vehicles: Planning for future built environment with shared mobility edited by Xu Zhang (ISBN 9780727764522) which explores the balance needed between passenger needs, sustainability objectives and healthy options with less cars.

Traffic congestion is stressful and is also expensive. EIU (2010) quotes congestion costs in a 2006 study as US$ 31 billion per year for New York. Dirks and Keeling (2009) estimate these costs worldwide as typically 1–3% GDP. In 2019 the UK economy lost £6.9 bn due to congestion equating to £894 per head. The Texas Transportation Institute estimated that in 2021 this cost was $78 bn for the US. This does not include the effects of mental frustration on people's health which impact their lives at work and at home which when added to the effects of pollution on health mean radical measures need to be taken to reduce the harmful impacts on health of our transport systems.

## 6 Adaptable Cities

Cities are long term and need to be adaptable to deal with the continual change in technology as well as society itself. The adaptability of places lies in their capacity to respond to forces of change in such a manner that the place remains positive in the eyes of the community who live in and use that place. Resilience in this sense requires viewing a place as a system of inter-connected parts, social, economic and environmental, which function in a synergistic fashion. Each part must be adequately robust in scale and quality so as not to be liable to quickly fail. Aligned together the parts must be complementary to each other so as to support the gradual adaptation of the place as a whole, to accommodate the pressures and effects of change. In this way the system is capable of cooperative-organising to better optimise the place to changed conditions and changing their composition to fit the changing patterns encountered (Lucas 2004).

The pre-requisites of such resilient adaptability are: that there is a belief shared by stakeholders in the value of the place; there is communication about and participation in the responses to changing conditions; and there is reliable, up-to-date information and analysis of place performance and of possible changes to allow the processes of place development to be intelligently informed. An adaptable city should be assessed by the success of our places on the basis of an area's ability to be 'change ready' and adaptable and capable of taking advantage of opportunities (McInroy and Longlands 2010).

## 7 Systems Approach to Planning and Designing Cities

Cities are a network of interactive systems. For planning and design one needs an overarching system that brings the various parts into a whole vision. Various conceptual frameworks are emerging that attempt to do this: Doughnut economics by Raworth (2017); the Flourish Model for health and wellbeing (Clements-Croome 2020); health and wellbeing in cities (see Fig. 2 and Barton and Grant 2006; Gatzweiler et al. 2017); and unhealthy neighbourhood syndrome (see Fig. 2 in AlWaer et al. 2021). All these frameworks aim to consider the environmental, the economic and the social value of design and planning in particular situations, and a balanced view of all factors is taken into account.

The THRIVES framework raises the importance of environmental degradation and structural barriers to health in urban policy and development. THRIVES stands for **T**owards **H**ealthy u**R**banism: **I**nclusi**V**e **E**quitable **S**ustainable and was developed by Pineo (2022).

THRIVES refocuses attention towards:

- awareness that health impacts often occur far away from new developments and sometimes many years after construction, so then requiring design teams to rethink of the impact on the spaces around the property;

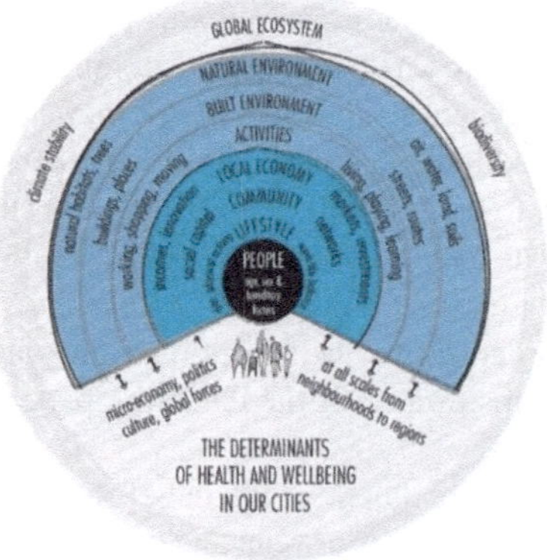


Health Map (Source: Barton and Grant 2006, redrawn By AlWaer, et al., 2021) .

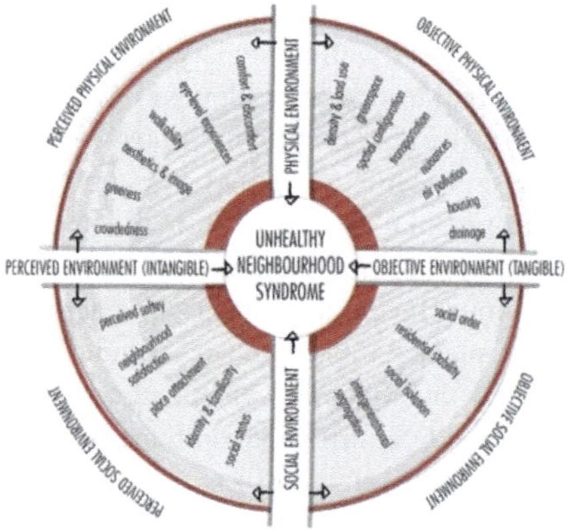


Conceptual Framework for understanding what contributes to Unhealthy Neighbourhood Syndrome (Source: AlWaer, et al., 2021, Adapted from Hajrasouliha et al. 2018).

**Fig. 2** Determinants of health and wellbeing in our cities and conceptual framework for understanding the factors that can make neighbourhoods unhealthy (adapted by AlWaer et al. 2021). *Source* Permission from Dr. Husam Alwaer

- structural barriers to health that are caused by urban design processes with outcomes that exclude groups in society, including: low income families, the disabled, elderly and ethnic minorities;
- the urgency of environmental degradation that is caused by urban environments, affecting health like extreme weather conditions (flooding, drought, fires, earthquakes, tsunamis) and air pollution from traffic and construction sites.

Community representatives and practitioners working in urban policy, planning, design, property development and public health can use THRIVES to reach a shared and collective understanding and so inform decisions (see www.healthyurbanism.net).

A recent UK Report by The Commission on Creating Healthy Cities concludes that we have to ensure old or new cities provide an environment in which cycling and walking are encouraged; there is a vibrant social environment and the infrastructure providing support services and mobility is effective (written by the Global Centre on Health and Urbanisation at Kellogg College Oxford University in collaboration with the Princes Foundation by authors Richards GC. Carpenter J. Heneghan C. Howard D. Healthy Cities Toolkit. 2022. https://www.healthycitiescommission.org/). The Report stresses the need for more affordable quality housing (Crisp 2020, 2021) Health is made at home, hospitals are for repair: Building a healthy and health creating society, Billericay with Salus Global Knowledge Exchange); planning cities for health and wellbeing (Pineo 2022); placemaking for healthy living; more public transport and less cars; less pollution and waste; effective energy and water strategies to meet climate change targets; access to more locally healthy food; and generous areas of accessible green space.

The WHO European Centre for Environment and Health in Bonn, Germany, was established in 1989. Their report on urban planning for resilience and health: key messages (14 June 2022 WHO/EURO:2022-5650-45415-64990) presents key messages on how to build forward better and apply environmental and infrastructural planning as an important pathway towards building urban resilience. It not only offers solutions in spatial planning using various tools, indicators and frameworks but also discusses the organisational processes necessary to achieve these objectives. In this way there will be a better quality of life for citizens and give an agenda for developing ecological cities.

## 8 Health and Wellbeing Cities

There are a number of practical things we can do to increase the probability of citizens experiencing an environment which is healthy and conducive to happiness.

- Ensure there is plenty of green space (2016; Gehl 2010).
- Space around houses and commercial buildings to be landscaped. Tree lined roadways as well as grass verges give pleasure and uplift one’s mood.

- Provide city feedback hubs for citizens to inform authorities about problems and let people work with professionals on the design, renovation and management of their city.
- Ensure easy to use and effective communication services for all especially regarding internet, police, fire and medical services. Decrease bureaucracy but increase effectiveness of services.
- Provide walking and cycling preferred ways of mobility but cater for the elderly and disabled people too. This approach to be complemented with affordable and frequent 24 h public transport system.
- Make cities attractive places of diversity for learning, cultural activities, job opportunities and a place where people are happy to live, work or visit.
- Cities should have a fair inclusive justice system.
- Cities should be welcoming for law abiding people of all races, genders and beliefs and viewed as places where humanity abounds.
- Cities are planned and operated to the changes necessary to comply with climate change issues regarding energy, waste, water and pollution.
- Cities must be resilient to deal with the ravages of flooding, fires, droughts, freak weather conditions and rising temperatures.
- Incorporate the social and spatial aspects of built environments into the decision-making of policy and planning (Zhong et al. 2022).

## 9 Conclusions

Cities require a systems approach for planning, design and management. It is also essential to set a vision for all the stakeholders so they feel a passion for the city project they are developing for the benefit of society. Reducing consumptions of energy and water, pollution and waste is essential but also cities must be accessible for everyone so amenities must be affordable. Continual feedback by monitoring enables conditions to be reviewed and improved.

This Chapter focuses on health and wellbeing because these are the roots of human energy needed for people to be active and feel happier. In summary liveable cities should not only have long-term resilience to counter rising temperatures, flooding, droughts, fires and disasters all of which cause severe mental stress and social unrest but rather should be joyful places that uplift the spirit and soul.

Architecture should emphasise the need for providing flourishing environments inside buildings and so achieve low infection risk, healthy and happier spaces for people to live and work in.

Let us remember what Churchill said:

> The further backward you look then the further forwards you can see.
>
> Quoted by the late Queen Elizabeth II at her Platinum Jubilee in June 2022.

## References

AlWaer H et al (2021) Unhealthy neighbourhood syndrome: a useful label for analysing and providing advice on urban design decision-making. Sustainability 13:6232. https://doi.org/10.3390/su13116232
Barton H, Grant M (2006) A health map for the local human habitat. J R Soc Promotion Health 126(6):252–253. ISSN 1466-4240 developed from the model by Dahlgren and Whitehead, 1991
Beatley T (2016) Handbook of biophilic city planning and design. Island Press
Carney M (2021) Value(s): building a better world for all. Harper Collins
Chua A et al (2004) The impact on mental well-being of the urban and physical environment: an assessment of the evidence. J Mental Health Promotion 2:8–17
Clements-Croome DJ (2020) Designing buildings for people. Crowood Press
Crisp N (2020) Health is made at home, hospitals are for repair: building a healthy and health creating society. Billericay with Salus Global Knowledge Exchange
Crisp N (2021) Health is made at home; hospitals are for repair
Deakin M, AlWaer H (2014) From intelligent to smart cities. Routledge, London
Dirks S, Keeling M (2009) A vision of smarter cities: how cities can lead the way into a prosperous and sustainable future. IBM Institute for Business Value, New York
EIU REPORTS (2010) Liveable Cities, the economist intelligence unit, also report liveanomics, commissioned by Philips: 2011
Ellin N (2006) Integral urbanism. Routledge, London
Forkenbrock DJ, Weisbrod GE (2001) NCHRP REPORT 456: guidebook assessing the social and economic effects of transportation projects. National Academy Press, Washington DC
Gatzweiler FW et al (2017) Lessons from complexity science for urban health and wellbeing. Cities Health 1(2):210–223. ISSN 237488349
Gehl J (2010) Cities for people. Island Press
Hagerty MR, Cummins RA, Ferriss AL, Land K, Michalos AC, Peterson M et al (2001) Quality of life indexes for national policy: review and agenda for research. Soc Indicators Res J 55:1–96
Hajrasoulia A et al (2018) Urban form and mental wellbeing: scoping a theoretical framework for action. J Urban Des Mental Health 5(10). ISSN 2398-4082
Hoornweg D, Pope K (2014) Ontario Institute of Technology Working Paper 4 by socioeco nomic pathways and regional distribution of the world's 101 largest cities
KPMG International (2013) Future state 2030: the global megatrends shaping governments, includes 9 global megatrends as analysed by Mowat Centre in Canada
Lucas C (2004) Complex adaptive systems—webs of delight, Version 4.83, May. Available at: http://www.calresco.org/lucas/cas.htm
Maslow A (1943) A theory of human motivation. Psychol Rev 50:370–396
Mcinroy N, Longlands S (2010) Productive local economies: creating resilient places. Centre for Local Economic Strategies, Manchester
Paskaleva KA (2011) The smart city: a nexus for open innovation? J Intell Build Int Intell Smart Cities 3(3):153–171
Pineo H (2022) Healthy urbanism: designing and planning equitable, sustainable and inclusive places. Palgrave Macmillan
Raworth K (2017) Doughnut economics: seven ways to think like a 21st century economist. Random House Business
Richards GC, Carpenter J, Heneghan C, Howard D (2022) Healthy cities toolkit. https://www.healthycitiescommission.org/
Rogers R (1998) Cities for a small planet. Faber and Faber: Icon Editions
Schumacher EF (1973) Small is beautiful; a study of economics as if people mattered. Blond and Briggs, London
UN (2018) Report on world urbanisation prospects. Department of Economic and Social Affairs, New York
Van der Ryn S, Cowan S (1996) Ecological design. Island Press, Washington DC

WEF (2022) Biodiversre cities by 2030: transforming cities relationship with nature, Insight Report, January 2022

Zhong J et al (2022) Role of built environments on physical activity and health promotion: a review and policy insights. Front Public Health. https://doi.org/10.3389/fpubh.2022.950348

# Mortality Risks and Burden of Heatwaves and High Temperature

**Jian Lei, Lu Zhou, and Haidong Kan**

**Abstract** Heatwaves and high temperatures are associated with increased mortality risks. Enhanced by the changing climate, the mortality risks and burden may keep rising in the future. With a nationwide disease surveillance point system in 272 cities in China, this study estimated city-specific and nationwide mortality risks of heatwaves and high temperatures using distributed lag nonlinear models and predicted the mortality burden in the future with global climate models from Coupled Model Intercomparison Project phase 6 (CMIP6). This study examined the population's vulnerability to heatwaves and high temperatures, including demographic and climatic modifications, highlighting the necessity for adaptation measures against heat stress to contribute to the reduction of the heat-related mortality burden.

**Keywords** Heatwave · High temperature · Mortality risk · Mortality burden · China

## 1 Background

### *1.1 Climate Change*

Global climate change may increase the frequency and intensity of extreme weather, especially the occurrence of heatwaves and high temperatures, which can directly affect human health by exposure to non-optimal ambient temperature (Gasparrini et al. 2017). In the context of global climate change, accumulating epidemiological evidence indicates that abnormal ambient temperatures could increase the risk of death from a wide range of cardiorespiratory diseases (Moghadamnia et al. 2017; Song et al. 2017). With accelerating climate change and population aging, the effects

J. Lei · L. Zhou · H. Kan (✉)
Key Lab of Public Health Safety of the Ministry of Education, NHC Key Lab of Health Technology Assessment, IRDR ICoE on Risk Interconnectivity and Governance on Weather/Climate Extremes Impact and Public Health, Department of Environmental Health, School of Public Health, Fudan University, Shanghai, China
e-mail: kanh@fudan.edu.cn

 
R. Yao (ed.), *Resilient Urban Environments*, Cities and Nature,
https://doi.org/10.1007/978-3-031-55482-7_4

of future heatwaves on human health in China are likely to increase continuously even under a low-emission scenario (Chen et al. 2022). A study warned that the increase in surface air temperature in China has been faster than the global rate, and intensive high-temperature and heatwave events are expected to occur in the future (Wang et al. 2019).

## 1.2 High Temperature

Climate change poses widespread and frequently aggressive threats to public health, and one primary impact is the mortality burden of high temperatures (Field et al. 2014). In several multicenter studies, researchers evaluated the exposure–response associations between temperature and total mortality and the estimated mortality burden of moderate and extreme non-optimum temperatures (Gasparrini et al. 2015; Guo et al. 2014). The impact of high temperatures has been well-documented (Chen et al. 2018; Dear and Wang 2015). Enhanced by the changing climate, the mortality risks and burden may keep rising in the future. It is estimated that additional warming from 1.5 to 2 °C will lead to more than 27.9 thousand additional heat-related deaths annually when considering all 831 million urban inhabitants in China (Wang et al. 2019). A recent study across Japan, South Korea, and China provided robust evidence for significant mortality risks and burden in association with night-time warming, which suggest a growing role of night-time warming in heat-related health effects in a changing climate (He et al. 2022). A better knowledge of how non-optimum temperatures affect population health is important not only to the scientific community but also to public health (Huang et al. 2012).

## 1.3 Heatwave

A heatwave is a duration of two or more consecutive days of hot weather (Yin et al. 2018) and is deemed to be one of the most lethal types of weather phenomenon. Although a heatwave is a pervasive natural hazard, it has become an increasingly severe public health issue in the context of accelerated global climate warming (Meehl and Tebaldi 2004). A study predicted that heatwaves will become more intense, more frequent, and longer lasting in the second half of the 21st century, eventually causing more pervasive health and socioeconomic consequences in the future (Field et al. 2014; Meehl and Tebaldi 2004). Although a number of epidemiological studies have evaluated the health impacts of heatwaves, the results were largely inconsistent and incomparable (Arbuthnott and Hajat 2017; Xu et al. 2016). Firstly, some studies in the USA, Europe, and developed Asian countries reported that heatwaves had added effects (i.e., additional risks due to sustained heat after adjusting for high temperature) on mortality (Anderson and Bell 2011; Hajat et al. 2006; Lee et al. 2016). Under the changing climate, the evaluations of heatwaves and high temperatures are significant for reducing the heat-related mortality burden and protecting vulnerable populations.

# 2 Heatwave and High Temperature in 272 Cities in China

## *2.1 The Data Source*

Daily mortality counts for each city were collected from China's Disease Surveillance Points System (DSPS). The DSPS monitored cause-specific mortality under strict quality control measures and showed good national and provincial representativeness (Liu et al. 2016). In brief, the DSPS included 272 cities in China and dispersed over all 31 provincial administrative regions accounting for > 80% of all cities at the prefecture level and above in China. The 272 cities were located in five climatic zones: the temperate monsoon zone (116 cities) is mainly characterized by high temperatures and rainy summers and cold and dry winters.

The subtropical monsoon zone (140 cities) mainly presents high temperatures and more rain in the summer, with mild temperatures and less rain in the winter.

The temperate continental zone (six cities) typically has a dry climate and scarce rainfall.

The alpine zone (five cities) covers the Qinghai-Tibet plateau region, represented by plenty of snow in the winter half year and a mild cool climate in the summer half year.

And the tropical monsoon zone (five cities) is characterized by high temperatures and rain throughout the year.

The causes of deaths evaluated in the present study were total non-accidental causes (briefed as "total", coded by the International Classification of Disease, 10th Revision: A00–R99).

The modeled daily mean temperature series from different climate-change scenarios were obtained to predict the future effects of non-optimal temperatures. Ten global climate models (GCMs) were chosen from the latest internationally coordinated Coupled Model Intercomparison Project Phase 6 (CMIP6, https://esgf-node.llnl.gov/search/cmip6/) (Eyring et al. 2015). We selected four climate-change scenarios, which considered socioeconomic development as updates of Representative Concentration Pathways (RCPs), making hypothetical future scenarios more reasonable and reliable (Hurtt et al. 2020). Specifically, the scenario with strict carbon emission reduction was selected, which is an update of the representative concentration pathway 1.9 based on shared socioeconomic pathways 1 (SSP119) and an update of the representative concentration pathway 2.6 based on shared socioeconomic pathways 1 (SSP126), and the scenario with certain carbon reduction, which is an update of the representative concentration pathway 4.5 based on SSP2 (SSP245) (O'Neill et al. 2016). In addition, the SSP585 scenario (with an assumption of less greenhouse-gas emission control) as the worst case was also evaluated, and mortality burden under the SSP585 scenario was compated with that under the SSP126 and SSP245 scenarios within the same period (2010–2099). Notably, SSP1–RCP1.9 scenario is a newly developed scenario combination, intending to achieve the Paris Agreement 1.5 °C warming goal (Hurtt et al. 2020; O'Neill et al. 2016).

The modeled daily temperature series were downscaled through bi-linear interpolation at a 1.0° × 1.0° spatial resolution and were linear interpolated (Gasparrini et al. 2017). The temperature series of each city from 2015 to 2099 were extracted by linking the city-specific coordinates with the corresponding grid cell. Notably, deviations might be inevitable when the modeled temperature series were applied to the temperature–mortality association evaluated from the observed temperature. Therefore, the modeled series were bias-corrected using the city-specific observations for all the GCMs and SSP-RCPs, respectively, producing a total of 10 × 4 datasets. In brief, the correcting factors obtained from the bias correction of the observed series and contemporaneous modeled series were then applied to the modeled series from 2015 to 2099 (Gasparrini et al. 2017). Generally, the modeled temperature series could be reproduced with minor model-observation differences, and the involved GCMs could well project daily temperatures in the future. A two-step analytic framework was used to statistically downscale and interpolate the temperature outputs from ten GCMs under these scenarios.

## 2.2 *High Temperature in China*

The geographical locations of 272 main Chinese cities and the warm-season (i.e., from May to September) mean temperature in 2013–2015 are displayed in Fig. 1. The descriptive statistics on non-accidental deaths and weather conditions in 272 main Chinese cities during the warm season from 2013 to 2015 are shown in Table 1. The mean temperature during the study period was 23.8 ± 3.6 °C and the relative humidity was 71.3 ± 12.6%. The annual heatwave days during the study were 16 ± 1.

Two-stage statistical analyses were used to fit the exposure–response associations and lag patterns between high temperature and total mortality in different climatic zones. Firstly, the association between temperature and mortality in each city was estimated by using a standard time-series regression model (Bhaskaran et al. 2013). To account for the nonlinear and delayed effects of temperature on total mortality, this stage analysis applied an over-dispersed generalized linear model in combination with a distributed lag nonlinear model (DLNM) (Gasparrini et al. 2015). Consistent with previous studies, the main model included a natural cubic B-spline of calendar day with 6 degrees of freedom per year to control for seasonality and long-term trend in mortality, a natural cubic B-spline of the present-day relative humidity with three degrees of freedom, and a categorical variable of the day of the week (Ma et al. 2014). The cross-basis function of daily temperature was built by the DLNM, including a

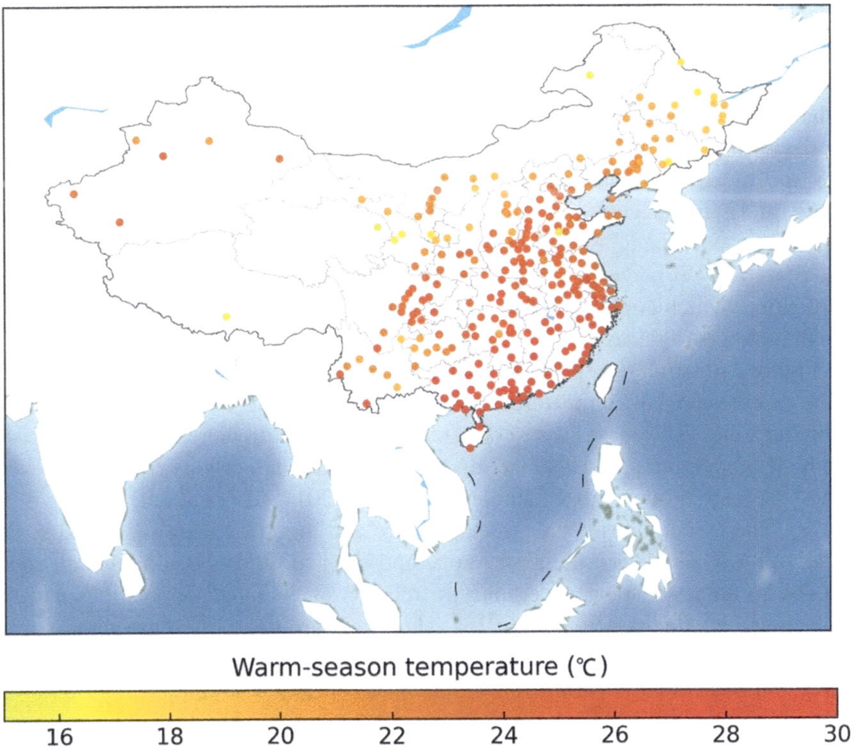


**Fig. 1** Maps of 272 main Chinese cities and warm-season mean temperature in 2013–2015

**Table 1** Descriptive statistics on daily non-accidental deaths and weather conditions in 272 main Chinese cities during the warm season, 2013–2015

| Variable | Mean | SD | Min | Median | Max |
|---|---|---|---|---|---|
| *Daily non-accidental deaths* | | | | | |
| Total | 15 | 5 | 2 | 11 | 156 |
| Male | 9 | 3 | 1 | 6 | 93 |
| Female | 6 | 3 | 1 | 5 | 63 |
| 5–64 years | 4 | 2 | 1 | 3 | 42 |
| 65–74 years | 3 | 2 | 0 | 3 | 36 |
| $\geq$ 75 years | 7 | 3 | 0 | 5 | 77 |
| $\leq$ 9 years of education | 14 | 5 | 2 | 10 | 149 |
| > 9 years of education | 1 | 1 | 0 | 1 | 11 |
| *Weather conditions* | | | | | |
| Temperature (°C) | 23.8 | 3.6 | 14.7 | 24.6 | 28.7 |
| Relative humidity (%) | 71.3 | 12.6 | 31.4 | 73.7 | 93.3 |
| Annual heatwave days | 16 | 1 | 11 | 16 | 19 |

*Note SD* standard deviation

quadratic B-spline with three internal knots placed at the 10th, 75th, and 90th centiles of city-specific temperature distributions, and a lag-response curve with a natural cubic B-spline with an intercept and three internal knots placed at equally spaced values in the log scale, with the maximum lag up to 10 days (Gasparrini et al. 2015; Guo et al. 2014). In the second stage, the best linear unbiased prediction (BLUP) of the city-specific association between temperature and mortality at both regional (climatic zones) and national levels was obtained using a recently developed multivariate meta-regression approach (Gasparrini and Armstrong 2013). The BLUP approach is made use of a trade-off between the city-specific association and the second stage pooled estimation, which could provide more precise estimations, especially in cities with small numbers of deaths.

Figure 2 illustrates the pooled cumulative exposure–response curves for the associations between daily temperature and total mortality in 272 main Chinese cities by climatic zones. The curves were consistently J shaped with increased mortality risks for high temperatures, and higher temperatures had larger mortality risks, which were also observed in other related studies (Guo et al. 2014; Ma et al. 2014; Song et al. 2017). The associations between temperature and mortality were weak and statistically non-significant in the alpine zone and tropical monsoon zone, which could be due to high statistical uncertainty in relation to the small populations and the few cities included (Chen et al. 2018). The minimum mortality temperatures were different across climatic zones (i.e., temperate monsoon zone: 23.5 °C, subtropical monsoon zone: 26.5 °C, temperate continental zone: 22.0 °C, alpine zone: 19.8 °C, tropical monsoon: 27.6 °C). It was lower in temperate continental zone and temperate monsoon zones, suggesting the potential adaptation ability across different climatic regions (Chen et al. 2018). This study provided robust evidence for the total mortality risks of high temperatures.

For the lag pattern of high temperature and total mortality, this study found that the effects of high temperatures appeared immediately, persisted for two or three days, and followed by a mortality displacement, which was consistent with previous studies (Anderson and Bell 2009; Guo et al. 2011). Most related studies also indicated that the effect of high temperature is instant and does not last for too long (Chen et al. 2018; Yin et al. 2018). Besides, the lag patterns of high temperature across different climatic regions were generally similar (Fig. 3).

## 2.3 *Heatwave in China*

The daily mean temperature was the main variable to define a heatwave according to previous studies (Guo et al. 2017b). Due to the lack of a universally accepted definition of heatwave, previous studies have proposed several definitions by using different heat thresholds (absolute or relative) and durations (different numbers of consecutive days) (Chen et al. 2015; Kent et al. 2014). The present analysis adopted city-specific relative thresholds (90th, 92.5th, 95th, and 97.5th percentiles) based on the statistical distributions of daily mean temperature during the study period,

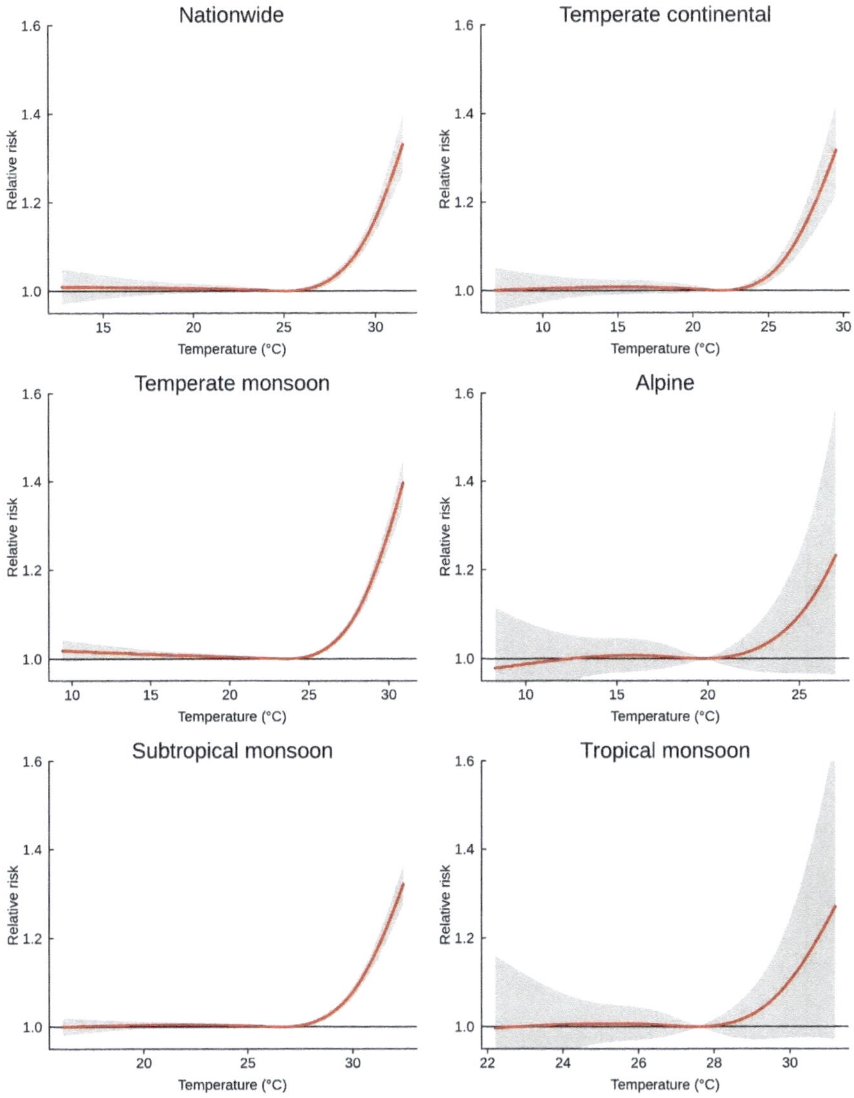


**Fig. 2** Exposure–response curves of total mortality associated with warm-season temperature over lag 0–10 days in different climatic zones

which allows for long-term local acclimatization (Guo et al. 2017b). Simultaneously, different durations of heatwaves were considered: two or more consecutive days, three or more consecutive days, and four or more consecutive days. Twelve heatwave definitions were used to comprise the above four thresholds and three durations

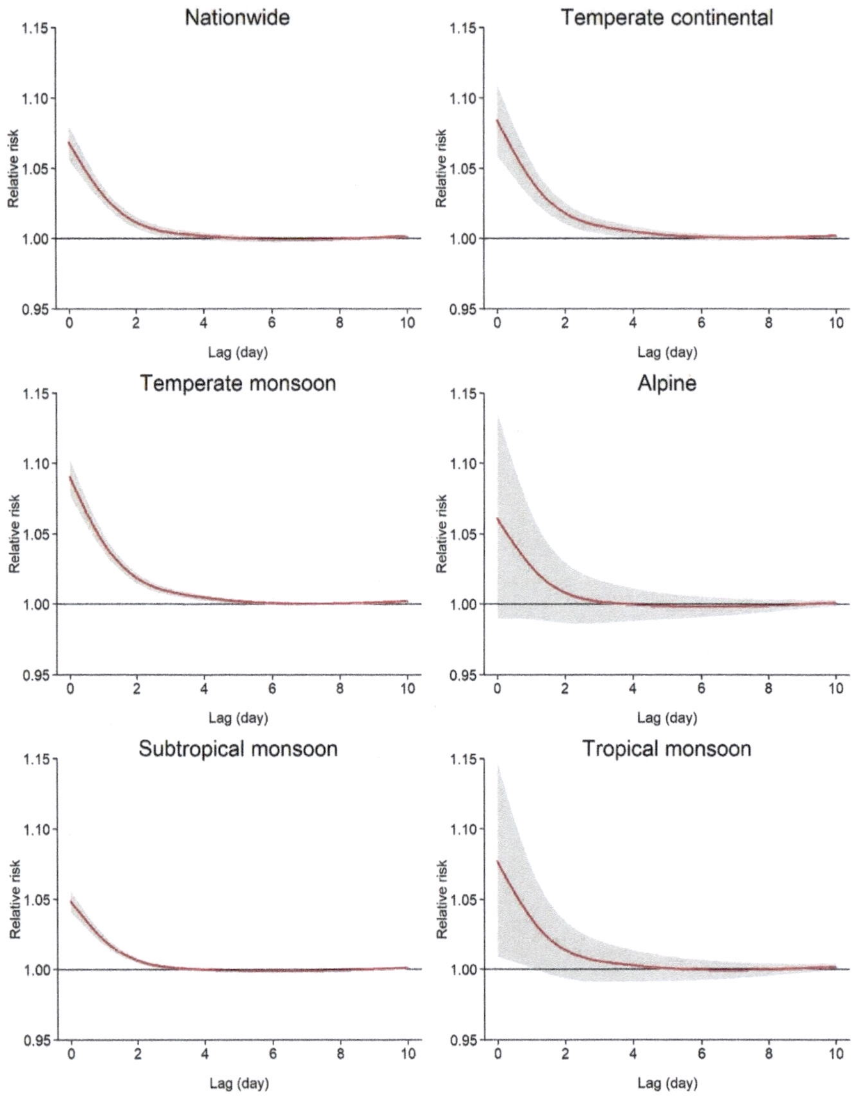


**Fig. 3** Pooled lag structures for the relative risks of total mortality associated with warm-season temperature over lag 0–10 days in different climatic zones

(Table 2). The days during the study period were categorized into heatwave or non-heatwave days according to the definitions of heatwave to allow for calculating a relative risk (RR) of total mortality.

As shown in Table 2, the annual-mean number of heatwave days for each city decreased substantially with more stringent definitions, while the number differed

**Table 2** Summary descriptive statistics on heatwave definitions and heatwave days in 272 Chinese cities, 2013–2015

| Name | Definition | Mean number of heatwave days per year | | | | |
|---|---|---|---|---|---|---|
| | | Mean | SD | Min | Median | Max |
| P90_ ≥ 2 d | > P90 with ≥ 2-day duration | 34 | 2 | 23 | 34 | 37 |
| P90_ ≥ 3 d | > P90 with ≥ 3-day duration | 29 | 3 | 20 | 30 | 35 |
| P90_ ≥ 4 d | > P90 with ≥ 4-day duration | 25 | 4 | 13 | 26 | 32 |
| P92.5_ ≥ 2 d | > P92.5 with ≥ 2-day duration | 25 | 2 | 17 | 25 | 29 |
| P92.5_ ≥ 3 d | > P92.5 with ≥ 3-day duration | 21 | 3 | 14 | 22 | 27 |
| P92.5_ ≥ 4 d | > P92.5 with ≥ 4-day duration | 18 | 4 | 9 | 18 | 25 |
| P95_ ≥ 2 d | > P95 with ≥ 2-day duration | 16 | 1 | 11 | 16 | 19 |
| P95_ ≥ 3 d | > P95 with ≥ 3-day duration | 13 | 2 | 6 | 14 | 18 |
| P95_ ≥ 4 d | > P95 with ≥ 4-day duration | 11 | 3 | 3 | 11 | 17 |
| P97.5_ ≥ 2 d | > P97.5 with ≥ 2-day duration | 8 | 1 | 5 | 8 | 10 |
| P97.5_ ≥ 3 d | > P97.5 with ≥ 3-day duration | 6 | 2 | 2 | 6 | 9 |
| P97.5_ ≥ 4 d | > P97.5 with ≥ 4-day duration | 4 | 2 | 0 | 4 | 8 |

slightly within each type of heatwave definition. This statistical analysis was conducted for heatwave definition of > P95 with ≥ 2-day duration.

A two-stage analytical framework was used to examine the heatwave–mortality associations, in which city-specific associations were estimated in the first stage and then regional- or national-average associations were obtained in the second stage. The statistical analysis was restricted to the warm season. In the first stage, the estimated RRs of cause-specific mortality were calculated by comparing the heatwave days to non-heatwave days in each city, using a time-series approach. Specifically, the associations between heatwaves and mortality were fit by a generalized additive model, with a quasi-Poisson link function to account for the overdispersion in the distribution of daily death counts. The generalized additive model was then linked with a distributed lag model to capture the lag effects of heatwaves (Chen et al. 2015; Lee et al. 2016).

A maximum lag of 10 days was determined according to previous studies that showed an immediate effect and possibly followed by a mortality displacement (Guo et al. 2014, 2017b). The long lag period was also supported by a previous study on the added health impacts of heatwaves (Saha et al. 2014). The lag-response curve was modeled with a natural cubic B-spline with four degrees of freedom (df) and two internal knots (plus an intercept) placed at equally spaced values in the log scale (Guo et al. 2017b). To allow for investigating the added effects of heatwaves, the basic time-series model introduced daily mean temperature using a distributed lag nonlinear model with a 6-df natural cubic B-spline for temperature effects and the same lag space as heatwaves. Other covariates in the basic model included: (1) a natural cubic B-spline of calendar day with 4 df per year to control for seasonal and long-term trends; (2) a natural cubic B-spline of the current-day relative humidity to adjust for its potentially nonlinear confounding effects; and (3) an indicator of the day of the week (Anderson and Bell 2011; Guo et al. 2017b). Then, the city-specific heatwave–mortality associations were reduced to the overall association that was cumulated throughout the lag period by reducing the number of parameters to be pooled in the meta-analysis (Gasparrini and Armstrong 2013). In the second stage, the pooled cumulative RRs were calculated using a univariate meta-regression model for each type of heatwaves and each cause of death at the national level or regional (climatic zones) levels (Gasparrini and Armstrong 2013).

Consistent with several studies of mortality and morbidity in Asia and the USA (Chen et al. 2015; Lee et al. 2016; Ma et al. 2015), this study found significant added effects of heatwaves on mortality. The results indicated that the effect of heatwave occurred at almost lag 1 day and could last for 4–5 days, which were also supported by several studies (Guo et al. 2017b; Hajat et al. 2006). The immediate effects of heatwaves suggested prompt and somewhat sustained actions in developing heatwave-warning systems. The lag patterns of heatwaves in the alpine and tropical monsoon regions were weak and statistically non-significant as the small populations and the few cities included in these regions (Chen et al. 2018), while the curves maintained the same tendency as the lag pattern nationwide.

The investigations of lag patterns and the identification of mortality displacement are important to fully understand the health impacts of heatwaves and to develop preventive public health policies. Mortality displacement is a phenomenon that excess deaths occurring during or just after hot weather brought forward by only a short time. Without consideration of the displacement effects, it is likely to overestimate the added effects of heatwave (Saha et al. 2014) (Fig. 4).

## 3 Population Vulnerability

Table 3 shows the population vulnerability by stratified analyses, which indicates that the effects of high temperature and heatwave were generally more prominent among the female population. One of the important contributions of the current study is to present the high vulnerability of the elderly, females, and less-educated people to the mortality risks associated with heatwaves and high temperature, which were consistent with previous heat-related studies (Chen et al. 2015; Kim and Kim 2017; Ma et al. 2015). The susceptibility of females may be caused by their relatively low ability to heat adaptation in relation to vulnerable physical conditions and disadvantaged socioeconomic status (Kim and Kim 2017). The elderly were susceptible because of their high prevalence of chronic cardiopulmonary diseases and weakened physiological capability of heat adaptation. The vulnerability of less-educated people may reflect the well-reported phenomenon that disadvantaged socioeconomic status was often companied by limited access to health care, poor nutritional conditions, and low prevalence of air conditioning. It remains to be determined why these subgroups were susceptible.

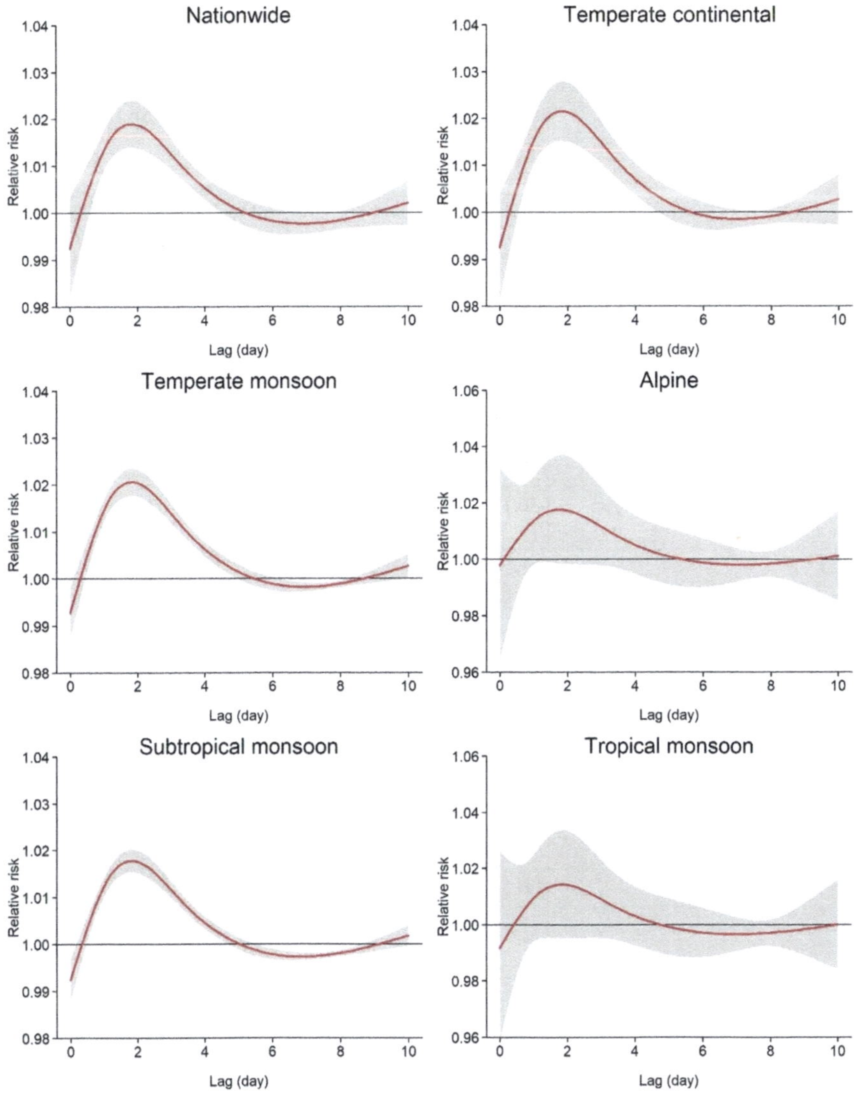


**Fig. 4** Pooled lag structures for the relative risks of total mortality associated with heatwaves over lag 0–10 days in different climatic zones

This nationwide analysis identified potential effect modification in terms of climatic characteristics, which were important to develop evidence-based health protection plans against abnormal weather or climate changes (Hajat et al. 2010). In general, the RRs in the temperate monsoon zone were similar to or higher than that in the subtropical monsoon zone, followed by the temperate continental zone. The RRs were the smallest and statistically insignificant in the alpine and tropical monsoon.

**Table 3** Relative risks of daily total deaths associated with warm-season temperatures and heatwaves in 272 main Chinese cities, classified by region, sex, age, and education level

| Variable | Warm-season temperature | | | Heatwave |
|---|---|---|---|---|
| | MMT (°C) | Extreme heat (°C) | RR (95% CI) | RR (95% CI) |
| Total | 25.0 | 30.5 | 1.213 (1.172, 1.254) | 1.022 (1.001, 1.043) |
| *Region* | | | | |
| Temperate monsoon | 23.5 | 29.6 | 1.242 (1.211, 1.275) | 1.030 (1.015, 1.045) |
| Subtropical monsoon | 26.5 | 31.7 | 1.234 (1.202, 1.268) | 1.016 (1.003, 1.029) |
| Temperate continental | 22.0 | 28.2 | 1.205 (1.141, 1.274) | 1.031 (0.998 1.064) |
| Alpine | 19.8 | 25.1 | 1.101 (0.966, 1.256) | 1.045 (0.950, 1.151) |
| Tropical monsoon | 27.6 | 30.1 | 1.114 (0.975, 1.274) | 1.019 (0.926, 1.117) |
| *Sex* | | | | |
| Male | 25.5 | 30.5 | 1.168 (1.129, 1.207) | 1.007 (0.982, 1.031) |
| Female | 25.0 | 30.5 | 1.274 (1.216, 1.334) | 1.045 (1.016, 1.076) |
| *Age* | | | | |
| 5–64 years | 26.5 | 30.5 | 1.093 (1.045, 1.144) | 0.992 (0.962, 1.024) |
| 65–74 years | 25.0 | 30.5 | 1.103 (1.054, 1.154) | 1.012 (0.979, 1.046) |
| ≥ 75 years | 24.5 | 30.5 | 1.330 (1.272, 1.391) | 1.042 (1.013, 1.073) |
| *Education* | | | | |
| ≤ 9 years | 25.0 | 30.5 | 1.216 (1.175, 1.259) | 1.021 (1.000, 1.043) |
| > 9 years | 25.0 | 30.5 | 1.106 (1.009, 1.211) | 1.027 (0.957, 1.101) |

*Note MMT* minimum mortality temperature; *RR* relative risk; *CI* confidence interval. Extreme heat denoted 99.0th percentile of warm-season temperature distributions

The findings highlighted the need for developing regional high-temperature and heatwave-warning systems in a large country like China. Also, future work would be designed to determine city- or region-specific threshold temperatures as well as significant heatwave duration, which were fundamentally important for developing local heatwave-prevention plans (Xu et al. 2016).

## 4 High Temperature and Heatwave in the Future

Figures 5, 6 and Tables 4, 5 show the predicted warm-season temperature and frequency of heatwave events under the SSP119, SSP126, SSP245, and SSP585 scenarios in different climatic zones. Generally, the warm-season temperatures in different climatic zones will keep growing in the coming future. Specifically, the nationwide warm-season temperature in the 2050s [25.0 °C (95%CI: 24.7 °C, 25.4 °C)] will increase by 1.1 °C, and the warm-season temperature in 2090s [25.1 °C (95%CI: 24.7 °C, 25.5 °C)] will increase 1.2 °C compared with the warm-season temperature in 2010s [23.9 °C (95%CI: 23.8 °C, 24.0 °C)] under the SSP126 scenario.

The heatwave days are predicted to increase in the different scenarios, especially in the SSP585 scenario (Tables 4 and 5). The nationwide annual heatwave days under SSP126 in the 2050s [31 (95%CI: 30, 33)] will increase by 15, and annual heatwave days in the 2090s [31 (95%CI: 28, 33)] will increase by 16 compared with that in the 2010s [15 (95%CI: 14, 15)]. Under the impact of climatic changes in the future, the effect of heatwaves should not be underestimated. Consistent with this prediction, abundant investigations also reported that the high temperatures and occurrences of heatwave days will keep increasing in the future (Guo et al. 2017a, 2018), which can help decision-makers in planning adaptation and mitigation strategies for climate change (Guo et al. 2018).

## 5 Mortality Burden of High Temperature and Heatwave in the Future

The attributable fractions associated with extreme heat were calculated by summing the subsets of days with relevant temperature ranges according to each city's specific percentile of temperature distributions (Gasparrini et al. 2015). Monte Carlo simulation was used to assess uncertainties in projection with 1000 samples produced from the BLUP coefficients and then generated results for each GCM (Gasparrini et al. 2017). The 95% empirical confidence intervals (eCIs) were defined as the 2.5th and 97.5th percentiles of empirical distribution across all samples and GCMs, accounting for uncertainties from exposure–response relationships and modeled temperatures.

The numbers of deaths attributable to heat during the warm season for each city were calculated by summing the subsets of modeled temperature series where ambient temperatures were higher than the city-specific MMT. For the projection, the heat-related deaths for each climate-change scenario in each city, and the attributable fraction was calculated as the proportion of heat-related deaths in the total number of deaths. Eventually, the attributable fractions were aggregated by country and region, decade, and scenario assuming no adaptation (Gasparrini et al. 2017).

Based on these findings, the mortality of high temperatures and heatwaves are likely to increase in the future, especially under the SSP245 and SSP585 scenario (Figs. 7, 8 and Tables 6, 7). For high temperature, the attributable fraction of

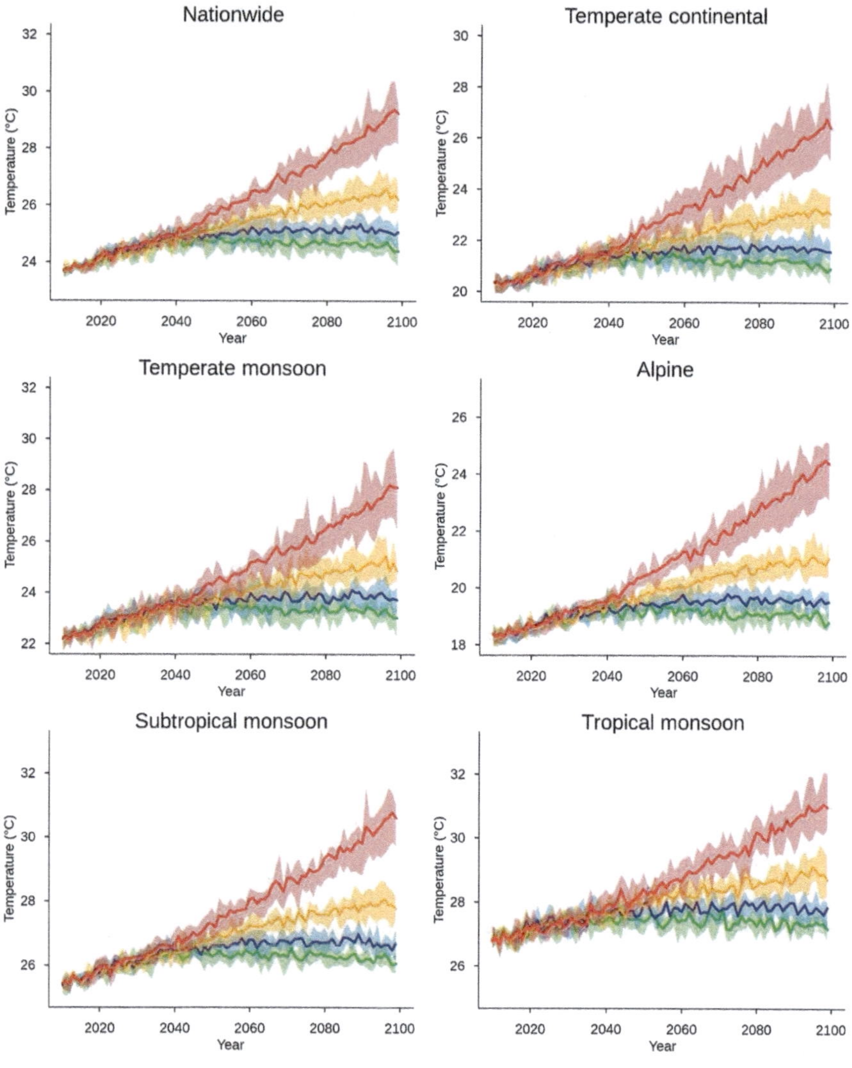


**Fig. 5** Projected warm-season temperature under the SSP119, SSP126, SSP245, and SSP585 scenarios in different climatic zones. Warm-season temperature across ten general circulation models under the SSP119, SSP126, SSP245, and SSP585 scenarios was presented in green, blue, orange, and red, respectively. SSP = shared socioeconomic pathway

high temperature under the SSP126 scenario was 1.1% (95%CI: 0.8%, 1.6%) in the 2010s. The attributable fraction of high temperature will be 2.4% (95%CI: 1.2%, 3.5%) in the 2050s and 2.7% (95%CI: 1.3%, 3.6%) in the 2090s. Generally, the mortality burden of high temperature is relatively small, but the increasing tendency should be paid necessary attention. For the mortality burden of heatwave under the SSP126 scenario, the percent changes of heatwave-related excess deaths will also

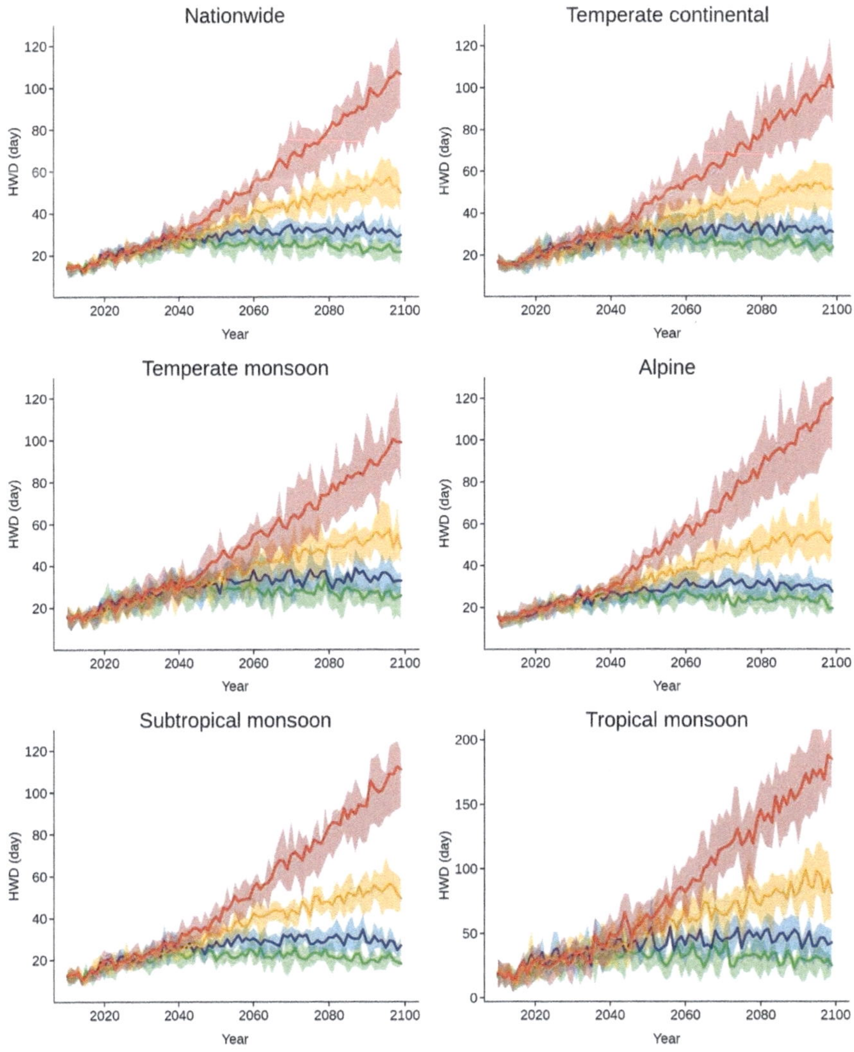


**Fig. 6** Projected annual heatwave days under the SSP119, SSP126, SSP245, and SSP585 scenarios in different climatic zones. Heatwave days across ten general circulation models under the SSP119, SSP126, SSP245, and SSP585 scenarios were presented in green, blue, orange, and red, respectively. SSP = shared socioeconomic pathway; HWD = heatwave days

increase in the future. This study highlights the significance of preventing climate-change policies to minimize increases in temperatures and mitigate the associated negative impacts on public health.

**Table 4** Projected future warm-season temperature under the SSP119, SSP126, SSP245, and SSP585 scenarios in different climatic zones

| Region | Period | SSP119 scenario | | SSP126 scenario | | SSP245 scenario | | SSP585 scenario | |
|---|---|---|---|---|---|---|---|---|---|
| | | Temperature | Δ | Temperature | Δ | Temperature | Δ | Temperature | Δ |
| Nationwide | 2010–19 | 23.8 (23.7, 23.9) | | 23.9 (23.8, 24.0) | | 23.9 (23.7, 23.9) | | 23.8 (23.7, 23.9) | |
| | 2050–59 | 24.7 (24.4, 25.0) | 0.9 | 25.0 (24.7, 25.4) | 1.1 | 25.3 (25.1, 25.4) | 1.4 | 25.9 (25.4, 26.1) | 2.1 |
| | 2090–99 | 24.5 (24.3, 24.8) | 0.7 | 25.1 (24.7, 25.5) | 1.2 | 26.4 (25.8, 26.9) | 2.5 | 28.9 (28.1, 29.6) | 5.1 |
| Temperate monsoon | 2010–19 | 22.3 (22.3, 22.4) | | 22.4 (22.3, 22.5) | | 22.4 (22.2, 22.4) | | 22.3 (22.1, 22.5) | |
| | 2050–59 | 23.4 (23.2, 23.9) | 1.1 | 23.6 (23.1, 24.2) | 1.2 | 23.9 (23.6, 24.0) | 1.5 | 24.6 (23.9, 24.6) | 2.2 |
| | 2090–99 | 23.2 (23.0, 23.5) | 0.9 | 23.8 (23.4, 24.3) | 1.4 | 25.1 (24.5, 25.9) | 2.7 | 27.7 (26.9, 28.5) | 5.4 |
| Subtropical monsoon | 2010–19 | 25.5 (25.4, 25.6) | | 25.6 (25.4, 25.7) | | 25.6 (25.5, 25.7) | | 25.6 (25.4, 25.6) | |
| | 2050–59 | 26.3 (26.1, 26.6) | 0.8 | 26.6 (26.6, 26.8) | 1.1 | 27.0 (26.8, 27.1) | 1.4 | 27.5 (27.2, 27.7) | 2 |
| | 2090–99 | 26.2 (25.9, 26.3) | 0.7 | 26.6 (26.4, 26.8) | 1.1 | 27.9 (27.4, 28.2) | 2.3 | 30.3 (29.5, 30.8) | 4.8 |
| Temperate continental | 2010–19 | 20.3 (20.3, 20.5) | | 20.4 (20.2, 20.5) | | 20.4 (20.3, 20.6) | | 20.4 (20.2, 20.7) | |
| | 2050–59 | 21.4 (21.0, 21.7) | 1.0 | 21.6 (20.9, 22.3) | 1.2 | 21.9 (21.5, 22.3) | 1.6 | 22.8 (22.2, 23.3) | 2.4 |
| | 2090–99 | 21.0 (20.9, 21.3) | 0.7 | 21.7 (21.3, 22.3) | 1.3 | 23.1 (22.6, 23.9) | 2.7 | 26.2 (25.1, 27.1) | 5.8 |
| Alpine | 2010–19 | 18.4 (18.3, 18.5) | | 18.4 (18.3, 18.5) | | 18.4 (18.3, 18.5) | | 18.4 (18.3, 18.6) | |
| | 2050–59 | 19.2 (18.9, 19.5) | 0.9 | 19.4 (19.1, 19.8) | 1 | 19.9 (19.6, 20.2) | 1.5 | 20.7 (20.4, 20.8) | 2.3 |
| | 2090–99 | 19.0 (18.8, 19.3) | 0.6 | 19.5 (19.3, 19.8) | 1.1 | 20.9 (20.6, 21.5) | 2.5 | 24.1 (22.9, 24.8) | 5.6 |
| Tropical monsoon | 2010–19 | 26.9 (26.7, 27.0) | | 26.9 (26.8, 27.1) | | 26.9 (26.8, 27.1) | | 26.9 (26.8, 27.2) | |
| | 2050–59 | 27.4 (27.3, 27.5) | 0.6 | 27.8 (27.6, 28.0) | 0.8 | 28.1 (28.0, 28.5) | 1.2 | 28.5 (28.3, 28.7) | 1.6 |
| | 2090–99 | 27.3 (27.0, 27.5) | 0.4 | 27.8 (27.5, 27.9) | 0.8 | 28.9 (28.6, 29.4) | 1.9 | 30.8 (30.3, 31.8) | 3.9 |

*Note SSP* shared socioeconomic pathway. The temperature was presented as the mean and 25th–75th range across ten general climate models. Δ denoted the projected warm-season temperature increase versus baseline period (2010–19)

**Table 5** Projected future heatwave days under the SSP119, SSP126, SSP245, and SSP585 scenarios in different climatic zones

| Region | Period | SSP119 scenario | | SSP126 scenario | | SSP245 scenario | | SSP585 scenario | |
|---|---|---|---|---|---|---|---|---|---|
| | | HWD | Δ | HWD | Δ | HWD | Δ | HWD | Δ |
| Nationwide | 2010–19 | 14 (13, 15) | | 15 (14, 15) | | 15 (14, 16) | | 15 (14, 16) | |
| | 2050–59 | 26 (23, 30) | 11 | 31 (30, 33) | 15 | 36 (33, 38) | 21 | 47 (39, 47) | 32 |
| | 2090–99 | 23 (20, 25) | 9 | 31 (28, 33) | 16 | 54 (46, 63) | 38 | 101 (83, 110) | 86 |
| Temperate monsoon | 2010–19 | 16 (12, 18) | | 17 (16, 18) | | 17 (15, 18) | | 16 (14, 18) | |
| | 2050–59 | 29 (24, 35) | 13 | 32 (29, 37) | 16 | 37 (32, 40) | 20 | 47 (38, 48) | 30 |
| | 2090–99 | 26 (21, 28) | 10 | 34 (29, 38) | 18 | 54 (46, 65) | 37 | 93 (82, 106) | 77 |
| Subtropical monsoon | 2010–19 | 13 (12, 13) | | 13 (12, 14) | | 14 (13, 15) | | 13 (12, 15) | |
| | 2050–59 | 22 (19, 26) | 10 | 29 (28, 32) | 16 | 35 (33, 38) | 21 | 46 (40, 48) | 33 |
| | 2090–99 | 20 (17, 24) | 8 | 28 (26, 32) | 15 | 53 (46, 59) | 39 | 105 (88, 113) | 91 |
| Temperate continental | 2010–19 | 16 (14, 17) | | 17 (15, 18) | | 17 (16, 17) | | 17 (16, 18) | |
| | 2050–59 | 27 (21, 30) | 11 | 31 (25, 37) | 14 | 35 (29, 41) | 19 | 49 (41, 56) | 32 |
| | 2090–99 | 24 (20, 27) | 8 | 31 (24, 36) | 15 | 53 (44, 63) | 36 | 97 (79, 107) | 80 |
| Alpine | 2010–19 | 15 (14, 17) | | 15 (14, 16) | | 15 (14, 16) | | 16 (15, 16) | |
| | 2050–59 | 25 (22, 28) | 10 | 29 (23, 34) | 14 | 36 (30, 40) | 20 | 49 (43, 52) | 33 |
| | 2090–99 | 23 (18, 26) | 7 | 30 (28, 33) | 14 | 54 (45, 62) | 38 | 110 (87, 121) | 94 |
| Tropical monsoon | 2010–19 | 18 (13, 20) | | 21 (18, 22) | | 20 (15, 24) | | 20 (15, 23) | |

(continued)

**Table 5** (continued)

| Region | Period | SSP119 scenario | | SSP126 scenario | | SSP245 scenario | | SSP585 scenario | |
|---|---|---|---|---|---|---|---|---|---|
| | | HWD | Δ | HWD | Δ | HWD | Δ | HWD | Δ |
| | 2050–59 | 32 (29, 39) | 14 | 44 (39, 50) | 23 | 56 (47, 69) | 36 | 72 (65, 83) | 53 |
| | 2090–99 | 29 (22, 36) | 11 | 44 (35, 49) | 23 | 89 (75, 113) | 69 | 173 (154, 203) | 153 |

*Note SSP* shared socioeconomic pathway; *HWD* heatwave days. Heatwave days were presented as the annual mean and 25th–75th range across ten general climate models. Δ denoted the projected increase in annual heatwave days versus baseline period (2010–19)

# 6 Conclusions

In summary, this analysis of data from 272 main Chinese cities provided ample evidence for the added mortality risks associated with high temperatures and heatwaves. Besides, this analysis characterized the exposure–response associations and lag patterns between ambient temperature and total mortality. The vulnerable population (elderly, female, less-educated populations) was identified. The mortality burdens of high temperatures and heatwave under the climate-change scenarios were estimated.

This study highlights the necessity for taking adaptation measures against heat stress and may contribute to the reduction of the heat-related mortality burden. The largest study from a developing country constituted the important scientific basis for designing heatwave-warning systems and robust rationales to mitigate the disease burden of heatwaves in a warming climate. Furthermore, it is recommended to tighten climate mitigation policies tailored to local conditions and enhance climate resilience technically and infrastructurally, especially for vulnerable populations.

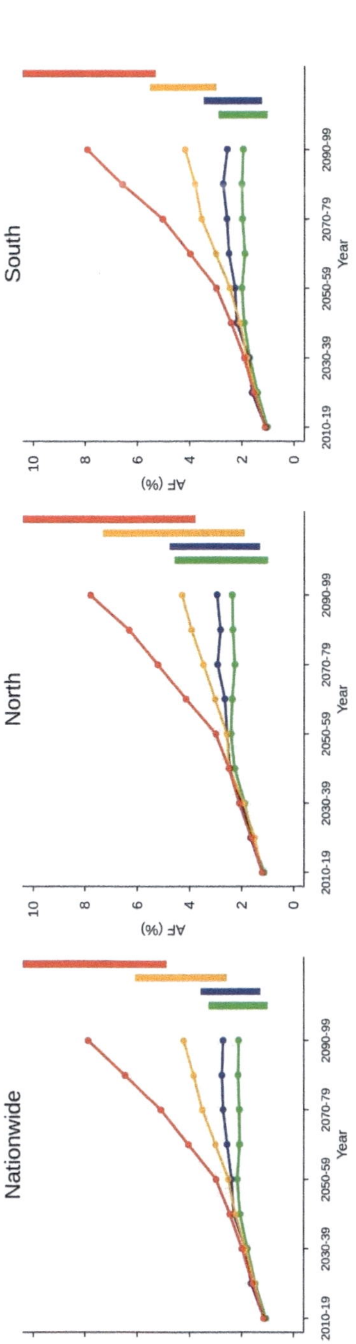


**Fig. 7** Mean heat-related attributable fractions (AFs) of total mortality under the SSP119, SSP126, SSP245, and SSP585 scenarios in different regions. AFs across ten general circulation models under the SSP119, SSP126, SSP245, and SSP585 scenarios were presented in green, blue, orange, and red, respectively. The bars represent 95% empirical confidential intervals in 2090–2099. SSP = shared socioeconomic pathway

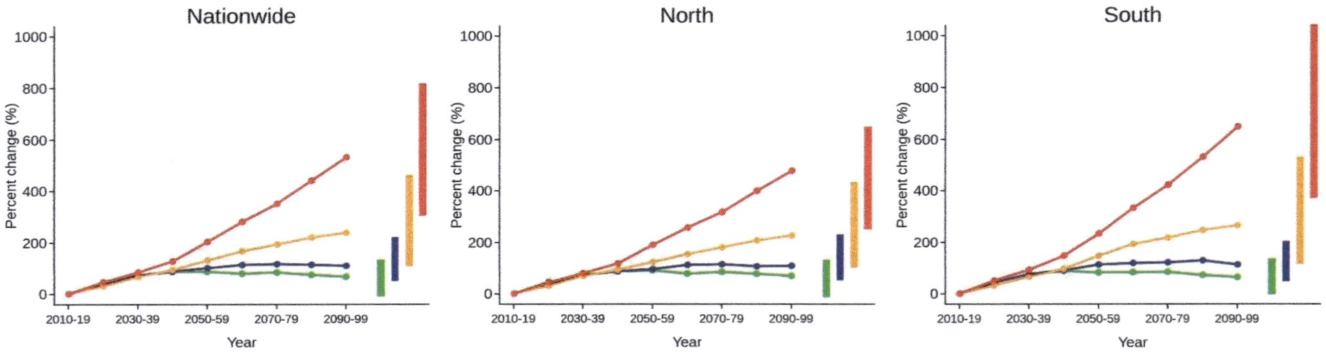


**Fig. 8** Mean percent change heatwave-related excess deaths under the SSP119, SSP126, SSP245, and SSP585 scenarios in different regions. Percent changes across ten general circulation models under the SSP119, SSP126, SSP245, and SSP585 scenarios were presented in green, blue, orange, and red, respectively. The bars represent 95% empirical confidential intervals in 2090–2099. SSP = shared socioeconomic pathway

**Table 6** Projected heat-related attributable fractions of total mortality under the SSP119, SSP126, SSP245, and SSP585 scenarios in 272 main Chinese cities

| Region | Period | SSP119 scenario | | SSP126 scenario | | SSP245 scenario | | SSP585 scenario | |
|---|---|---|---|---|---|---|---|---|---|
| | | AF | Δ | AF | Δ | AF | Δ | AF | Δ |
| Nationwide | 2010–19 | 1.1 (0.7, 1.7) | | 1.1 (0.8, 1.6) | | 1.2 (0.8, 1.6) | | 1.1 (0.7, 1.7) | |
| | 2050–59 | 2.2 (1.1, 3.1) | 1.1 (0.2, 1.7) | 2.4 (1.2, 3.5) | 1.3 (0.4, 2.3) | 2.5 (1.7, 3.4) | 1.3 (0.8, 2.2) | 3.0 (2.1, 4.1) | 1.8 (0.9, 2.8) |
| | 2090–99 | 2.1 (1.0, 3.3) | 1.0 (0.2, 2.1) | 2.7 (1.3, 3.6) | 1.6 (0.4, 2.5) | 4.2 (2.6, 6.1) | 3.1 (1.6, 5.0) | 7.9 (4.9, 13.8) | 6.7 (3.9, 12.7) |
| North | 2010–19 | 1.1 (0.6, 2.1) | | 1.2 (0.7, 2.0) | | 1.2 (0.7, 2.1) | | 1.2 (0.7, 2.4) | |
| | 2050–59 | 2.4 (0.7, 4.2) | 1.3 (0.0, 2.3) | 2.6 (1.0, 3.9) | 1.4 (0.2, 2.4) | 2.6 (1.6, 4.2) | 1.4 (0.7, 3.0) | 3.0 (1.6, 4.7) | 1.8 (0.5, 3.4) |
| | 2090–99 | 2.4 (1.0, 4.6) | 1.2 (0.3, 3.3) | 2.9 (1.3, 4.8) | 1.7 (0.6, 3.6) | 4.3 (1.9, 7.3) | 3.1 (1.1, 5.7) | 7.8 (3.8, 14.3) | 6.6 (3.0, 12.9) |
| South | 2010–19 | 1.0 (0.6, 1.7) | | 1.1 (0.7, 1.5) | | 1.1 (0.8, 1.4) | | 1.1 (0.7, 1.6) | |
| | 2050–59 | 2.0 (1.2, 2.7) | 1.0 (0.2, 1.6) | 2.3 (1.3, 3.6) | 1.2 (0.4, 2.3) | 2.5 (1.6, 3.3) | 1.3 (0.7, 2.1) | 3.0 (1.9, 4.0) | 1.9 (0.9, 2.8) |
| | 2090–99 | 1.9 (1.0, 2.9) | 0.9 (0.1, 1.7) | 2.6 (1.2, 3.5) | 1.5 (0.3, 2.5) | 4.2 (3.0, 5.5) | 3.1 (1.8, 4.6) | 7.9 (5.3, 13.5) | 6.8 (4.2, 12.6) |

*Note SSP* shared socioeconomic pathway; *AF* attributable fraction. AF was presented as the mean and 95% empirical confidential intervals. Δ denoted the projected increase in AF versus baseline period (2010–2019)

**Table 7** Projected percent change heatwave-related excess deaths under the SSP119, SSP126, SSP245, and SSP585 scenarios in 272 main Chinese cities, with an assumption of non-adaptation

| Region | Period | SSP119 scenario | SSP126 scenario | SSP245 scenario | SSP585 scenario |
|---|---|---|---|---|---|
| Nationwide | 2030–39 | 72.4 (17.6, 120.9) | 72.8 (20.2, 136.9) | 67.1 (19.4, 161.4) | 83.6 (16.9, 171.5) |
| | 2050–59 | 87.4 (14.9, 181.8) | 100.9 (41.2, 173.5) | 129.8 (70.2, 251.9) | 202.5 (116.4, 388.1) |
| | 2070–79 | 84.1 (− 3.4, 148.9) | 116.2 (30.1, 215.5) | 192.0 (104.9, 333.1) | 349.2 (216.3, 595.4) |
| | 2090–99 | 67.2 (− 8.9, 132.5) | 109.4 (50.7, 221.4) | 237.6 (106.7, 461.4) | 530.1 (303.9, 816.4) |
| North | 2030–39 | 73.0 (13.6, 128.6) | 71.1 (9.8, 138.8) | 68.2 (10.3, 187.7) | 78.8 (1.2, 178.8) |
| | 2050–59 | 91.2 (17.2, 211.1) | 95.2 (30.7, 182.7) | 121.5 (73.7, 243.6) | 189.1 (91.4, 348.3) |
| | 2070–79 | 84.7 (− 2.9, 169.0) | 114.0 (31.4, 229.9) | 179.3 (88.8, 309.3) | 316.3 (185.2, 495.3) |
| | 2090–99 | 69.8 (− 12.4, 131.1) | 108.5 (52.2, 230.4) | 225.4 (100.9, 431.1) | 475.8 (249.0, 647.6) |
| South | 2030–39 | 74.0 (23.6, 143.0) | 76.4 (35.4, 134.0) | 64.7 (29.9, 119.9) | 92.0 (44.3, 153.2) |
| | 2050–59 | 82.9 (10.8, 150.6) | 112.7 (36.4, 181.0) | 145.9 (57.2, 270.3) | 232.0 (137.9, 490.3) |
| | 2070–79 | 84.7 (− 4.7, 140.0) | 120.9 (27.2, 208.0) | 216.2 (126.4, 385.5) | 420.8 (269.7, 846.5) |
| | 2090–99 | 64.3 (− 2.5, 136.2) | 112.1 (47.7, 203.7) | 263.3 (114.7, 528.0) | 647.6 (370.1, 1227.1) |

*Note SSP* shared socioeconomic pathway. Percent changes were presented as the mean and 95% empirical confidential intervals versus baseline period (2010–19)

## References

Anderson GB, Bell ML (2009) Weather-related mortality: how heat, cold, and heat waves affect mortality in the United States. Epidemiology 20:205–213. https://doi.org/10.1097/EDE.0b013e318190ee08

Anderson GB, Bell ML (2011) Heat waves in the United States: mortality risk during heat waves and effect modification by heat wave characteristics in 43 U.S. communities. Environ Health Perspect 119:210–218. https://doi.org/10.1289/ehp.1002313

Arbuthnott KG, Hajat S (2017) The health effects of hotter summers and heat waves in the population of the United Kingdom: a review of the evidence. Environ Health 16:119. https://doi.org/10.1186/s12940-017-0322-5

Bhaskaran K, Gasparrini A, Hajat S, Smeeth L, Armstrong B (2013) Time series regression studies in environmental epidemiology. Int J Epidemiol 42:1187–1195. https://doi.org/10.1093/ije/dyt092

Chen K, Bi J, Chen J, Chen X, Huang L, Zhou L (2015) Influence of heat wave definitions to the added effect of heat waves on daily mortality in Nanjing, China. Sci Total Environ 506–507:18–25. https://doi.org/10.1016/j.scitotenv.2014.10.092
Chen R et al (2018) Association between ambient temperature and mortality risk and burden: time series study in 272 main Chinese cities. BMJ 363:k4306. https://doi.org/10.1136/bmj.k4306
Chen H et al (2022) Projections of heatwave-attributable mortality under climate change and future population scenarios in China. Lancet Reg Health West Pac 28:100582. https://doi.org/10.1016/j.lanwpc.2022.100582
Dear K, Wang Z (2015) Climate and health: mortality attributable to heat and cold. Lancet 386:320–322. https://doi.org/10.1016/S0140-6736(15)60897-2
Eyring V, Bony S, Meehl GA, Senior C, Stevens B, Stouffer RJ, Taylor KE (2015) Overview of the Coupled Model Intercomparison Project Phase 6 (CMIP6) experimental design and organisation. Geosci Model Dev 8:10539–10583. https://doi.org/10.5194/gmdd-8-10539-2015
Field CB, Barros VR, Dokken DJ, Mach KJ, Mastrandrea MD (2014) Climate change 2014 impacts, adaptation, and vulnerability, vol 1. Cambridge University Press, Cambridge. https://doi.org/10.1017/cbo9781107415379
Gasparrini A, Armstrong B (2013) Reducing and meta-analysing estimates from distributed lag non-linear models. BMC Med Res Methodol 13:1. https://doi.org/10.1186/1471-2288-13-1
Gasparrini A et al (2015) Mortality risk attributable to high and low ambient temperature: a multi-country observational study. Lancet 386:369–375. https://doi.org/10.1016/s0140-6736(14)62114-0
Gasparrini A et al (2017) Projections of temperature-related excess mortality under climate change scenarios. Lancet Planet Health 1:e360–e367. https://doi.org/10.1016/S2542-5196(17)30156-0
Guo Y, Barnett AG, Pan X, Yu W, Tong S (2011) The impact of temperature on mortality in Tianjin, China: a case-crossover design with a distributed lag nonlinear model. Environ Health Perspect 119:1719–1725. https://doi.org/10.1289/ehp.1103598
Guo Y et al (2014) Global variation in the effects of ambient temperature on mortality: a systematic evaluation. Epidemiology 25:781–789. https://doi.org/10.1097/EDE.0000000000000165
Guo XJ, Huang JB, Luo Y, Zhao ZC, Xu Y (2017a) Projection of heat waves over China for eight different global warming targets using 12 CMIP5 models. Theor Appl Climatol 128:507–522. https://doi.org/10.1007/s00704-015-1718-1
Guo Y et al (2017b) Heat wave and mortality: a multicountry, multicommunity study. Environ Health Perspect 125:087006. https://doi.org/10.1289/EHP1026
Guo Y et al (2018) Quantifying excess deaths related to heatwaves under climate change scenarios: a multicountry time series modelling study. PLoS Med 15:e1002629. https://doi.org/10.1371/journal.pmed.1002629
Hajat S et al (2006) Impact of high temperatures on mortality: is there an added heat wave effect? Epidemiology 17:632–638. https://doi.org/10.1097/01.ede.0000239688.70829.63
Hajat S, O'Connor M, Kosatsky T (2010) Health effects of hot weather: from awareness of risk factors to effective health protection. Lancet 375:856–863. https://doi.org/10.1016/S0140-6736(09)61711-6
He C et al (2022) The effects of night-time warming on mortality burden under future climate change scenarios: a modelling study. Lancet Planet Health 6:e648–e657. https://doi.org/10.1016/S2542-5196(22)00139-5
Huang CR, Barnett AG, Wang XM, Tong SL (2012) The impact of temperature on years of life lost in Brisbane, Australia. Nat Clim Change 2:265–270. https://doi.org/10.1038/Nclimate1369
Hurtt GC et al (2020) Harmonization of global land use change and management for the period 850–2100 (LUH2) for CMIP6. Geosci Model Dev 13:5425–5464. https://doi.org/10.5194/gmd-13-5425-2020
Kent ST, McClure LA, Zaitchik BF, Smith TT, Gohlke JM (2014) Heat waves and health outcomes in Alabama (USA): the importance of heat wave definition. Environ Health Perspect 122:151–158. https://doi.org/10.1289/ehp.1307262

Kim EJ, Kim H (2017) Effect modification of individual- and regional-scale characteristics on heat wave-related mortality rates between 2009 and 2012 in Seoul, South Korea. Sci Total Environ 595:141–148. https://doi.org/10.1016/j.scitotenv.2017.03.248

Lee WK, Lee HA, Lim YH, Park H (2016) Added effect of heat wave on mortality in Seoul, Korea. Int J Biometeorol 60:719–726. https://doi.org/10.1007/s00484-015-1067-x

Liu S et al (2016) An integrated national mortality surveillance system for death registration and mortality surveillance, China. Bull World Health Organ 94:46–57. https://doi.org/10.2471/BLT.15.153148

Ma W, Chen R, Kan H (2014) Temperature-related mortality in 17 large Chinese cities: how heat and cold affect mortality in China. Environ Res 134:127–133. https://doi.org/10.1016/j.envres.2014.07.007

Ma W et al (2015) The short-term effect of heat waves on mortality and its modifiers in China: an analysis from 66 communities. Environ Int 75:103–109. https://doi.org/10.1016/j.envint.2014.11.004

Meehl GA, Tebaldi C (2004) More intense, more frequent, and longer lasting heat waves in the 21st century. Science 305:994–997. https://doi.org/10.1126/science.1098704

Moghadamnia MT, Ardalan A, Mesdaghinia A, Keshtkar A, Naddafi K, Yekaninejad MS (2017) Ambient temperature and cardiovascular mortality: a systematic review and meta-analysis. PeerJ 5:e3574. https://doi.org/10.7717/peerj.3574

O'Neill BC et al (2016) The Scenario Model Intercomparison Project (ScenarioMIP) for CMIP6. Geosci Model Dev 9:3461–3482. https://doi.org/10.5194/gmd-9-3461-2016

Saha MV, Davis RE, Hondula DM (2014) Mortality displacement as a function of heat event strength in 7 US cities. Am J Epidemiol 179:467–474. https://doi.org/10.1093/aje/kwt264

Song X et al (2017) Impact of ambient temperature on morbidity and mortality: an overview of reviews. Sci Total Environ 586:241–254. https://doi.org/10.1016/j.scitotenv.2017.01.212

Wang Y et al (2019) Tens of thousands additional deaths annually in cities of China between 1.5 °C and 2.0 °C warming. Nat Commun 10:3376. https://doi.org/10.1038/s41467-019-11283-w

Xu Z, FitzGerald G, Guo Y, Jalaludin B, Tong S (2016) Impact of heatwave on mortality under different heatwave definitions: a systematic review and meta-analysis. Environ Int 89–90:193–203. https://doi.org/10.1016/j.envint.2016.02.007

Yin P et al (2018) The added effects of heatwaves on cause-specific mortality: a nationwide analysis in 272 Chinese cities. Environ Int 121:898–905. https://doi.org/10.1016/j.envint.2018.10.016

# Modelling Urban Microclimates

**Runming Yao and Wenbo Wang**

**Abstract** A microclimate refers to a distinctive climate of a small-scale area, such as a garden, park, valley, or part of a city. Urban microclimates are both complex and dynamic. Modelling urban microclimate can provide robust measures for analyzing urban physical parameters including wind, temperature and solar radiation in complex urban areas at different scales, assessing outdoor thermal environments and performing parametric studies for successful urban planning and design of a liveable urban environment for citizens. This chapter introduces a robust and fast-running tool that can be used to simulate urban solar radiation, surface temperatures and air temperature. A case study of the implementation of this tool is demonstrated.

**Keywords** Microclimate · Urban heat island · Thermal comfort · Geographic information system

## 1 Introduction

A microclimate refers to a distinctive climate of a small-scale area, such as a garden, park, valley, or part of a city (Met Office 2019). An *urban microclimate*, for example, can therefore be considered as the climate of a specific area within a city or a town, which accommodates human activities. An urban microclimate is formed by the local weather, such as solar radiation, air temperature, humidity, wind speed and direction, and the area's physical parameters such as the form of onsite building formations; surface materials and vegetation; as well as human activities. Studies of urban microclimates had drawn increased attention in academic research as well as urban planning practice. A better understanding of the city's urban microclimates can support urban planners and building designers for better urban and building design. It

R. Yao (✉) · W. Wang
Chongqing University, Chongqing, China
e-mail: r.yao@cqu.edu.cn; r.yao@reading.ac.uk

University of Reading, Reading, UK



R. Yao (ed.), *Resilient Urban Environments*, Cities and Nature,
https://doi.org/10.1007/978-3-031-55482-7_5

can help environmental scientists assess health and well-being. It plays an important role in the resilience of cities.

Computer modelling has been developed as a tool for the simulation and prediction of urban microclimates, which has been widely conducted in many aspects including assessment of the effectiveness of green infrastructure, urban ventilation design, assessment of outdoor thermal comfort, potentials of utilization of solar energy and daylight, energy performance analysis of buildings in the urban context and the feasibility of the deployment of renewable energy in buildings. For instance, a widely used Computational Fluid Dynamic (CFD)-based tool called ENVI-met simulates surface–plant–air interactions to assess the impact of the built environment on the local microclimate (Bruse and Fleer 1998). It is typically applied within the fields of urban climatology, architecture, building design, or environmental planning. However, ENVI-met is often computationally expensive to simulate. Many more simple but robust urban microclimate models have been developed (Robinson et al. 2007; Yao et al. 2011; Huang et al. 2020).

This chapter updates the previously published Chapter *Urban Microclimate and Simulation* by Yao (2013) with the authors' newly developed fast and robust tools for the simulation of urban microclimates based on previous studies (Yao et al. 2011). The tools can be used for the first-cut calculations providing interactive design feedback quickly to test the viability of an idea and compare multiple options at the early planning stage by urban planners, architects, building designers and environment assessors who need to make quick comparisons of the impact of different urban parameters on urban microclimates and urban heat island (UHI) mitigation strategies. It attempts to facilitate sustainable climate-responsive urban and building design to make the city resilient.

## 2 Factors Affecting Urban Microclimates

In addition to the local climates, the principal factors that affect urban microclimates include:

(1) Urban Forms

A city is formed by numerous buildings with different heights, orientations and shapes, which are set in various patterns. Martin and March (1972) examined the six archetypal urban forms named pavilions, slab, terraced, terrace courts, pavilion courts and courts.

Urban forms describe the geometrical characteristics of urban spaces. A street canyon is a valley-like space, often created by adjacent buildings. Deep street canyons with higher than average 'height-to-width' ratios (aspect ratio) reflect and absorb both short-wave and long-wave radiation. Thus, heat is magnified during the daytime causing higher temperatures and cooling effects are diminished during the night. Buildings' different shapes and heights affect the distribution of wind and turbulence in urban streets. It, therefore, has a significant impact on urban microclimates.

(2) Building and Surface Materials

Buildings, together with hard non-porous surface materials such as concrete, brick and asphalt, have a large heat capacity and high thermal conductivity. This means that they can store heat received from both solar radiation and other heat sources in an urban area. A material's radiative properties, such as its albedo and emissivity, also have detrimental effects, as does its colour. Dark surfaces tend to absorb electromagnetic radiation and warm up.

(3) Anthropogenic Heat Sources

Heat emitted from transportation (e.g. exhaust fumes), industrial processes, heating and air conditioning systems and other human activities can substantially contribute to heat gains within an urban space.

(4) Local Atmospheric Pollution

Atmospheric pollution such as smoke, chemical and dust particles, as well as global greenhouse gas emissions, absorbs and scatters solar radiation. Unfortunately, it also traps more long-wave radiation which exacerbates the greenhouse effect and increases temperatures.

(5) Urban Wind

Urban wind patterns tend to be more turbulent. Hence, the resulting wind speeds are slower than average, because of the urban form, their settings and their surface 'roughness'. Heat and pollution dissipation can suffer as a consequence of reduced convection.

(6) Evapotranspiration

Lower levels of vegetation (e.g. trees and plants) reduce the potential for evapotranspiration, that is the evaporation and plant transpiration, from the earth's surface into the atmosphere. For urban hydrological microclimates, urban green and blue (e.g. river, lake, pond and fountain) infrastructure affects soil moisture dynamics, surface runoff and the overlying atmosphere through water infiltration and evapotranspiration (Koster 2015; Stavropulos-Laffaille et al. 2018). This has a significant impact on urban latent heat fluxes and contributes directly to urban near-surface climates.

Vegetation and water feature could directly impact the urban climate by reducing surface and local air temperatures or indirectly modify it by reducing heat exchange with occupied spaces thereby reducing mechanical cooling loads or any other anthropogenic heat patterns linked to rising temperatures (Gunawardena et al. 2017).

Differences in underlying surface characteristics and the human activities intensity between cities and suburbs cause variations in urban energy balance, which forms the basis of urban microclimate change (as shown in Fig. 1). Thus, the research on the interaction between different urban constituent elements and urban climate could effectively guide the migration strategies of the UHI and its negative effects.

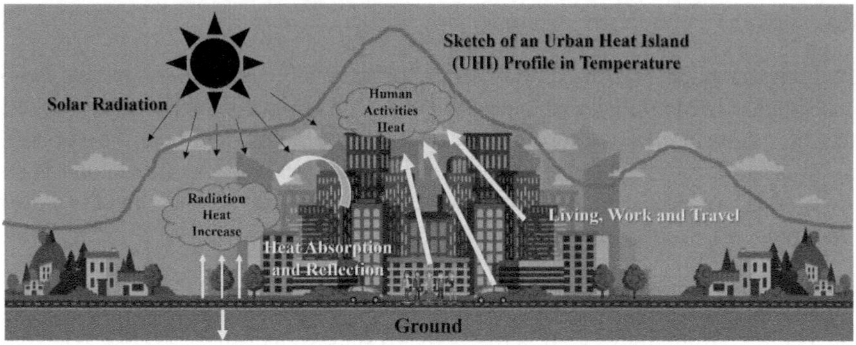


**Fig. 1** Graph showing the relationship between temperature and geographical form

# 3 Modelling Urban Microclimates

Modelling urban microclimates poses a challenge owing to the complexities in defining boundary conditions appropriate to the various environmental parameters. In a like manner, modelling such phenomena requires detailed descriptions of the urban landscape which results in heavy demands placed on input and computing capabilities. In addition, the spatial and temporal nature of dynamic interactions between a building and its urban environment creates further complexities.

As one of the fundamental tools for understanding urban climate, there are a number of urban climate simulations' tool serving different purposes. These range from complex models used to predict real-world outcomes to highly abstract models used only to guide research. Generally, we can distinguish between climate numerical models by: (1) the scale and dimensions and (2) the physical processes included and their description of the urban surface.

## *3.1 The Scale of Urban Climate Models*

The complex combination of structures, materials and human activity features on urban surfaces makes initial microclimate changes difficult to interpret. Their complexity and spatial variability extend from a few metres (facades or canyons) to tens of kilometres (city-wide). Therefore, scale issues should be initially considered in urban climate measurement, modelling and the application of scenario designs.

(a) Street Scale

A structure formed by the arrangement of a street and its flanking buildings is often referred to as an urban canyon or street canyon. Its biggest feature lies in its unique geometry. Canyons are described by their 2D cross-sections, expressed as the ratio of the height of buildings near the street to the width of the street (i.e. H/W, where

H is the height of the building and W is the width of the street). H/W is also known as the canyon aspect ratio. Its value affects many characteristics of an urban climate, including solar potential, shading and trapping, outdoor thermal comfort, wind corridors and the spread of vehicular pollutants. Since it is most clearly defined in dense central areas of cities, the canyon aspect ratio is the most basic unit in urban microclimate simulations along with the Urban Canyon Model (UCM) which is the most commonly used tool to characterize urban microclimates (Masson 2000). It is worth mentioning that the airflow situation in UCM is decoupled and then coupled with the calculation of the energy budget model.

(b) Block Scale

A city block unit consists of a road network, usually with multiple, simply structured, adjacent street canyons. Due to the various arrangements of the road network, blocks may be regular shapes such as triangles, squares, or circles or irregular shapes such as crescents and five-pointed stars. The combination of these different blocks produces a variety of climates and is large enough to generate localized breeze systems (such as parks) or clouds (such as factory complexes). However, block units can be difficult to identify in all cities.

(c) Neighbourhood Scale

Unlike block scales, neighbourhood scales are broader in scope and contain more complex surface features including urban planning information on land-use areas, such as industrial, commercial, residential, public, and undeveloped land. The numerical model of neighbourhood scales was developed based on the interaction of a building with its surroundings at the surface layer. In principle, such models include physical quantities such as solar radiation, heat exchange and convection on the building and land surfaces. In many numerical models for microclimate simulation, the CFD technique, which deals with the equations governing the flow (Navier–Stokes equations), is applied to describe the airflow patterns around and inside buildings. In general, Microclimate Climate Models (MCMs) are widely used by architects and building scientists to study the effects of different urban elements at neighbourhood scales (e.g. building orientation, street canyon aspect ratios, surface materials, vegetation coverage, pedestrian comfort and urban ventilation) (Haghighat and Mirzaei 2011; Tominaga et al. 2015).

(d) City Scale

Investigation of the large-scale UHI variation within a city is broadly adopted in urban climatology and meteorology fields. The impact of urban-scale policies to mitigate the UHI, e.g. urban ventilation, pollution dispersion management and greening, is mostly analyzed using meso-scale model (MM) tools (Lemonsu and Masson 2002). The developed models are based on the equations governing fluid dynamics with equally important models such as radiation, cloud cover and soil integrated into the calculations. As the major limitation, the meso-scale models are applied to very coarse cells, implying a weak resolution on the surface layer to observe interactions between buildings and their environment.

Predictive models at various scales often serve different purposes. In contrast, microclimate numerical models are the most conducive to exploring the relationship between different urban constituent elements and urban climate as well as building-energy-consumption related to near-surface climate. Therefore, microclimate modelling typically focuses on the development of numerical models from the neighbourhood scale to the street scale.

## 3.2 Modelling Process

Since the horizontal and vertical dimensions are approximately equal between the micro- and local-scale modelling domains, capturing the climate impacts of urban surface characteristics (e.g. land covers, street canyon structures and surface materials) is the focus of microclimate modelling. In such microclimate models, it can be simply distinguished as models that focus on momentum fluxes (airflow) or energy fluxes (energy balance).

Models based on the urban surface energy balance (SEB) equations are one of the most common ways to build up numerical models for microclimate. The energy balance equation reveals the main processes of energy exchange in an urban environment. By calculating and modelling the energy term of the SEB equation, it can effectively predict the thermal climate of urban surfaces. It integrates the urban surface climate modelling process more effectively, enabling fast and brief simulations of the near-surface urban thermal climate. The formulation of the volumetric energy balance represents the response of the whole urban environmental system, as shown in Fig. 2. The balance of such urban surface energy can be expressed as (Oke et al. 2017):

$$Q^* + Q_{\mathrm{F}} = Q_{\mathrm{H}} + Q_{\mathrm{E}} + \Delta Q_{\mathrm{s}} + \Delta Q_{\mathrm{A}} + S, \tag{1}$$

where

$Q^*$ is net all-wave radiation,
$Q_{\mathrm{F}}$ is the anthropogenic heat fluxes w/m$^2$,
$Q_{\mathrm{H}}$ is the turbulent sensible heat fluxes w/m$^2$,
$Q_{\mathrm{E}}$ is the turbulent latent heat fluxes w/m$^2$,
$\Delta Q_{\mathrm{s}}$ is the heat storage which represents the net change in heat storage of all elements of the city w/m$^2$,
$\Delta Q_A$ is the net horizontal advection energy w/m$^2$,
$S$ is all other energy sources and sinks.

The most commonly used SEB model is the Urban Canyon Model (UCM) concept developed by Oke and colleagues, which involves an urban street and a road bounded

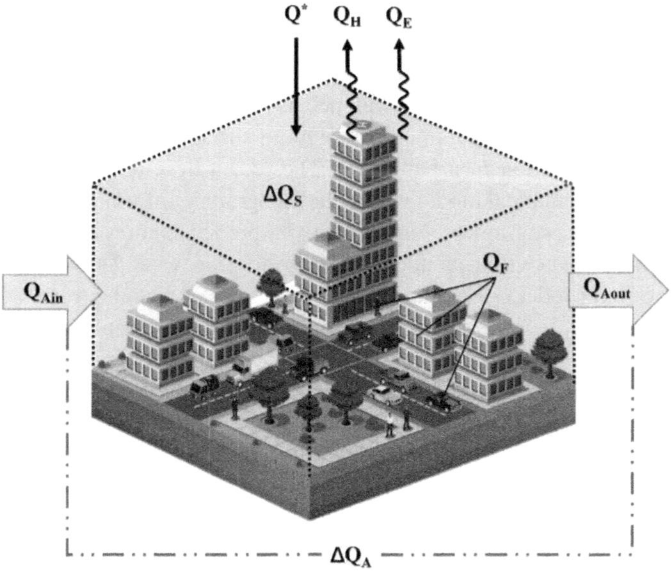


**Fig. 2** Schematic representation of the fluxes in the urban surface energy balance. Arrows are drawn in the direction considered to be positive in the corresponding flux

by two walls (Oke 1979). This method simply averages the urban geometry such as the height, width and impervious surface of the building, thereby effectively representing the urban surface with minimal computational cost. UCM usually takes two different approaches: (a) the Single-Layer Urban Canopy Model (SLUCM) and (b) the Multi-Layer Urban Canopy Model (MLUCM). The SLUCM is based on simplified urban geometry, which is as close as possible to the actual urban surface. In this model, the building is considered to have a 3D shape, and the scheme calculates separate energy budgets for roofs, walls and roads and then aggregates at the canopy level. Specifically, the turbulent sensible heat flux from the roof to the air, the heat energy flux from the building wall to the canopy and the surface-to-ground flux were calculated based on the temperature difference at different locations. These fluxes are then calculated together to give the energy flux transferring to the air. The other method is known as the MLUCM. Unlike SLUCM, this method treats the canopy as a single layer but subdivides it into many sub-layers. Therefore, it is more representative of the atmosphere in urban canyons than the SLUCM and slab models. Chen et al. (2011) assert that this representation can better simulate the mechanical and thermodynamic structure of the urban roughness sub-layer and that of the urban boundary layer. Since it can simulate the most observed urban features (e.g. UHI and enhanced inversion layers above urban surfaces), this method is adopted for temperature and wind profiles within the urban canopy layer (Chen et al. 2011).

Some numerical models based on the SEB formulation also seek a more efficient solution for high-resolution urban microclimate simulation. One such model is Urban

Microclimate simulation model (UMsim) (Yao et al 2011). The model applies a nodal network method to couple thermal and airflow sub-models to simulate urban microclimates. The coupled thermal and airflow model includes heat balance, mass balance and pressure balance sub-models. Developed as a software programme, it can simulate thermal environmental parameters including surface temperatures, air temperatures and solar radiation in urban areas. The recent development of the UMsim integrates the mathematical model within the Geographic Information System (GIS) platform. One of the programme's significantly improved features is its ability to store large quantities of data within an integrated sub-model in raster data format including Digital Elevation Model (DEM), building usage types and land covers. UMsim has been also enhanced by integrating open-source GIS data within the programme to facilitate the simulation of multiple-scale urban microclimates. A new numerical simulation framework based on raster data operations provides a fast computational method for the analysis of large volumes of data. Particularly useful is UMsim's capability of assessing the impact of urban-built forms on microclimates in order to help create sustainable urban designs during the master-planning stage.

## 3.3 *Framework of the Model Structure*

The main role of UMsim based on SEB is to simulate the energy exchange between different urban elements. This simulation process is achieved by establishing a link between a set of properties (e.g. temperature, material heat capacity, etc.) and a physical process (e.g. heat flux density). Therefore, when building a numerical model, it is essential to determine the necessary property data in addition to constructing a reasonable physical process. In other words, the structure of this type of model consists of two main components. One is to construct the property data used for the simulation as boundary conditions. These sets of property data act as the input data for the numerical model. For urban climate models, there are usually two types of input data: meteorological files used to determine climate boundary conditions and urban surface properties' files used to characterize urban morphology. The second is the connection of hypothetical physical processes by means of numerical simulations. These physical processes often involve coupling multiple heat exchange processes such as airflow, solar radiation, heat conduction and anthropogenic heat. Finally, outdoor thermal climate-related parameters are often used as output data, such as air temperature, surface temperature, wind speed and wind direction. Thus, the UMsim has followed the same model structure as shown in Fig. 3 to create a numerical simulation framework. The following briefly describes the structure of the UMsim model in terms of (1) the input data and (2) the physical model.

(1) **Input Data**

For urban climate models, there are typically two types of input data: meteorological files used to determine climate boundary conditions and urban surface property files used to characterize urban morphology. As shown in Fig. 3, the model constructed for

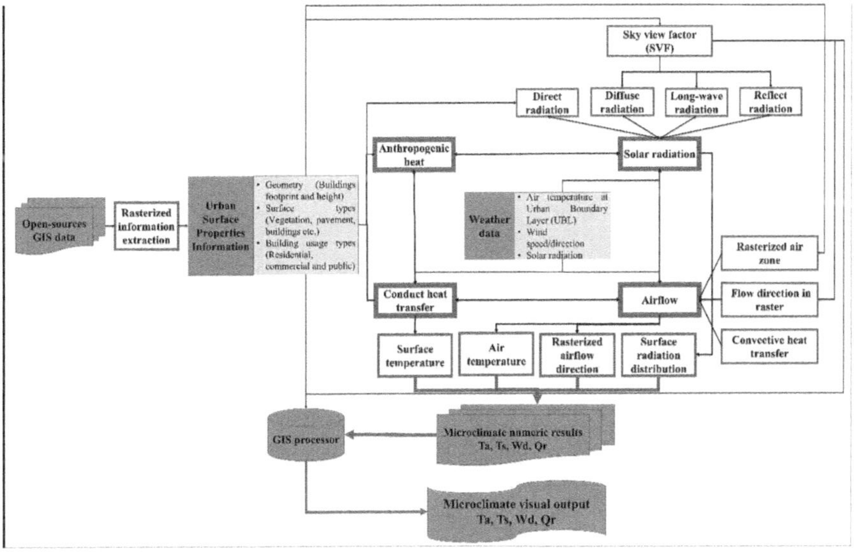


**Fig. 3** Brief framework of the proposed numerical method for urban climate simulations based on raster operations

UMsim also requires input of climate boundary data and data describing the surface properties of the city (obtained with the help of open-source geographic information).

### Weather Data

Since the volumetric energy model assumed in Eq. 1 is used to predict energy and mass exchange at the urban surface, the boundary conditions of the urban surface climate model are determined by the state of the atmosphere. In general terms, the initial meteorological data ultimately required are the temperature, wind speed, wind direction and global solar radiation at the air boundary layer (ABL).

### Urban Surface Properties

In addition, the combination of the shape and orientation of buildings, roads and other man-made structural surfaces affects the mechanical, thermal and radiative properties of the atmosphere. Accurate and more detailed descriptions of urban information can have a positive effect on urban climate modelling. The updated UMsim model therefore constructs a dataset of urban surface features based on a Geographic Information System (GIS) raster data structure. This raster representation of urban surface attributes not only for high-resolution descriptions, but also for storing multidimensional content. This also provides baseline data for climate models based on the analysis of raster operations. Table 1 briefly summarizes the urban surface properties obtained and required by the UMsim model.

### (2) Numerical Modelling

**Table 1** Input urban surface property types for the UMsim model

| General classification | Subcategory for each category |
|---|---|
| Land surface classification | Buildings<br>Roads<br>Trees<br>Grasses<br>Bare soil<br>Waterbody |
| Buildings' geometry | Height of buildings<br>Footprint of buildings |
| Buildings' use types | Public<br>Residential<br>Commercial |

Having identified the data for the properties being studied, it is intuitively important to structure the simulation process according to these properties. This is because the numerical model includes only the processes associated with these attributes. At the same time, the assumptions and coupling patterns of physical processes are different for climate models used for many different purposes. As shown in Fig. 3, the UMsim model constructs four main sub-physical process models on thermal climate simulation based on the above input data.

#### Radiation Budget

Solar radiation plays a major role in outdoor comfort and surface energy balance. The net total radiation in an urban microclimate is affected by the geometry (i.e. the location, orientation and shape of the building) (Shishegar 2013). The total net solar radiation received on the surface is:

$$Q_{\mathrm{R}} = K\downarrow - K\uparrow + L\downarrow - L\uparrow, \tag{2}$$

where

$Q_{\mathrm{R}}$ is net radiation,
$K\downarrow$ is incoming direct and diffuse short-wave radiation,
$K\uparrow$ is reflected short-wave radiation,
$L\downarrow$ is incoming long-wave radiation from the atmosphere,
$L\uparrow$ is outgoing long-wave radiation from the surface.

Additionally, the net radiation energy absorbed by a canyon surface could be expressed as:

$$Q_{\mathrm{R}} = (1-\alpha)(Q_{\mathrm{BH}} + Q_{\mathrm{D}} + Q_{\mathrm{env}}) + Q_{\mathrm{L}}, \tag{3}$$

where

$Q_{\mathrm{R}}$ is net radiation,
$\alpha$ is the surface albedo,

$Q_{BH}$ is the direct beam radiation on a horizontal surface,
$Q_D$ is diffuse radiation,
$Q_{env}$ is the reflection of radiation from the surrounding urban environment,
$Q_L$ is surface-absorbed long-wave radiation.

The UMsim model provides a fast calculation of the surface radiation distribution based on raster operations with the help of Eq. 3. This simulation method is characterized by the fact that the texture of the urban interior can be represented by DEM data (a type of raster data).

**Anthropogenic Heat**

Anthropogenic heat flux usually depends on the population density of the city. Furthermore, the seasonal temperature variability also tends to affect heat flux from human activities. For instance, in some hot areas, more energy is used for ventilation; and in other areas under cold winter conditions, more energy is used for heating, especially at high altitudes. At the same time, anthropogenic heat flux has a huge impact on the modelling of urban energy flux, as it accounts for at least 10% of urban surface energy (Grimmond 1992). However, the energy released by human activities such as space heating, transportation, air conditioning and industrial waste is not easily available. Meanwhile, the bottom-up approach is one of the most commonly used methods of anthropogenic heat estimation. It explores the typical elements of an urban ecosystem, usually considering vehicles, people and buildings. The composition of the anthropogenic heat ($Q_F$) can be written as in the following equation:

$$Q_F = Q_{FV} + Q_{FB} + Q_{FM}, \tag{4}$$

where

$Q_{FV}$ is heat emissions from vehicles,
$Q_{FB}$ is heat emissions from buildings,
$Q_{FM}$ is heat emissions from human metabolism.

The UMsim model also provides an efficient anthropogenic heat estimate with the help of the above equation by combining open-source geographic information data (information on building types, population and traffic conditions).

**Airflow**

Since raster data consist of image elements, each image element has its own space. In order to quickly estimate the airflow direction, the UMsim model proposes a scheme for setting up airflow zones based on raster cells. In other words, each image element represents an airflow zone (see Fig. 4). According to the wind velocity and direction of the input weather data, the velocity along and coordinate components are calculated:

$$U_X = U \cdot \cos \alpha_1, \tag{5a}$$

**Fig. 4** Schematic diagram of air zone establishment

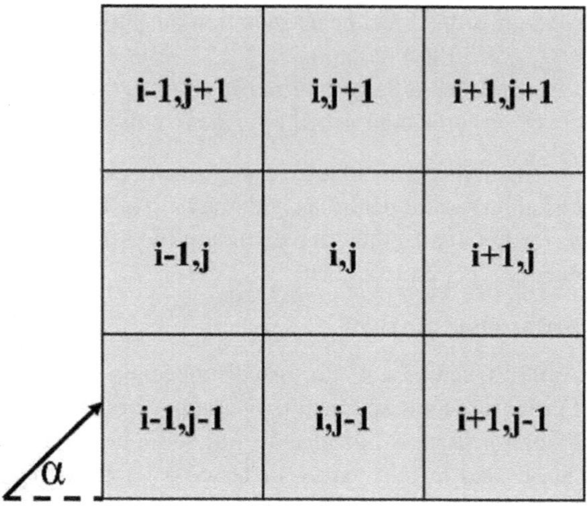


$$U_Y = U \cdot \sin \alpha_1, \tag{5b}$$

where $\alpha_1$ is angel between wind direction and $X$ coordinate.

As the air mass moves, the energy contained in the air flows from one air zone to the adjacent air zone. The amount of energy flow is therefore determined on the basis of the mass flow rate and is calculated by substituting it into the mass balance equation.

## Conductive Heat Transfer

Heat conduction can be seen as the interaction between particles in a substance. It is manifested as the transfer of energy from particles with higher energy to particles with lower energy. The quantitative calculation of heat conduction through a material can use the heat conduction rate equation represented by Fourier's law. It can be expressed on a one-dimensional flat wall as:

$$q = -\lambda \frac{\partial T}{\partial x}, \tag{6a}$$

$$\rho c \frac{\partial T}{\partial t} = \frac{\partial q}{\partial x}, \tag{6b}$$

where

$q$ is the heat flux density in the x direction,
$\lambda$ is the physical property thermal conductivity (W $m^{-1}$ $K^{-1}$),
$T$ is the temperature (K),
$\rho$ is the physical property density (kg $m^{-3}$),
$c$ is the physical property specific heat capacity (J $kg^{-1}$ $K^{-1}$),

$t$ is the time step (s).

In the simple case where the boundary conditions are strictly defined, the analytical method can obtain the exact solution. However, numerical methods are required in real-world situations where boundary conditions vary non-uniformly or where material properties vary with time (e.g. moisture in the soil). The UMsim, on the other hand, uses raster elements to create different nodes with attribute raster data for heat transfer calculations.

## *3.4 Urban Microclimates*

Numerical urban climate modelling conditions are easily controlled and repeatable, which reduces the human and material costs of urban climate research. However, the modelling results must be experimentally verified to prove its validity. Due to the complexity of urban and building thermal climate issues, numerical simulation has become an important aid in the field of urban environmental planning with the development of computer technology in recent years. In developing planning strategies, quantitative design tools for microclimate assessment purposes can better support sustainable urban environmental design, as traditional design processes are mostly based on experience rather than science (Evins 2013; Czmoch and Pękala 2014). Modelling techniques can provide the expected spatial distributions of wind speed and direction, air and surface temperatures and scalar fields inside complex urban areas at the district scale, which can help sustainable environment design associated with outdoor thermal climate. Urban climate models also provide useful information on the microclimates affecting cities in the context of climate change. This information will assist in the development of policies for adaptation and mitigation of climate change. In addition, UMsim aims to provide an intuitive quantification of information that can be used to explore options for better urban function, sustainability and environmental planning in the context of rapid urban construction. For this purpose, it can provide the user or researchers with the following microclimate result parameters related to the urban-built environment.

**Solar**

The quantity of solar energy received by the building envelope and at ground level is significantly influenced by the urban form, in particular, building layout, height and surface properties which are specified by a reflectance coefficient (determined by material, colour and texture). It is reported that the temperature difference between indoor and outdoor surfaces can reach 11 °C in some urban canyons (Georgakis and Santamouris 2006); an urban canyon is being 'a canyon-like channel' formed by buildings (e.g. Michigan Avenue, Chicago).

To no surprise, the availability of solar radiation can have a significant impact on energy-efficient building design and occupant well-being. During the summer day, the intensity of solar radiation received by the urban surface is positively correlated with the intensity of the urban heat island. As one expects, estimating the quantity of radiation received by urban surface is extremely helpful when planning efficient urban environments. The efficacy of solar radiation is defined as: 'the ratio of the illuminance produced by radiation, to the radiation itself'; or in other words, how well visible light is produced. There are three luminous efficacies: (1) global radiation; (2) direct/beam radiation; and (3) diffuse radiation (De Rosa et al. 2008). Luminous efficacy is a useful parameter in the calculation of daylight illuminance for places with solar irradiance data (Li et al. 2008). For this purpose, UMsim is an effective software model for calculating the total irradiance or illuminance value of any surface within an urban DEM model.

**Wind**

Urban form, characterized by building density, size, height, orientation and layout, significantly impacts on the local wind pattern and creates considerable variation in wind speeds around a building complex. Notably, there are numerous implications of urban form on the design and management of natural and hybrid ventilation and cooling strategies; this is a fundamental challenge within industry when designing effective ventilation strategies. This challenge partly exists, because such strategies are commonly based on generalized weather station wind data that have a propensity to overestimate the availability for natural ventilation (i.e. strategies are designed using higher wind speeds than those that occur in reality). Consequently, this results in underperforming cross, displacement and stack ventilation systems that may have to be corrected retrospectively. Subsequent correction procedures are often disruptive, costly, and can cause excessive carbon emissions relative to the initial design assessment. Therefore, in order to complete the simulation more efficiently, UMsim uses a simplified airflow algorithm. It only provides a simple predicted airflow direction result within the city corridor.

**Air Temperature**

Air temperature is a major determinant of urban comfort. As one might expect, air temperature in and around a building has a huge influence on its energy demand. Temperature is determined by a multitude of factors and as with other parameters can be calculated. The updated UMsim model can quickly simulate the outdoor thermal climate in the environment of the building group by coupling all the energy models. In the specific solution process, first a grid digital model of the building group is established according to the planning map or actual geographic information. The respective data on building layout, height, building materials, greening and other factors are stored as raster data. It implements the mutual call of raster analysis technology and the mathematical model to complete the analysis of the outdoor thermal process for the entire building complex.

In general, UMsim is very suitable for rapid urban microclimate simulation to support urban design in both urban construction and renewal as it facilitates early urban design evaluation. It can calculate the urban air temperature, ground surface temperatures, solar radiation falling on the ground, etc. A simple case study demonstrates its practicality in the following section.

# 4 Case Examples

The UMsim model allows rapid assessment of urban thermal climate environments at the neighbourhood scale. It can be seen as an intermediate model between the Urban Canopy Model (UCM) and the CFD-based microclimate model. The numerical model structure is designed for a more effective and practical assessment of the design of urban environments in relation to microclimate at an early planning stage. In current urban developments, data on outdoor microclimatic features including air temperature and surface temperature are desired for outdoor thermal comfort assessment and also for the design of low-energy buildings. The newly improved UMsim can provide microclimate data to engineers and building designers.

## 4.1 *Basic Information for the Study Area*

In order to demonstrate more intuitively the function and role of the proposed model in specific application scenarios, several microclimate simulations in different neighbourhood contexts are performed in this section. Three 1 km * 1 km areas have been chosen for this case study, using London (51° N, 0° E) as an example. Figure 5 shows the specific locations of three selected regions in London which are numbered 1, 2 and 3, respectively: Area 1 is west of central London, Area 2 in north London and Area 3 northeast of London. The three areas were selected with regard to two main indicators of urban planning: Building Volume Density (BVD) and Fractional Vegetation Cover (FVC), both of which directly affect the urban thermal environment (Isa et al. 2020; Qi et al. 2022).

The specific calculation process for these two indicators is as follows. Firstly, the BVD indicator can be obtained with the help of Formula 7 as below.

$$\mathrm{BVD} = \frac{\sum_{n=1}^{N}(C_n * H_n)}{S_\mathrm{L}}, \tag{7}$$

where $C_n$ is the area of building $n$, $H_n$ is the height of building $n$, $S_\mathrm{L}$ is the total area of land and $N$ is the total number of buildings.

According to this formula, the BVD mainly reflects the ratio of the building volume to the corresponding land area. Thus, the BVD can be obtained by zonal statistical analysis (a basic raster operation in GIS) of Digital Elevation Model (DEM) (1 m resolution) data for the whole London area. DEM data are produced using the open-source data provided by the UK Environment Agency (2015). The results in Fig. 6a show the BVD values for the whole of London for a partition of 1 km * 1 km. Secondly, the FVC index is calculated using the Normalized Difference Vegetation Index (NDVI) according to Eq. 8a. NDVI is a standardized index used to generate images showing the amount of vegetation. This index is usually obtained from remote sensing images in the red and near-infrared red (NIR) bands according to Eq. 8b

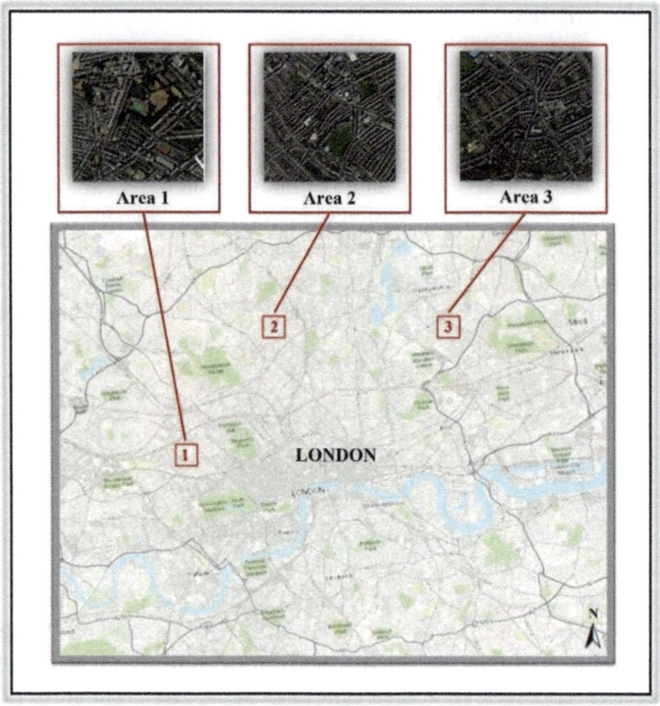


**Fig. 5** Three selected case site locations in London

acquired by Landsat 8 (Jeevalakshmi et al. 2016). As the calculation of FVC and NDVI is well established and has been discussed in numerous studies (e.g. Song et al. 2017; Anees et al. 2022), it will not be repeated in this section. Similarly, zonal statistics for the FVC values for London give the average value of a 1 km * 1 km grid as shown in Fig. 6b.

$$FVC = \frac{NDVI - NDVI_{min}}{NDVI_{max} - NDVI_{min}}, \tag{8a}$$

$$NDVI = \frac{NIR - R}{NIR + R}, \tag{8b}$$

where $NDVI_{min}$ is the normalized value of pure soil cells, $NDVI_{max}$ is the normalized value of pure vegetation cells, NIR is the cell value for NIR band images and R is the cell value for red band images.

After filtering based on the BVD and FVC values calculated in Fig. 6, the three areas shown in Fig. 5 were selected. The urban morphology parameters for the three selected regions are shown in Table 2. Whilst Areas 1 and 3 have similar vegetation cover, Area 1 has a higher building volume than Area 3. Conversely, the building

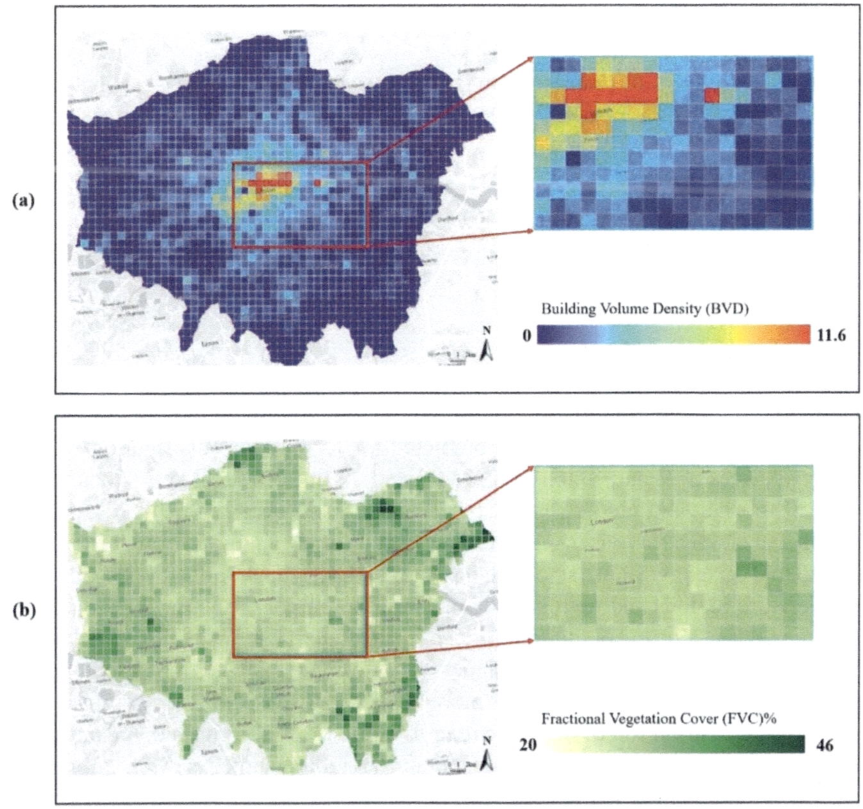


**Fig. 6** BVD and FVC values for London with 1 km * 1 km resolutions. **a** BVD values calculated based on London DEM data. **b** FVC values calculated based on London NDVI values for July 21, 2020

volumes in Areas 2 and 3 are almost identical, but vegetation cover is much lower in Area 2 than in Area 3. Once the simulation areas have been determined, the environment is modelled in raster data format. The numerical model UMsim is then used to obtain the corresponding outdoor thermal climate predictions. A typical summer day (July 21, 2020) is selected as the meteorological data input for this case study simulation. The selection of the meteorological data time refers to the date of the remote sensing data used for the NDVI calculation.

## 4.2 *Estimation of Outdoor Thermal Climates*

Hourly air temperatures were calculated for the three selected 1 km * 1 km areas using the model developed for this study. The results of the model can be used for

**Table 2** Basic morphology indicators of the three selected case sites

| Area number | Fractional vegetation cover (FVC) % | Building volume density (BVD) | Building coverage ratio (BCR) % | Number of buildings | Mean height of buildings (m) | Centre coordinates |
|---|---|---|---|---|---|---|
| Area 1 | 36 | 2.86 | 23 | 365 | 10 | 0.19° E, 51.52° N |
| Area 2 | 25 | 1.76 | 25 | 521 | 6.2 | 0.01° E, 51.56° N |
| Area 3 | 38 | 1.82 | 25 | 480 | 7.5 | 0.12° E, 51.57° N |

the cross-sectional detection of outdoor thermal characteristics in different building environments. Figure 7 shows the predicted air temperature distributions for the selected Areas 1, 2 and 3 at 10.00, 12.00 and 14.00 h, respectively. Focusing on the daily variation in predicted air temperature for Area 1 in Fig. 7, its results show significant daily variations in temperature in urban street valleys in both the temporal and spatial domains. Such mapping helps to quickly screen the area for possible high-temperature regions and optimize them. Thus, the output of the model can quickly provide high-resolution outdoor thermal climate quantification data for a neighbourhood area. This will help urban planners to assess outdoor thermal climate conditions in existing urban environments more visually and efficiently. By comparing the temperature distribution of the three regions at a uniform given time, it can be seen that the area occupied by the high-temperature region is larger in Area 2 compared to Area 1 and Area 3. Combined with the parameters in Table 2, it is known that Area 2 represents a lower building and less green area. The research of Perini and Magliocco (2014) shows that shorter buildings and sparser vegetation cover will lead to an increase in outdoor heat effects. This makes it clear that the model's simulation results are informative. Therefore, based on the simulation results of Fig. 7, it is recommended that the environment of Area 2 could be improved with additional shading measures or more green areas in the hot areas of mapping to improve the summer heat situation. To further explore the effect of different neighbourhood environments on the outdoor thermal climate, the time-by-time results for the three selected areas were averaged.

Figure 8 shows the simulated average air temperatures for the three selected areas for the whole day. The maximum air temperature was around 28 °C at 4.00 p.m. The air temperature at 10.00 a.m. local time was around 10 °C lower compared to the maximum temperature. From the results of the linear plot, it can be seen that the temperature in Area 2 is quite high during the day. However, at night Area 1 shows a relatively high-temperature trend. This could be attributed to its high building volume. At the same time, the simulation results of this case study show a consistent trend with the results of other London thermal climate distribution studies (Taylor et al. 2015; Schneider dos Santos 2020).

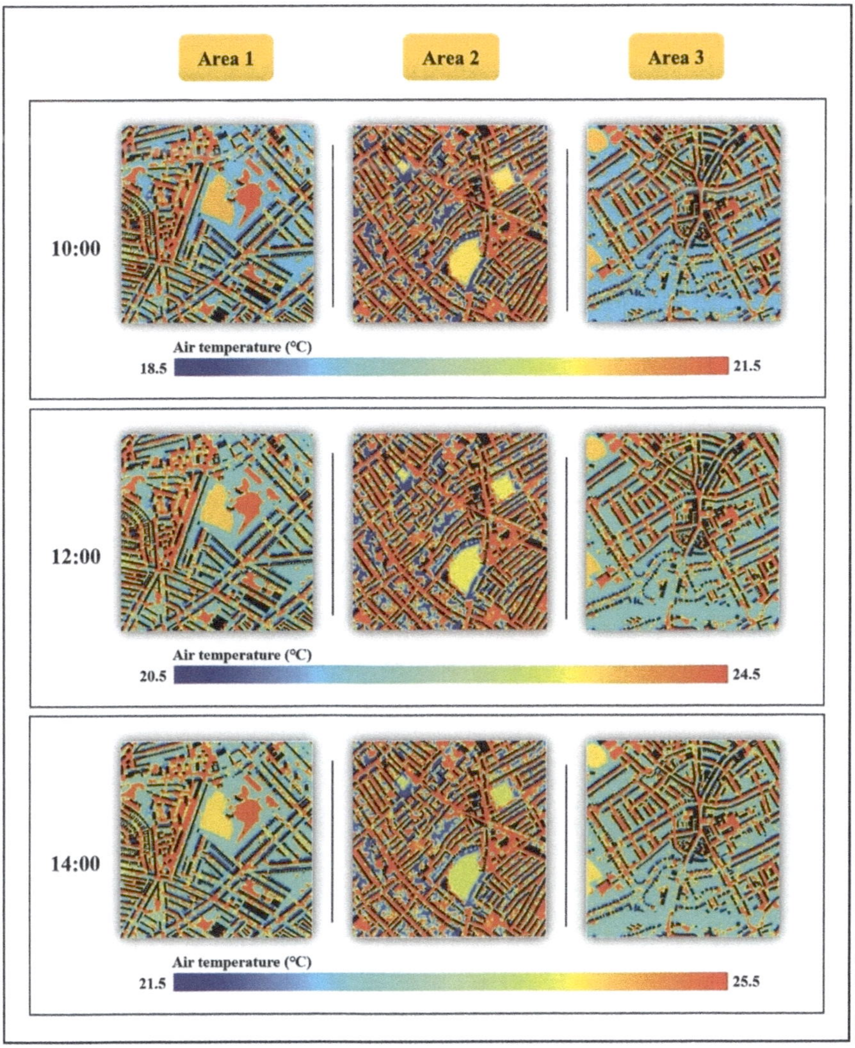


**Fig. 7** Predicted air temperature for three selected areas in London at 10.00, 12.00 and 14.00 h, respectively, on July 21, 2020. (The black parts are buildings)

In addition, as a prerequisite for the calculation of outdoor temperatures, the developed model can also provide maps of the distribution of solar radiation received by the urban surface. The distribution of solar radiation received by the surface in the Case 1 area at different times of the day is shown in Fig. 9.

The final calculation time for each area is around 4 h for an output condition with a resolution of 1 m * 1 m. The entire simulation was done with a laptop with 8 GB of RAM and a 2.20 GHz 8th generation i7 CPU.

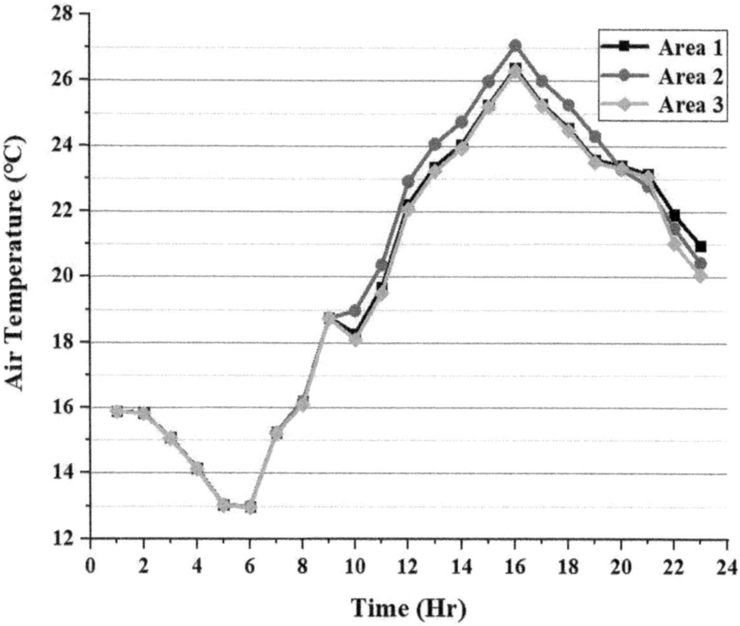


**Fig. 8** Average predicted air temperature for three selected areas in London on July 21, 2020

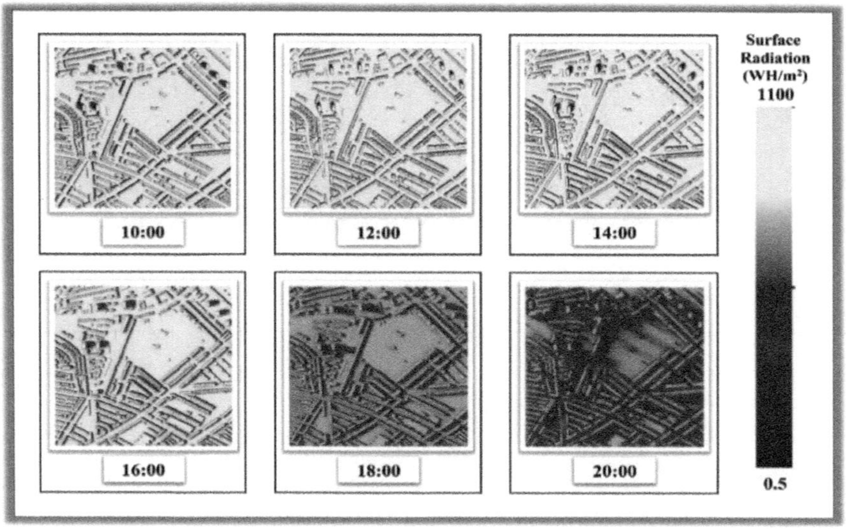


**Fig. 9** Daily variation of surface solar radiation for the Case 1 area in London on July 21, 2020

This is a new method for rapid, high-resolution, neighbourhood-scale, outdoor thermal climate estimation. It suggests that the proposed model is an effective method for rapidly estimating the outdoor thermal climate at the neighbourhood scale for different environmental patterns.

# 5 Summary

This chapter provides a brief introduction to the concept of urban microclimate and the modelling process. It also presents a newly updated raster data format-based UMsim model for real urban environment assessment with the help of the case study. The advantages of this updated approach to urban climate modelling in practice are now evident through this simple case illustration. It presents the following advantages from the pre-modelling to the output of the post-results:

- The UMsim model allows the rapid creation of environmental models for climate simulation from existing building design models. This is mainly due to the compatibility of existing GIS data platforms for a wide range of data and the proposed representation of the physical model of the environment based on the raster data format in GIS. This allows for effective savings in upfront model processing time in the evaluation of real projects. This can increase the efficiency for the assessment of the project.
- The improved raster data operation-based urban climate simulation solution provides reliable and efficient results. It enables a rapid high-resolution assessment of the outdoor thermal climate at a neighbourhood scale. This provides planners and designers with visual and quantitative data for assessing the urban thermal environment in the early planning stages. It not only provides relevant thermal climate simulation data but also provides relevant building design indices for reference. This is all thanks to the excellent regional statistical power of raster data operations.
- The flexible data clipping and combining capabilities of the raster data also provide additional possibilities for the creation of environmental models. They allow the integration of the status model with the planning model in the modelling of the actual environment. This provides more comprehensive information for urban climate modelling and also a more comprehensive response to the possible interaction of the planning model with existing scenarios.

In general, the UMsim is a robust and practical numerical simulation model for the urban microclimate. It can be used in a variety of scenarios to assist in the assessment of urban planning/building design projects related to the outdoor thermal environment. The case study presented briefly demonstrates some of the functions that can be used in the planning and design phase.

## References


Agency E (2015) The environment agency national LIDAR Programme. https://www.data.gov.uk/dataset/f0db0249-f17b-4036-9e65-309148c97ce4/national-lidar-programme#licence-info. Accessed 21 March 2023

Anees SA, Zhang X, Shakeel M et al (2022) Estimation of fractional vegetation cover dynamics based on satellite remote sensing in Pakistan: a comprehensive study on the FVC and its drivers. J King Saud Univ—Sci 34:101848. https://doi.org/10.1016/j.jksus.2022.101848

Bruse M, Fleer H (1998) Simulating surface-plant-air interactions inside urban environments with a three dimensional numerical model. Environ Model Softw 13:373–384. https://doi.org/10.1016/S1364-8152(98)00042-5

Chen F, Kusaka H, Bornstein R et al (2011) The integrated WRF/urban modelling system: development, evaluation, and applications to urban environmental problems. Int J Climatol 31:273–288. https://doi.org/10.1002/joc.2158

Czmoch I, Pękala A (2014) Traditional design versus BIM based design. Procedia Eng 91:210–215. https://doi.org/10.1016/j.proeng.2014.12.048

De Rosa A, Ferraro V, Kaliakatsos D, Marinelli V (2008) Simplified correlations of global, direct and diffuse luminous efficacy on horizontal and vertical surfaces. Energy Build 40:1991–2001. https://doi.org/10.1016/j.enbuild.2008.04.018

Evins R (2013) A review of computational optimisation methods applied to sustainable building design. Renew Sustain Energy Rev 22:230–245. https://doi.org/10.1016/j.rser.2013.02.004

Georgakis C, Santamouris M (2006) Experimental investigation of air flow and temperature distribution in deep urban canyons for natural ventilation purposes. Energy Build 38:367–376. https://doi.org/10.1016/j.enbuild.2005.07.009

Grimmond CSB (1992) The suburban energy balance: methodological considerations and results for a mid-latitude west coast city under winter and spring conditions. Int J Climatol 12:481–497. https://doi.org/10.1002/joc.3370120506

Gunawardena KR, Wells MJ, Kershaw T (2017) Utilising green and bluespace to mitigate urban heat island intensity. Sci Total Environ 584–585:1040–1055. https://doi.org/10.1016/j.scitotenv.2017.01.158

Haghighat F, Mirzaei PA (2011) Impact of non-uniform urban surface temperature on pollution dispersion in urban areas. Build Simul 4:227–244. https://doi.org/10.1007/s12273-011-0035-6

Huang J, Jones P, Zhang A et al (2020) Urban Building Energy and Climate (UrBEC) simulation: example application and field evaluation in Sai Ying Pun, Hong Kong. Energy Build 207:109580. https://doi.org/10.1016/j.enbuild.2019.109580

Isa NA, Salleh SA, Naim Wan Mohd WM et al (2020) Building volume effects on ambient temperature in the Kuala Lumpur City. IOP Conf Ser Earth Environ Sci 489.https://doi.org/10.1088/1755-1315/489/1/012011

Jeevalakshmi D, Reddy SN, Manikiam B (2016) Land cover classification based on NDVI using LANDSAT8 time series: a case study Tirupati region. In: International conference on communication and signal processing (ICCSP) (ICCSP 2016) 560056, pp 1332–1335. https://doi.org/10.1109/ICCSP.2016.7754369

Koster R (2015) “Efficiency space”: a framework for evaluating joint evaporation and runoff behavior. Bull Am Meteorol Soc 96:393–396. https://doi.org/10.1175/BAMS-D-14-00056.1

Lemonsu A, Masson V (2002) Simulation of a summer urban breeze over Paris. Boundary-Layer Meteorol 104:463–490. https://doi.org/10.1023/A:1016509614936

Li DHW, Lam TNT, Chu VWC (2008) Relationship between the total solar radiation on tilted surfaces and the sunshine hours in Hong Kong. Sol Energy 82:1220–1228. https://doi.org/10.1016/j.solener.2008.06.002

Martin L, March L (1972) Urban space and structures, 1st edn. Cambridge University Press, London

Masson V (2000) A physically-based scheme for the urban energy budget in atmospheric models. Boundary-Layer Meteorol 94:357–397. https://doi.org/10.1023/A:1002463829265

Met Office (2019) The national meteorological library and archive fact sheet 14—microclimates. https://www.metoffice.gov.uk/research/library-and-archive/publications/factsheets. Accessed 21 March 2023
Oke TR (1979) Review of urban climatology 1973–1976, 1st edn. Organization, Secretariat of the World Meteotological, Geneva
Oke TR, Mills G, Christen A, Voogt JA (2017) Urban climates. Cambridge University Press, Cambridge
Perini K, Magliocco A (2014) Effects of vegetation, urban density, building height, and atmospheric conditions on local temperatures and thermal comfort. Urban Urban Green 13:495–506. https://doi.org/10.1016/j.ufug.2014.03.003
Qi Y, Li H, Pang Z et al (2022) A case study of the relationship between vegetation coverage and urban heat island in a coastal city by applying digital twins. Front Plant Sci 13:1–15. https://doi.org/10.3389/fpls.2022.861768
Robinson D, Campbell N, Gaiser W et al (2007) SUNtool—a new modelling paradigm for simulating and optimising urban sustainability. Sol Energy 81:1196–1211. https://doi.org/10.1016/j.solener.2007.06.002
Schneider dos Santos R (2020) Estimating spatio-temporal air temperature in London (UK) using machine learning and earth observation satellite data. Int J Appl Earth Obs Geoinf 88:102066. https://doi.org/10.1016/j.jag.2020.102066
Shishegar N (2013) Street design and urban microclimate: analyzing the effects of street geometry and orientation on airflow and solar access in urban canyons. J Clean Energy Technol 1:52–56. https://doi.org/10.7763/jocet.2013.v1.13
Song W, Mu X, Ruan G et al (2017) Estimating fractional vegetation cover and the vegetation index of bare soil and highly dense vegetation with a physically based method. Int J Appl Earth Obs Geoinf 58:168–176. https://doi.org/10.1016/j.jag.2017.01.015
Stavropulos-Laffaille X, Chancibault K, Brun JM et al (2018) Improvements to the hydrological processes of the Town Energy Balance model (TEB-Veg, SURFEX v7.3) for urban modelling and impact assessment. Geosci Model Dev 11:4175–4194. https://doi.org/10.5194/gmd-11-4175-2018
Taylor J, Wilkinson P, Davies M et al (2015) Mapping the effects of urban heat island, housing, and age on excess heat-related mortality in London. Urban Clim 14:517–528. https://doi.org/10.1016/j.uclim.2015.08.001
Tominaga Y, Akabayashi S, Kitahara T, Arinami Y (2015) Air flow around isolated gable-roof buildings with different roof pitches: wind tunnel experiments and CFD simulations. Build Environ 84:204–213. https://doi.org/10.1016/j.buildenv.2014.11.012
Yao R (2013) Design and management of sustainable built environments, 1st edn. Springer, London
Yao R, Luo Q, Li B (2011) A simplified mathematical model for urban microclimate simulation. Build Environ 46:253–265. https://doi.org/10.1016/j.buildenv.2010.07.019

# Predictive Modelling in Urban Environments

**Fangxin Fang**

**Abstract** Urban problems (e.g. turbulent flows, chemical reactions, radiation, heat exchange and interaction between the physical processes) typically have important dynamics that operate over a range of length scales. In this chapter, we first introduce multiscale adaptive mesh numerical methods for the dispersion of urban pollutants at city, building and street scales down to personal scales. The use of adaptive unstructured meshes in urban environment modelling can provide accurate results since the mesh is dynamically adapted according to the evolving physical features. We also present new numerical techniques such as machine learning (ML), reduced-order modelling (ROM) and data assimilation (DA), for real-time operational modelling and uncertainty analysis. The combination of ROM, ML and DA enables a rapid and accurate modelling response in emergencies. The multiscale adaptive mesh and rapid response modelling capability will be demonstrated in realistic cases—London (UK) and Denmark.

**Keywords** Adaptive meshes · Machine learning · Data assimilation · Urban environment

## 1 Introduction

As urbanization and city-living increase dramatically worldwide, an environment that allows a healthy and productive lifestyle for all ages is a priority. Thus, generating healthy urban environments necessitates their rational design and management for daily operation and emergency response, especially considering their perpetually evolving nature due to climate change. Now, more than ever, there is a pressing need for a roadmap to ensure that decisions can be taken to allow the resilient development of cities.

Air quality is closely linked to transport systems, energy production, naturally ventilated buildings—avoiding massively energy consuming, and polluting, heating/

F. Fang (✉)
Department of Earth Science and Engineering, Imperial College London, London SW7 2BP, UK
e-mail: f.fang@imperial.ac.uk

 
R. Yao (ed.), *Resilient Urban Environments*, Cities and Nature,
https://doi.org/10.1007/978-3-031-55482-7_6

air conditioning, green infrastructure and urban design. Managing and controlling these factors is a complex task that needs to both maximize efficiency and reduce costs. These challenges highlight the demand for strategies, innovative methods and measures to design and manage cities, taking into account an ever-changing urban landscape and people's lifestyle—all the while making cities safer, more pleasant, healthier and productive places in which to live and work.

Models are powerful tools for understanding pollutant and carbon lifecycles and atmospheric processes, making predictions, uncertainty quantification and optimal control/design for resilient cities. Further, the model provides an objective and measurable approach to evaluate various environmental and planning choices. This includes making decisions for emergency response in the short term, as well as for transportation and the built environment in the long term. The complex pollutant and atmospheric physical processes act over a wide range of spatial and temporal scales (Carmichael et al. 2008). It thus requires multiscale models for accurately assessing the pollutant/carbon emissions from human activity and the impact of green and blue infrastructures.

It is expected that future model tools will have the capability of addressing: (i) effective ventilation of the urban environment at city and street scales; (ii) thermal and wet environmental conditions; (iii) impact of architectural design; (iv) integrated vehicle traffic, cyclist and pedestrian pathways; (v) incorporation of local power and thermal energy plants; (vi) emergency response; (vii) transport of pollutants; (viii) tree, greening and water system placement; and (ix) transfer of critical information (via clear visual interfaces) on pollution levels, winds and temperatures to city planners, administrators, emergency services and the general public. This is a complex modelling problem that can only be satisfactorily treated through an innovative and fully-integrated model framework. Key to conditioning and evaluating such a modelling system will be the capacity to provide both simulated and in-situ measured observations into the modelling framework.

This chapter introduces multiscale atmospheric and air quality models and advanced data-driven models (machine learning and data assimilation) in urban environments.

## 2 Multiscale Urban Atmospheric and Air Quality Models in Urban Environments

The grand challenge in modelling complex physical phenomena is to predict their characteristics and evolution with adequate accuracy and reliability. This has remained an open scientific problem for decades, unsurprisingly given the extreme disparity in the length and time scales involved, spanning several orders of magnitude. For example, urban modelling involves both large and smaller scale features such as boundary currents, mixing, chemical interactions and transport, overflows and mesoscale eddies. Dynamic mesh adaptation is the only effective way to encompass different scales (e.g. local, urban, regional, in a unified modelling system to better capture the interactions among the processes relevant at each scale). Over traditional structured mesh models, the use of adaptive unstructured meshes provides several profound and widely acknowledged advantages (Piggott et al. 2009). These include the ability to conform accurately and efficiently to complex domain geometries (e.g. bathymetry, a complex cityscape) and the ability to dynamically adapt mesh resolution to improve the accuracy of model results, to capture detailed dynamics or to follow the physical processes. Here, we introduce an advanced adaptive mesh fluid flow model, Fluidity, a finite element-based flow simulator developed at Imperial College and incorporating an anisotropic mesh adaptivity capability as well as the large eddy simulation methodology (https://github.com/FluidityProject/fluidity). Fluidity runs efficiently on modern parallel computing hardware and has been shown to scale well on supercomputing facilities. It achieves this through hybrid open-MP/MPI parallelization, which when combined with the adaptive resolution makes it ideal for spatially-distributed modelling of city-scale problems.

## 2.1 *Multiscale Urban Atmospheric Modelling*

Atmospheric variability has a critical impact on urban air quality through its controlling effect on local meteorological conditions and, in turn, greatly influences airborne pollutant dispersion. Moreover, the understanding of the relationship of pollutant concentrations with the prevailing circulation, both synoptic and local scale, is considered crucial for the forecasting of air pollution episodes. Advances of adaptive mesh physical modelling and data-driven approaches enable to model complex multiscale physical processes, including deep and shallow convection, stratiform clouds and precipitation, microphysics, surface fluxes, planetary boundary layer processes, turbulence and diffusion and radiation cloud-radiation interaction:

- Detailed atmospheric dynamic codes.
- Coupled modelling of aerosol cloud microphysics.
- Modelling techniques for thermodynamics.
- Consistent surface energy fluxes and boundary conditions.
- Coupling of atmospheric and chemical modelling.

Through the use of high-resolution earth observing satellites, a retrospective assessment of atmospheric conditions typical of the region can be constructed, capturing the frequency of events such as dust storms, air pollution and other atmospheric phenomena (e.g. rainfall) that impact the urban. To inform synthetic experiments examining interactions of the city-scape with its surrounding environment, atmospheric profiles of temperature, humidity and even air quality can be collected via UAVs through vertical soundings at key locations throughout the study domain. For larger-scale spatial assessments, high-resolution satellite data will be coupled with in-situ data to provide insights into surface-air temperature gradients throughout the year. Through combining both the UAVs and satellite data, a thorough assessment of the expected urban thermal environment will be explored.

## 2.2 Multiscale Air Quality Modelling

Air quality modelling systems mainly involve three components: emissions, meteorology and chemical transport (Zheng et al. 2020). Once air pollutants are emitted into the atmosphere, the dynamic and chemical processes would transport and transform them continuously. The interactions between these processes involve a wide range of spatial scales. The initial transformation of emissions from urban and industrial centres or dispersion of plumes from large power plant stacks occurs on relatively small scales, but would be engaged to much larger scales after long-distance transport (Zheng et al. 2015). In air quality modelling it is challenging to accurately represent the nonlinear photochemical processes from a very fine scale (a few metres, near point sources) to a large scale (several kilometres) simultaneously (Karamchandani et al. 2011).

The large-scale grid models are unable to catch the sharp change in pollutant concentrations, thus leading to an artificial dilution of stack emissions and incorrect chemical reaction rates. The artificial dilution of the emissions in the grid models results in a short lifetime of pollutants (Kim et al. 2017). To overcome this issue, most of existing air quality models (e.g. CMAQ, CAMx and WRF-Chem) have incorporated subgrid-scale modules (Karamchandani et al. 2011).

Over embedded subgrid scale modelling and static nested grid techniques, adaptive mesh modelling may be the best approach to resolve multiscale dynamic and chemical processes in a consistent way (Zheng et al. 2020). In dynamic mesh adaptations, mesh resolution can be changed automatically to improve the accuracy of model results, to capture detailed dynamics or follow the chemical evolution of plumes.

As an example of using adaptive meshes, the fluid flow model (Fluidity) mentioned above has been used for the study of COVID-19 spatial–temporal spreading in London (Zheng et al. 2021): how long can the virus spread for given certain meteorological conditions? The virus transport equation is:

$$\frac{\partial C}{\partial t} + \nabla \cdot (\mathbf{u}C) - \nabla \cdot (\overline{\kappa} \nabla C) = S + D, \tag{6.1}$$

where $C$ is the mass concentration of the virus, $\mathbf{u} = (u, v, w)^T$ is the velocity vector to be resolved using the three-dimensional (3D) Navier–Stokes equations with the large eddy simulation (LES), $t$ is the time, $\overline{\kappa}$ is the tensor of turbulent diffusivity, $S$ represents the source term of virus, and $D$ is the decay of virus.

The virus exponential decay formulation (van Doremalen et al. 2020) is given as follows:

$$C = C_0 \cdot e^{-\lambda t}, \tag{6.2}$$

where $C_0$ is the initial virus concentration and $\lambda$ denotes the decay rate. The decay term in Eq. (6.1) can thus expressed as $D = -\lambda C$. Studies have shown that the decay of airborne viruses is sensitive to the metrological conditions (wind velocity, ambient humidity and temperature).

Once viruses are emitted into the air, the dynamic and transmission processes involve a wide range of spatial scales. An artificial dilution of viruses may lead to a shorter lifetime in existing fixed grid models if the resolution of grids is not high enough. It has been proved (Zheng et al. 2015, 2020) that mesh adaptivity is the most efficient and effective approach for resolving multiscale dynamic processes. The mesh is adapted with respect to the dynamic flow features and virus concentrations in time and space. Using the adaptive mesh, the detailed flow dynamics and the temporal and spatial evolution of COVID-19 viruses during the spreading process can be captured, especially the local turbulent flows around buildings.

Figure 6.1 shows the locations of virus emissions, where the high mesh resolution is adopted around the virus source locations. In this study, the mesh is dynamically adapted with respect to both the wind velocity field and virus concentration. The spatial distribution of viruses is plotted in Fig. 6.2. We can see that the concentration of viruses decreases rapidly by a factor of 2–3 near the sources. The majority of the viruses remain trapped within a short radical distance of less than 50 m from the release location.

## 2.3 Green–Blue Infrastructure and Land Surface Thermal Dynamics Modelling

Modelling the turbulent exchanges of mass, momentum and dispersion is essential in determining environmental conditions. The thermal structure is particularly important in determining the state of temperature stratification that, in turn, has a major bearing on transport processes within the atmospheric boundary and canopy layers. The complex heat balance within the canopy needs to be treated in detail because of its impact on thermal comfort, building service heating and cooling demands and, most importantly, the enhancement or suppression of canopy ventilation through buoyancy effects. A land surface and tree/green model has recently been implemented within Fluidity (Fluidity-Urban) along with a 3D complex urban geometry tool which can be used for generating high-quality urban geometric meshes and identifying the land types (water, tree, roof, wall, greening, etc.). The use of dynamically adaptive meshes allows the mesh to change throughout the simulation enabling the resolution to focus on fields of interest and areas where important physical processes need to be captured.

Fluidity-Urban has been applied to the London Road case (Fig. 6.3). The computational domain is a 1.2 km (west–east) × 1.0 km (south-north) rectangular area

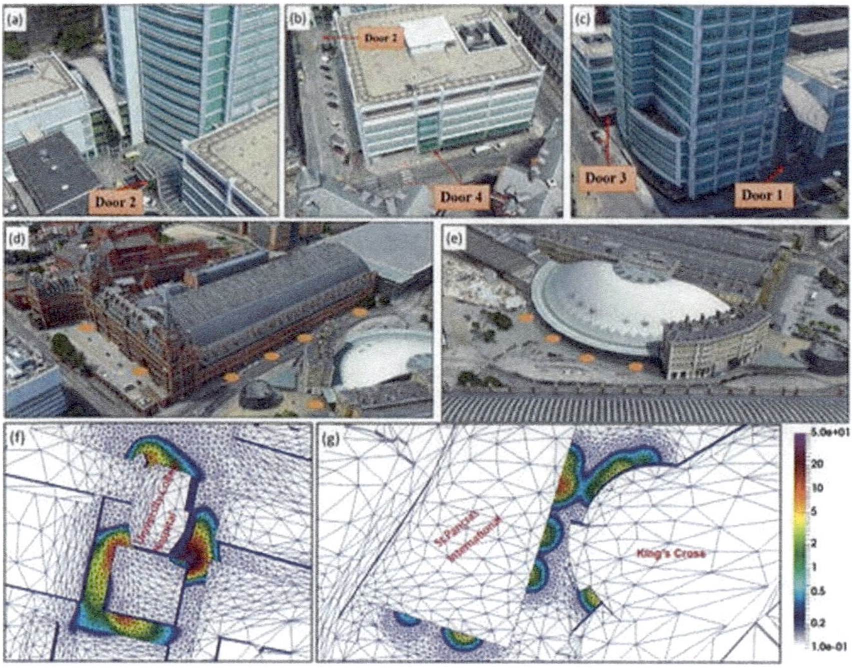


**Fig. 6.1** Release location of viruses in **a–c** doors 1–4 in University College Hospital (UCH), **d** the St Pancras International and **e** King's cross stations, as well as the corresponding initial adaptive meshes and virus concentration (unit: copies/m$^3$) around the source locations **f** and **g** railway stations (Zheng et al. 2021)

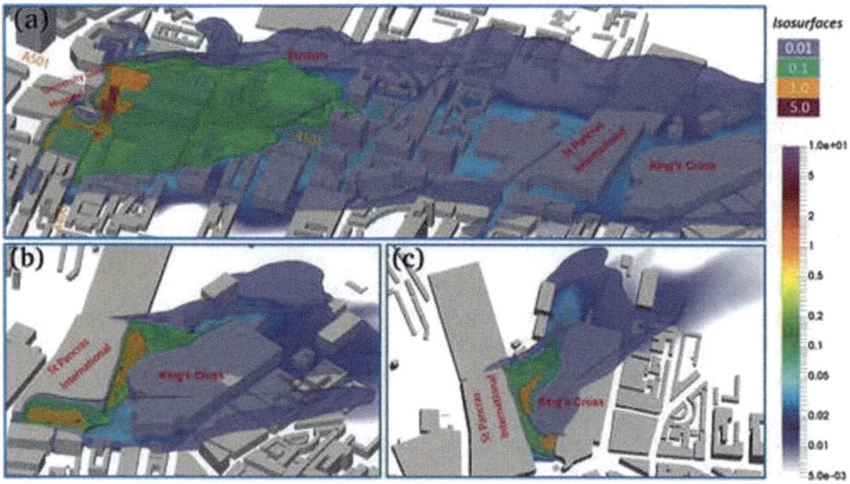


**Fig. 6.2** 3D iso-surface of the virus concentration (copies/m$^3$), where the viruses are released from **a** UCH, **b** the St Pancras International and **c** King's cross station where the colour bar represents the value of the virus concentration (unit: copies/m$^3$) (Zheng et al. 2021)

around the London Road with a height of 1000 m to capture the dynamical characteristics of the atmospheric boundary layer. In this domain, there are 322 buildings, 37 green spaces and 1631 trees. The buildings have different shapes and heights, with the tallest building up to 74 m. The tree canopy is approximately represented as a cylinder with a diameter $d_0 = 8$ m and height $h_0 = 16$ m, based on the tree species around this region. Here we set up the geometries, green infrastructure (GI) and associated surface fluxes along with the formation of a database containing the physics associated with different plant species. In this study, we mainly focus on (i) the detailed dynamics of, e.g. heat, moisture, radiation, mass transfer mechanisms and turbulence; (ii) the impact of trees and green parts on the urban environment. The impact of greening on temperature is shown in Fig. 6.4. The difference of temperatures with and without trees is plotted in Fig. 6.4. We can clearly see the impact of green infrastructures—the temperature is decreased by ~ 6° near the green parks.

**Fig. 6.3** Study site: London Road. Left: Google map; right: 3D geometry for modelling, where greening represents grass and trees (green dots) (provided by Dr. X. Wu and supported by the UK EPSRC MAGIC and INHALE projects)

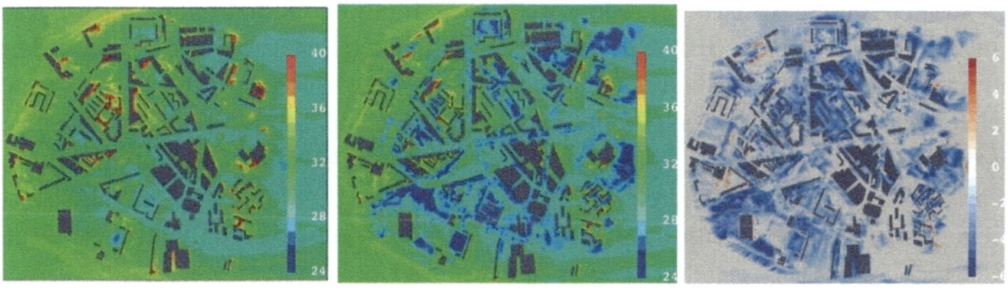


**Fig. 6.4** Impact of greening on temperature. The temperature results with (left) and without (middle) trees, as well as the differences (right) of temperature with and without trees (provided by Dr. X. Wu and supported by the UK EPSRC MAGIC and INHALE projects)

# 3 Predictive Modelling: Data-Driven Modelling—Machine Learning and Data Assimilation

## 3.1 Data Assimilation

Data assimilation techniques incorporate information from experiments and observations to reduce uncertainties in numerical prediction (Kalnay 2003). To achieve a predictive capability, numerical modelling often needs to be linked with experiments/monitoring. Data assimilation is an efficient means of producing a best estimate model solution by fitting a numerical simulation to observational data over both space and time. The technique also facilitates the estimation of the error sources caused by uncertainties in the models. Data assimilation is achieved by minimizing a cost function that reconciles the misfits between the data and dynamical modelling results, as well as the covariances specifying spatial and temporal correlations of errors (Fang et al. 2017; Du et al. 2016). A variety of approaches have been used to facilitate data assimilation and include statistical interpolation methods, nudging data assimilation, Kalman filter (Stratonovich 1959), ensemble Kalman filter (Evensen 2009), particle filter (Fearnhead and Künsch 2018) and variational (adjoint) methods (Cacuci et al. 1980). Data assimilation techniques are widely used in operational modelling.

To speed up the assimilation process, reduced-order modelling (ROM) approaches have been introduced to data assimilation. For example, ROM approaches are used for improving data assimilation under the framework of an EnKF (He et al. 2011). In the work of Farrell and Ioannou (2001), the balanced truncation-based ROM was used to obtain the state estimators. To reduce the effort required to implement the data assimilation codes, reduced-order approaches have been introduced into 4D variational data assimilation systems (Fang et al. 2013). An approximate linear reduced model can be obtained by projecting the forward model onto a reduced subspace. Compared to the classical variational method, the adjoint of the tangent linear model is replaced by the adjoint of a linear reduced forward model. The minimisation process of the related cost function is carried out in reduced subspace and hence reduces the computational cost albeit subject to limitations. More recently, data assimilation techniques have been integrated with machines learning methods (Farias et al. 2021). This increases the reliability of prediction and reduces error by including information with a physical meaning from observed data (Buizza et al. 2022).

Optimal monitoring location is the key in data assimilation, which will maximally improve the accuracy of numerical solutions at forecast times typical for the considered models. Targeted observations represent a subcategory under adaptive observations where data collection is optimized to improve a particular forecast aspect (e.g. energy, health impact). The targeting approach incorporates dynamical information from numerical model predictions to identify when, where and what types of observations would provide the greatest improvement to specific model.

forecasts at a future time (Daescu and Camichael 2003). Such targeted observations are important as they will allow the most effective use of available monitoring resources.

## 3.2 *Machine Learning-Based Data-Driven Modelling*

### 3.2.1 Machine Leaning Model and an Example of Urban Flooding Forecasting

Flooding and resilience are interconnected in the context of natural disasters and climate change. Resilience refers to the ability of a system or community to withstand and recover from shocks and stresses, while flooding is a common natural disaster that can cause significant damage to infrastructure, property and human lives. Building resilience to flooding involves physical infrastructure improvements, social and economic factors such as community preparedness and emergency response planning. With climate change increasing the frequency and severity of extreme weather events, including floods, it is important to address both the immediate impacts of flooding and underlying drivers of climate change through sustainable development practices and reducing greenhouse gas emissions.

Efficient and accurate real-time forecasting of flooding is critical to the provision of effective early warning. Traditional numerical models require a high computational cost associated with running large-scale numerical simulations. Nowadays, due to advancement in satellite sensors and availability of various observation data networks, more attentions have been given to big data solution for flooding forecast. Big data technologies, machine learning techniques, real-time weather monitor and evaluation are desirable for future smart cities. Recently, researchers have exploited hybrid efficient and accurate models, for example, a variety of machine learning algorithms (Weyn et al. 2020; Wang et al. 2018).

Machine learning approaches have achieved notable success in modelling ordered sequences and data with spatial context in related scientific fields (Brunton et al. 2020; Cheng et al. 2020a, b). Machine learning-based methods are powerful in simulating nonlinear and complex systems (Pawar et al. 2021; Cheng et al. 2022). The deep learning methods can identify the main dynamic patterns from historical datasets and reveal the intricate structures (Cheng et al. 2021; Wiewel et al. 2019). Recent research efforts in machine learning have been made to predict the spatiotemporal physical dynamics (Hu et al. 2019; Cheng et al. 2022; Wang et al. 2020).

As an example, the long lead-time predictive capability of the machine learning model (deep convolutional generative adversarial network—DCGAN) has been demonstrated in a flood event in the northeastern part of Greve, Denmark (Cheng et al. 2021). As shown in Fig. 6.5, Greve is a coastal city connected to the Baltic Sea and potentially affected by flood disasters induced by extreme sea level events along its coast. An extreme seawater level event with a maximum value 3.08 m is used as the input boundary condition along the coastal line. No normal flow boundary

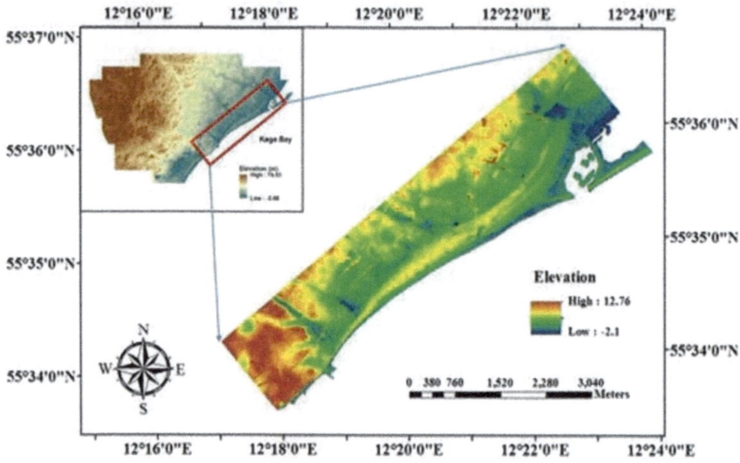


**Fig. 6.5** Study area in Greve, Municipality of Denmark (published in Cheng et al. 2021)

conditions were applied to other boundaries. The total simulation time was 24 h with a time step $\Delta t = 1$ s. Figure 6.6 shows the spatial distribution of the predicted water depth at time $t = 15.5$, 17.9 and 22.8 h over the areas marked with rectangles in Fig. 6.5. We can see that the results from the machine learning model (DCGAN) are in a good agreement with the original high fidelity model (MIKE 3 FM). The underlying flow pattern is captured well by the predictive DCGAN, especially in complex topographic features, including buildings and channels.

### 3.2.2 Hybrid Machine Learning and Physical Modelling

As well known, purely data-driven models are strongly dependent on the availability (locations and period) of monitoring data. The predictive accuracy is usually restricted due to a lack of sufficiently dense observations. Importantly, big data-driven (machine learning) modelling has its limitation due to a lack of physical knowledge (Kang et al. 2018). This limitation has hampered its wider applications to problems where the spatial–temporal dynamics are important, e.g. atmospheric, green and chemical processes. More recently, a hybrid physical and generative adversarial network (GAN) model was developed for real-time ozone forecasting with a high spatial resolution (Cheng et al. 2022). In the work of Cheng et al. (2022), the training and validation datasets originated from the reanalysis data which optimally combines the numerical solutions from the physical air quality model (here, the Nested Air Quality Prediction Modelling System—NAQPMS) and observed data in monitoring stations. In hybrid physical and machine learning modelling, physical modelling results from the governing equations can be used to improve understanding of complex ozone transport processes and provide the details of the spatial distribution of ozone concentrations, while monitoring data can be used for correcting

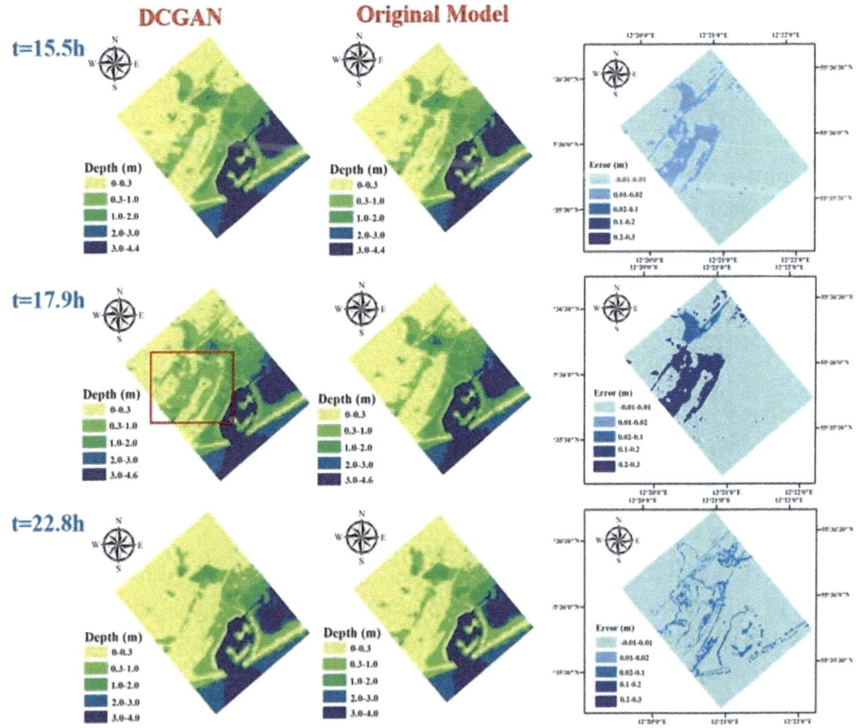


**Fig. 6.6** Comparison of the spatial distribution of predictive water depth obtained from the DCGAN (left column), the original high fidelity model (middle column) and the difference (right column) at time steps $t = 15.5$, 17.9 and 22.8 h, where the high fidelity model is MIKE 3 FM, a software tool for modelling unsteady three-dimensional free surface flows (published in Cheng et al. 2021)

physical modelling results, thus improving predictive accuracy. Recently, people have paid more attention to physics-informed neural networks (PINNs) (Karniadakis et al. 2021), which are a type of universal function approximators that can embed the knowledge of any physical laws (represented by Partial Differential Equations—PDEs) in the learning process (Raissi et al. 2019). Such hybrid physical-informed machine learning combination is in an optimal way that physical dynamic understanding can be provided from the governing equations, while data learning may find some patterns which is not expected in physical modelling. With advances of AI and physical modelling approaches, we will benefit from the hybrid approach in many ways:

- Uncertainty quantification: Data-driven approaches can used for determining uncertain parameters and replace subgrid empirical models in physical modelling.
- Improved accuracy of predictions: Assimilation of data into physical modelling can alleviate the misfit between data and modelling results. Big data can also be used for validation of physical models.

- Improved understanding of flow dynamics by considering dynamic knowledge from physical modelling and extracting unexpected flow patterns from big datasets.
- Speedup of simulations: Introducing physical modelling results to data training will help build a promising data-driven model which can then be used for spatiotemporal prediction where the CPU time is significantly reduced in comparison with complex physical modelling.

## 4 Summary

This chapter introduced recent development of advanced numerical techniques in urban environmental modelling, including 3D dynamic unstructured meshes, machine learning techniques, data assimilation and reduced-order modelling for accuracy and efficiency of simulations. Examples of realistic applications are presented.

Adaptive mesh is a technique used to dynamically adjust the mesh size and resolution of a model based on the local variation of the parameters being simulated. This approach allows for a more accurate representation of complex geometries and phenomena, while reducing computational costs. In urban modelling, adaptive mesh can be applied to represent different levels of detail in various parts of the city, such as high-density areas or open spaces, and to capture the impact of various urban design interventions.

Machine learning is a data-driven approach that uses algorithms and statistical models to automatically identify patterns in large datasets and make predictions based on them. Machine learning can be applied to urban modelling to analyse the behaviour and interactions of various urban systems, such as transportation, energy and water, and to predict their future performance under different scenarios. Machine learning can also be used to optimize urban design and planning decisions by evaluating the impact of various interventions and predicting their outcomes.

Data assimilation is a technique used to combine observational data with model simulations in order to obtain more accurate estimates of system behaviour. In urban modelling, data assimilation can be used to incorporate real-time data from sensors, cameras and other sources into models of traffic flow, air quality and other urban systems. This allows for more accurate predictions of system behaviour and more effective management of urban systems.

Reduced order modelling is a technique used to simplify complex models by reducing the number of parameters and equations required to describe a system. This approach allows for faster simulations and more efficient optimization of system design. In urban modelling, reduced-order modelling can be used to represent complex urban systems, such as the interactions between buildings and the urban environment, in a more computationally efficient way.

When combined with adaptive mesh and machine learning, data assimilation and reduced-order modelling can enable even more accurate and efficient urban modelling. Data assimilation can provide more accurate inputs for machine learning models, while reduced-order modelling can make simulations more computationally efficient, allowing for larger and more complex models to be used. Overall, these approaches can help planners and decision-makers make more informed and effective decisions about urban design and management.

## References


Brunton SL, Noack BR, Koumoutsakos P (2020) Machine learning for fluid mechanics. Annu Rev Fluid Mech 52:477–508

Buizza C, Casas CQ, Nadler P, Mack J, Marrone S, Titus Z, Le Cornec C, Heylen E, Dur T, Ruiz LB, Heaney C (2022) Data learning: integrating data assimilation and machine learning. J Comput Sci 58:101525

Carmichael GR, Sandu A, Chai T, Daescu DN, Constantinescu EM, Tang Y (2008) Predicting air quality: improvements through advanced methods to integrate models and measurements. J Comput Phys 227(7):3540–3571

Cacuci DG, Weber CF, Oblow EM, Marable JH (1980) Sensitivity theory for general systems of nonlinear equations. Nucl Sci Eng 75(1):88–110

Cheng M, Fang F, Pain CC, Navon IM (2020a) Data-driven modelling of nonlinear spatio-temporal fluid flows using a deep convolutional generative adversarial network. Comput Methods Appl Mech Eng 365:113000

Cheng M, Fang F, Kinouchi T, Navon IM, Pain CC (2020b) Long lead-time daily and monthly streamflow forecasting using machine learning methods. J Hydrol 590:125376

Cheng M, Fang F, Navon IM, Pain CC (2021) A real-time flow forecasting with deep convolutional generative adversarial network: application to flooding event in Denmark. Phys Fluids 33(5):056602

Cheng M, Fang F, Navon IM, Zheng J, Tang X, Zhu J, Pain CC (2022) Spatio-temporal hourly and daily ozone forecasting in China using a hybrid machine learning model: autoencoder and generative adversarial networks. J Adv Model Earth Syst 14(3):e2021MS002806

Cohen AJ, Brauer M, Burnett R, Anderson HR, Frostad J, Estep K, Balakrishnan K, Brunekreef B, Dandona L, Dandona R et al (2017) Estimates and 25-year trends of the global burden of disease attributable to ambient air pollution: an analysis of data from the Global Burden of Diseases Study 2015. Lancet 389:1907–1918

Daescu DN, Carmichael GR (2003) An adjoint sensitivity method for the adaptive location of the observations in air quality modeling. J Atmos Sci 60:434–450

Daescu DN, Navon IM (2008) A dual-weighted approach to order reduction in 4DVAR data assimilation. Mon Weather Rev 136(3):1026–1041

Du J, Zhu J, Fang F, Pain CC, Navon IM (2016) Ensemble data assimilation applied to an adaptive mesh ocean model. Int J Numer Meth Fluids 82(12):997–1009

Evensen G (2009) Data assimilation: the ensemble Kalman filter, vol 2. Springer, Berlin

Fang F, Pain CC, Navon IM, Cacuci DG, Chen X (2013) The independent set perturbation method for efficient computation of sensitivities with applications to data assimilation and a finite element shallow water model. Comput Fluids 76:33–49

Fang F, Pain CC, Navon IM, Xiao D (2017) An efficient goal-based reduced order model approach for targeted adaptive observations. Int J Numer Meth Fluids 83(3):263–275

Farias SV, Saotome O, Campos Velho HF, Shiguemori EH (2021) A damage detection method using neural network optimized by multiple particle collision algorithm. J Sens 2021

Farrell BF, Ioannou PJ (2001) State estimation using a reduced-order Kalman filter. J Atmos Sci 58(23):3666–3680
Fearnhead P, Künsch HR (2018) Particle filters and data assimilation. Annu Rev Stat Appl 5:421–449
Hu R, Fang F, Pain CC, Navon IM (2019) Rapid spatio-temporal flood prediction and uncertainty quantification using a deep learning method. J Hydrol 575:911–920
Kang GK, Gao JZ, Chiao S, Lu S, Xie G (2018) Air quality prediction: big data and machine learning approaches. Int J Environ Sci Dev 9(1):8–16
Karamchandani P, Vijayaraghavan K, Yarwood G (2011) Sub-grid scale plume modeling. Atmosphere 2(3):389–406
Karniadakis GE, Kevrekidis IG, Lu L et al (2021) Physics-informed machine learning. Nat Rev Phys 3:422–440. https://doi.org/10.1038/s42254-021-00314-5
Kim YH, Kim HS, Song CH (2017) Development of a reactive plume model for the consideration of power-plant plume photochemistry and its applications. Environ Sci Technol 51(3):1477–1487
Pawar S, San O, Aksoylu B, Rasheed A, Kvamsdal T (2021) Physics guided machine learning using simplified theories. Phys Fluids 33(1):011701
Piggott MD, Farrell PE, Wilson CR, Gorman GJ, Pain CC (2009) Anisotropic mesh adaptivity for multi-scale ocean modelling. Philos Trans R Soc A Math Phys Eng Sci 367:4591–5461
Raissi M, Perdikaris P, Karniadakis GE (2019) Physics-informed neural networks: a deep learning framework for solving forward and inverse problems involving nonlinear partial differential equations. J Comput Phys 378:686–707
Stratonovich RL (1959) Optimum nonlinear systems which bring about a separation of a signal with constant parameters from noise. Radiofizika 2(6):892–901
Wang Z, Xiao D, Fang F, Govindan R, Pain CC, Guo Y (2018) Model identification of reduced order fluid dynamics systems using deep learning. Int J Numer Meth Fluids 86(4):255–268
Wang HW, Li XB, Wang D, Zhao J, Peng ZR (2020) Regional prediction of ground-level ozone using a hybrid sequence-to-sequence deep learning approach. J Clean Prod 253:119841
Weyn JA, Durran DR, Caruana R (2020) Improving data-driven global weather prediction using deep convolutional neural networks on a cubed sphere. J Adv Model Earth Syst 12(9):e2020MS002109
Wiewel S, Becher M, Thuerey N (2019) Latent space physics: towards learning the temporal evolution of fluid flow. Comput Graph Forum 38(2):71–82
Xiao D, Heaney CE, Mottet L, Fang F, Lin W, Navon IM, Guo Y, Matar OK, Robins AG, Pain CC (2019) A reduced order model for turbulent flows in the urban environment using machine learning. Build Environ 148:323–337
Zheng J, Zhu J, Wang Z, Fang F, Pain CC, Xiang J (2015) Towards a new multiscale air quality transport model using the fully unstructured anisotropic adaptive mesh technology of Fluidity (version 4.1.9). Geosci Model Dev 8(10):3421–3440
Zheng J, Fang F, Wang Z, Zhu J, Li J, Xiao H, Pain CC (2020) A new anisotropic adaptive mesh photochemical model for ozone formation in power plant plumes. Atmos Environ 229:117431
Zheng J, Wu X, Fang F, Li J, Wang Z, Xiao H, Zhu J, Pain C, Linden P, Xiang B (2021) Numerical study of COVID-19 spatial–temporal spreading in London. Phys Fluids 33(4):046605

**Fangxin Fang** is a senior research fellow at Imperial College, and executive manager of data assimilation laboratory at the Data Science Institute, Imperial. She leads the research on advanced computational tools and data science technologies that can help us to manage a safe, comfortable, and healthy environment. She has published over 90 journal papers.

# Urban Design to Reduce Air Pollution Exposure in High-rise and High-density Cities

**Shuo-Jun Mei, Zhiwen Luo, and Jian Hang**

**Abstract** High-rise and high-density urban design has been associated with severe air pollution and negative impacts on urban pedestrians' health. However, proper urban design in such cities has the potential to effectively enhance outdoor ventilation and reduce air pollution exposure while maintaining land use efficiency. By better understanding the ventilation mechanisms and air pollution exposure patterns in high-rise and high-density cities, we proposed six design principles that can improve urban ventilation in high-density cities and therefore reduce air pollution exposure. They include (a) "lift-up" buildings, (b) creating height variability, (c) taking advantage of open space and central high-rise buildings, (d) creating wind paths, (e) avoiding projecting obstructions in the street canyon, and (f) taking advantage of thermal buoyancy. By incorporating these climate-sensitive design strategies, high-density cities can potentially reduce both indoor and outdoor exposure to air pollution, leading to improved sustainability and resilience.

**Keywords** Air pollution exposure · Design strategies · High-rise and high-density cities · Urban ventilation · Indoor-outdoor coupling

## 1 Introduction

The largest and fastest-growing cities in the world are now situated in Asia and Africa. According to the United Nations Human Settlements Programme (UN-HABITAT 2010), nearly 75% of urbanites will reside in these regions by the year 2050. High-density cities exhibit potential for sustainability, as they can reduce energy consumption for transportation and other activities by maximizing facility utilization rates (Kammen and Sunter 2016). However, inhabitants of high-rise and high-density

S.-J. Mei · J. Hang
School of Atmospheric Sciences, Sun Yat-Sen University, Southern Marine Science and Engineering Guangdong Laboratory (Zhuhai), Zhuhai, China

Z. Luo (✉)
Welsh School of Architecture, Cardiff University, Cardiff, UK
e-mail: LuoZ18@cardiff.ac.uk



R. Yao (ed.), *Resilient Urban Environments*, Cities and Nature,
https://doi.org/10.1007/978-3-031-55482-7_7

cities are at greater risk of severe environmental hazards, such as air pollution. Typically, high-density cities have high pollutant emission rates, attributable to high traffic volume per unit area. For example, Hong Kong has 275 vehicles per kilometre of roadway in 2002 (Hong Kong Traffic Department 2002). In addition, high-density urban areas possess low pollutant dispersion potential, since arrays of high-rise buildings generate strong resistance to approaching winds (Hang et al. 2012b; Mei et al. 2017; Yuan et al. 2014). Finally, the high population density results in a high intake fraction (Luo et al. 2010), defined as the fraction of total emissions inhaled by a receptor population. Zhou and Levy (2008) demonstrated that vehicular intake fractions were as large as 3000 ppm in a typical deep street canyon in midtown Manhattan, New York, due to its high population density and poor ventilation. To establish a healthy urban environment and mitigate health risks associated with urban residence, improving air quality in high-density cities is imperative.

In high-density cities, vehicle exhausts make a major contribution to air pollution. Motor vehicle exhausts contain a variety of air pollutants (e.g., particulate matter, benzene, carbon monoxide, nitric oxide and polycyclic aromatic hydrocarbons), which can have detrimental effects human health (Chan and Yao 2008). In Hong Kong, the chronic effects of air pollution on human health cause at least 1600 deaths, 6.8 million family doctor visits, 64,000 hospital bed-days and economic losses of 20 billion Hong Kong dollars per year (Wong et al. 2001). A study that surveyed 29,290 participants aged 50 years and above shows a linear positive exposure-response association between ambient PM2.5 and chronic obstructive pulmonary disease (Lin et al. 2018a). Exposure to PM2.5 is found directly related to an increase in overall cardiovascular mortality (Lin et al. 2018b).

The impact of air pollution can be estimated by the intake fraction, which is defined as the fraction of total emissions that are inhaled by a receptor population (Bennett et al. 2002). The intake fraction is a better indicator of air pollution impact than pollutant concentration as it describes the source-to-receptor relationship for an air pollution source. For a specific pollutant, the intake fraction (iF) can be expressed as the ratio of two terms: (a) the aggregate inhalation intake of a pollutant by an exposed population that can be attributed to emissions from a particular source or source class, divided by (b) the total emissions from that source (Marshall et al. 2003). Another exposure indicator, the personal intake fraction ($P_{IF}$), is also widely used to quantitatively evaluate the exposure to traffic-related air pollution for indoor and outdoor environments (Hang et al. 2020). Taking advantage of these mathematic indices, the urban pollutant dispersion can be more precisely evaluated. Additionally, the mitigation strategies can be better evaluated by these indices.

# 2 Ventilation Mechanism in High-rise and High-density Cities

The mitigation of air pollution in high-density cities can be achieved through effective urban ventilation. The performance of urban ventilation is determined by various factors, such as geographic and topographic features, meteorological conditions, and urban morphology (Luo et al. 2011). To quantitatively evaluate the performance of urban ventilation and its correlation with influencing factors, numerous studies have been conducted in recent years.

## 2.1 *Urban Ventilation Concept*

The concept of "urban ventilation" or "city breathability" is the capacity of built area in replacing polluted air with ambient fresh air (Hang et al. 2015, 2012a). The wind blowing from rural areas is considered a source of clean air and plays a crucial role in improving the air quality in urban regions.

The efficiency of urban ventilation can be assessed quantitatively using various indices, including the age of air (Hang and Li 2011), exchange velocity (Mei and Yuan 2021a), purging flow rate (Mei et al. 2017) and net escape velocity (Hang et al. 2015), air exchange rate (Mei et al. 2019). The local mean age of air ($\overline{\tau_p}$) refers to the duration for rural air to reach a specific point after entering the city and indicates the efficiency of rural air distribution in urban regions. Figure 7.1 illustrates the age of air concept in a city. The exchange velocity ($U_E$) represents the average velocity of mass transfer between the in-canopy and above-canopy flows. The purging flow rate (PFR) is defined as the ratio of pollutant generates rate to the average pollutant concentration in the urban canopy layer, defined in Fig. 7.1, or pedestrian level. The net escape velocity (NEV) is the velocity corresponding to the purging flow rate. The air exchange rate (ACH) represents the number of air exchanges between urban canopy and above air per hour. These ventilation indices enable the quantitative assessment of urban ventilation and assist in developing effective strategies for mitigating urban air pollution.

## 2.2 *Urban Morphology Influences on Urban Ventilation*

Changing the urban morphology is a practical approach to mitigate air pollution in high-density cities because many other factors are not controllable. By adopting these indices mentioned above, the efficiency of changing urban morphology in improving urban ventilation can be estimated.

Reducing building density is an effective approach to improving the urban ventilation (Hang et al. 2015; Mei et al. 2017), although it may reduce land use efficiency. For

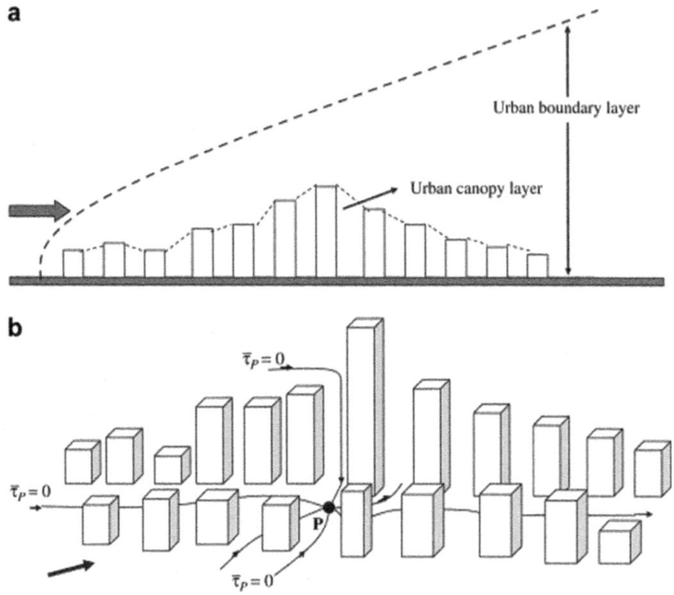


**Fig. 7.1** **a** Definition of urban canopy layer and the illustration of the urban wind environment and **b** illustration of the concept of age of air in the urban canopy layer by resetting the age of air to a reference value such as zero for air entering the urban canopy layer (Hang et al. 2009a)

example, the personal intake fraction increases from 5.6–5.8 ppm to 43.6–120.8 ppm when increasing the aspect ratio ($H/W$, ratio between building height to street width) from 3 to 5, while it only increases from 3.8–4.3 ppm to 5.6–5.8 ppm when increasing the $H/W$ from 1.0 to 3.0 (Zhang et al. 2019). This indicates that high-density cities can avoid severe pollutant accumulation by limiting the aspect ratio.

Studies have shown that increasing building height variations can significantly improve urban ventilation without sacrificing land use efficiency. For instance, Lin et al. (2014) found that increasing the standard deviation of building height variations from 0 to 83.3% led to an increase in air change rate (ACH) from 1.7 to 7.5. This increase in ACH resulted in better air circulation and reduced air pollution levels in the city.

The direction of wind is another significant factor that impacts urban ventilation efficiency. When incoming winds are parallel to the street axis, they can efficiently remove traffic pollutants (Hang et al. 2010). On the other hand, when incoming winds are perpendicular to the street axis, street canyon ventilation becomes worse, leading to high pollutant accumulation (Mei et al. 2019). Designing the urban space with street axis slightly tilting to the prevailing wind from perpendicular direction can improve urban ventilation (Hang et al. 2018a).

Creating open space in high-density areas is another effective measure to increase urban ventilation efficiency (Wang et al. 2017). The shape of the city also has a significant impact on urban ventilation efficiency. Studies have shown that round cities have better urban ventilation than square ones (Hang et al. 2009c, 2009b). These measures can improve urban ventilation without losing land use efficiency and are therefore suitable for reducing air pollution in high-density cities.

## 2.3 Thermal Buoyancy Influences on Urban Ventilation

Apart from the urban morphology, the thermal characteristics of urban structure can also change the urban ventilation performance. The urban surfaces, which can be heated by solar radiation, can alter the pollutant dispersion and ventilation performance of street canyons through buoyancy-driven ventilation (Liu et al. 2021b, 2021a). The interplay between wind-driven flow and buoyancy-driven flow can significantly change the flow pattern inside the street canyon; see Fig. 7.2. For example, one vortex in the street canyon can be split into multi-vortex when a windward wall is heated (Mei et al. 2016). Hence, the pollutant dispersion capacity is significantly reduced. For other heating surfaces, the effect of buoyancy on pollutant dispersion varies. Ground heating can significantly enhance mean flow, turbulence fluctuation, and pollutant flux inside the street canyons (Li et al. 2010). For leeward wall heating, the vortex rotation is accelerated by buoyancy and therefore pollution dispersion rate is enhanced (Mei and Yuan 2022). The effect of buoyancy can be classified as "assisting condition" in which buoyancy-driven flow has the same direction as that of wind-driven vortex and "opposing condition" in which buoyancy-driven flow has the opposite direction to that of wind-driven vortex (Cai 2012).

Several parameters were proposed to quantify the relative strength between ambient wind and buoyancy: Froude number (Fr) and Richardson number (Ri) (Ruck 1993). These two parameters can be regarded as equivalent as they are reciprocal to

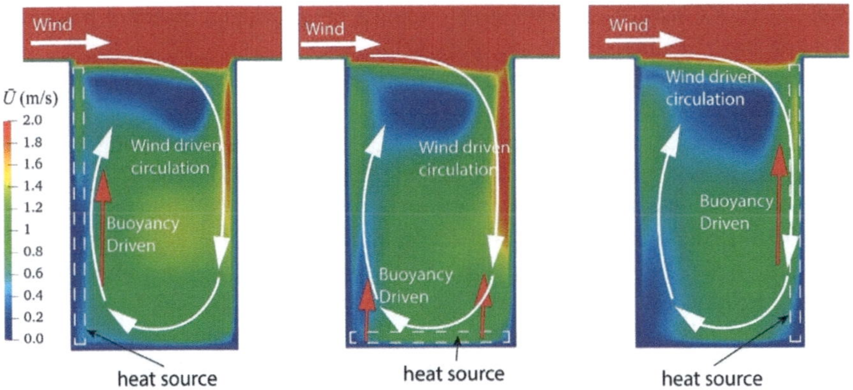


**Fig. 7.2** Mechanism of the interaction between the wind-driven circulation and buoyancy force, from (Mei and Yuan 2021a). The dashed white box marks the location of heat emission. The white arrows represent the wind-driven flow direction, and the red arrows represent the buoyancy-driven direction

each other, i.e. $\mathrm{Ri} = 1/\mathrm{Fr}$. The large absolute value of Ri number indicates a strong buoyancy effect, and negative value indicates positive buoyancy, and naturally, the small value of Fr number shows strong buoyancy effect. The definition of Richardson number was frequently written as $\mathrm{Ri} = \mathrm{Gr}/\mathrm{Re}^2$, where Gr is the Grashof number and Re is the Reynolds number. Their definitions are as follows,

$$\mathrm{Re} = \frac{u_0 H}{\nu} \tag{1}$$

$$\mathrm{Gr} = \frac{g\beta H^3 (T_w - T_\infty)}{\nu^2} \tag{2}$$

where $u_0$: ambient wind speed, $g$: gravitational acceleration, $H$: characteristic length (e.g., building heights), $\beta$: thermal expansion coefficient, $T_w$: surface temperature, $T_\infty$: ambient air temperature, $\nu$: air kinematic viscosity.

The airflow pattern at the urban canopy layer is found strongly related to these dimensionless parameters. For example, the wind tunnel experiment with a heated cubic block shows the reattachment lengths behind the building decreased with Ri when $\mathrm{Ri} < -0.2$, due to strong buoyancy effect, and unchanged when $\mathrm{Ri} > -0.2$ (Ruck 1993). A CFD simulation of street canyon shows that the single vortex flow was transferred into two vortices flow when $\mathrm{Ri} < -2$ under heated windward wall conditions (Xie et al. 2005). For a heated ground surface, two counter-rotating flow structures appeared when $\mathrm{Ri} < -4$. The simulation of detected one or two stable counter-rotating vortices is at high Fr numbers ($\approx 10^1$) and lower vortex-dominated flow at low Fr numbers ($\approx 10^{-1}$) (Oliveira Panão et al. 2009). The effect of buoyancy on urban ventilation also depends on the local geometry and wind direction. A numerical simulation shows that parallel wind and wall heating get better urban ventilation than oblique wind and isothermal cases (Hang et al. 2015). A field measurement Scaled Outdoor Measurement of Urban Climate and Health (SOMUCH) shows that narrower streets attain a more evident impact of buoyancy effect on street ventilation (Chen et al. 2017).

Buoyancy plays a more important role in urban ventilation when there is a lack of ambient wind. Instead, the vertical plume flows dominate the street canyon. Under an extremely low wind scenario, the heated surfaces will generate thermal plumes under that condition, with warm air moving upward (Pham et al. 2005). The field measurement conducted in a high-rise building found the maximum wall-flow velocity was about 0.76–1.33 m/s under very weak background wind (Fan et al. 2016). The closely spaced thermal plumes attract each other and generally merged into a single plume. In that condition, the air exchange velocity was found directly related to the wall temperature in these simulations (Mei et al. 2018). A recent study indicates that the airflow at the pedestrian level under a pure buoyancy-driven condition can reach about 1.5–4 m/s in a real urban neighbourhood (Mei and Yuan 2021b). This level of wind speed can be regarded as an effective source of urban ventilation. On a larger scale, the thermal buoyancy also drives thermal circulations, such as sea–land breeze, country breeze and slope wind. These thermal circulations dominate the urban ventilation under a calm conditions (Wang et al. 2020). In Hong Kong, the slope wind was found to dominate the ventilation when the building is lower than 60 m. For higher buildings, the building surface heating also plays an important role in urban ventilation (Luo et al. 2010).

## 3 Indoor–Outdoor Air Pollution Coupling in Cities

Urban areas are heavily impacted by outdoor air pollutants, which can penetrate indoor spaces through various means such as doors, windows, building cracks, and ventilation duct systems. According to a study conducted in London, indoor exposure contributes to 83% of the total population exposure concentration, making it the largest contributor (Kazakos et al. 2020). Due to spending most of their time indoors, individuals residing in near-road buildings may experience 6–9 times higher vehicular pollutant exposure compared to those at pedestrian level (Hang et al. 2017). As such, it is important to consider both indoor and outdoor exposure to outdoor pollution.

In high-density cities, the natural ventilation of buildings is driven by street canyon airflow, as shown in Fig. 7.3. This creates a coupling effect between indoor and outdoor ventilation, where the natural ventilation airflow changes the street canyon airflow, and vice versa. This coupling effect can significantly impact the prediction of indoor air pollution exposure, particularly since the wind inside high-density cities tends to be weak, single-sided ventilation is commonly adopted, and the building geometry and thermal characteristics vary from building to building. Additionally, the natural ventilation airflow between rooms can interact with each other, further complicating the overall exposure evaluation. Therefore, when evaluating overall exposure in high-density cities, it is important to consider the coupling effect of indoor and outdoor ventilation.

The main influence factors of indoor exposure in high-density cities include the ventilation mode, room location, building opening, street canyon geometry (Yuan et al. 2022), wind speed and direction (Dai et al. 2020), indoor–outdoor temperature differences (Hu et al. 2022), emission rate and emission location (Hang et al. 2018b). The numerical simulations coupling indoor and outdoor ventilation (Hu et al. 2022; Dai et al. 2022; Huang et al. 2021; Yang et al. 2020; Gough et al. 2020; Hang et al. 2016) show that the main characteristics of indoor–outdoor dispersion include:

(a) Indoor exposure in upwind rooms with single-side ventilation is high. The traffic pollutants are transported to the upwind wall and then penetrate the indoor space.
(b) Indoor exposure in low-elevation rooms is high. The traffic pollutants in deep street canyons are trapped near ground level. At that level, the high concentration pollutants in the outdoor space can easily penetrate indoor space.
(c) Traffic pollutants can diffuse to higher floors in cross-ventilation modes as the vortex flow inside the street canyon is destructed.

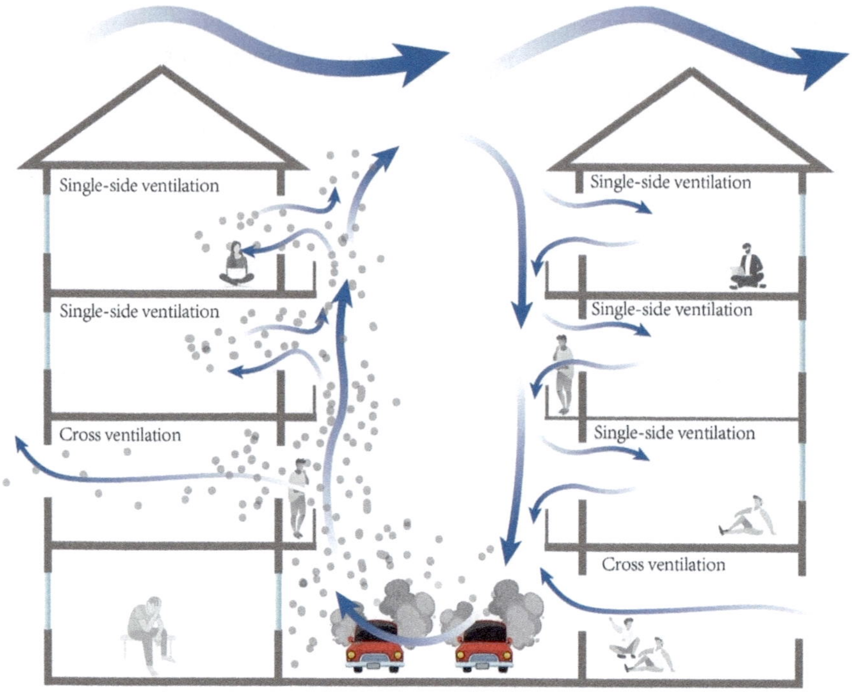


**Fig. 7.3** Schematic explanation of the traffic pollutant dispersion in high-density cities with natural ventilation

(d) When indoor–outdoor temperature difference exits, cross-ventilation rooms have a smaller air pollution exposure. The personal intake fraction can decrease by as much as 66% compared to single-side ventilation.
(e) Ground heating reduces indoor people's exposure by increasing outdoor ventilation.

1. **Design strategies to improve outdoor ventilation and air quality**

Since urban morphology has a significant impact on outdoor ventilation, it is possible to mitigate air pollution in high-density cities via proper urban design. Previous studies have shown that the following approaches can improve outdoor ventilation and reduce air pollution without lost land use efficiency.

(a) **"Lift-up" buildings**

The "lift-up" buildings provide extra breezeways at near ground level, which can effectively remove the traffic pollutants and prevent the development of vortex flow in street canyon (Sha et al. 2018), as shown in Fig. 7.4.

(b) **Creating height variability**

The building arrays with height variability experience better air quality than building arrays with uniform height, as shown in Fig. 7.5. The vortex flow pattern between buildings can be distributed by the height variation, and then the outdoor ventilation rate can be improved (Chen et al. 2017).

(c) **Open space and central high-rise buildings**

The simulation results (Li et al. 2022) show that open space increases the spatially-averaged velocity ratio for the whole area by 0.40–2.27%. Central open space and high-rise buildings decrease $P_{IF}$ under all four wind directions, by 6.56–16.08% and 9.59–24.70%, respectively. Another simulation (Li et al. 2022) shows that central high-rise building (2H) can increase wind speed by 4.73–23.36% and decrease the CO concentration by 4.39–23.00%. As shown in Fig. 7.6, the open space and high-rise buildings can improve the surrounding ventilation.

(d) **Creating wind paths**

Making the street direction parallel to prevailing wind direction can effectively decrease the pollutant concentration. When the street is perpendicular to the prevailing wind, street canyon and vortex flow forms, resulting in a low dispersion rate and high pollutant concentration (Mei et al. 2019), shown as in Fig. 7.7.

(e) **Avoiding projecting obstructions in the street canyon**

The advertisement boards weaken urban wind and increase pedestrian pollutant exposure (Lin et al. 2019). The impact of these obstructions depends on their shape. For example, the vertical projections show less blocking effect compared to horizontal projections (Ng 2009), see Fig. 7.8.

(f) **Taking advantage of thermal buoyancy**

Thermal buoyancy has the potential to improve the outdoor ventilation rate, but it may also have negative impacts on the thermal environment if the heat sources are not well arranged. The main principle of taking advantage of thermal buoyancy is to avoid increasing the air temperature in the street canyon. One approach to achieving this is by using a heated upwind wall to enhance outdoor ventilation, where the wind-driven and buoyancy-driven airflow are in the same direction, as shown in Fig. 7.9a. In such a scenario, the heat from the upwind wall can be efficiently transported out of the street canyon, without heating the pedestrian level air. Under no wind conditions, as depicted in Fig. 7.9b, buoyancy plays a dominant role in ventilating the street canyon, with the updraft buoyant airflow entraining the air from near ground level. In this case, the "Lift-up" buildings prove to be much more effective in promoting outdoor ventilation, as they facilitate air entry into the street canyon via the pedestrian level.

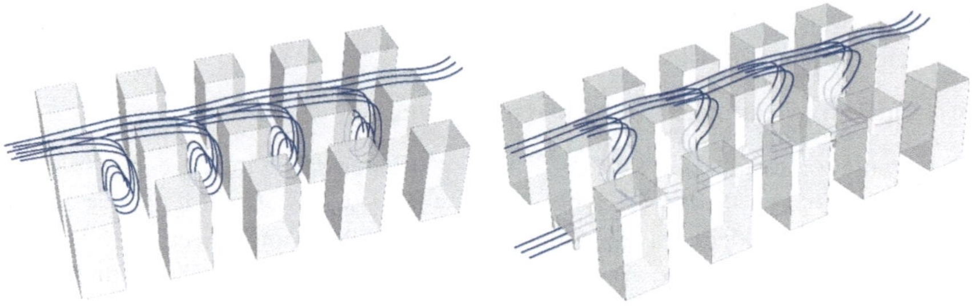

**Fig. 7.4** Schematic illustration of airflow pattern in typical building arrays and in "lift-up" building arrays

**Fig. 7.5** Schematic illustration of airflow pattern in building arrays with height variability

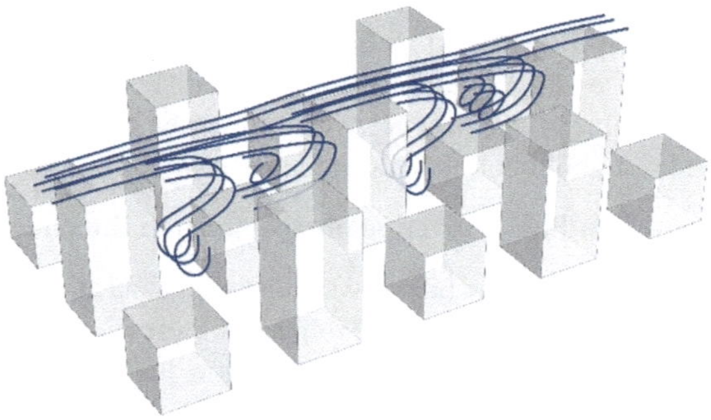

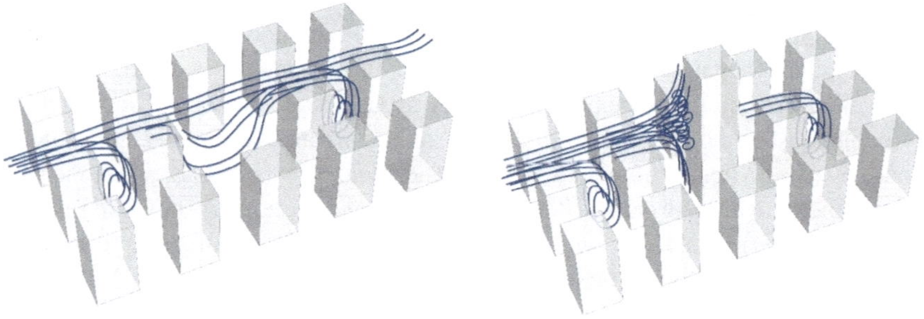

**Fig. 7.6** Schematic illustration of airflow pattern in building arrays with open space (left) and central high-rise building (right)

## 4 Conclusions

High-density design has both positive and negative impacts on urban living. While it reduces energy usage for traffic, it also brings severe air pollution, which significantly damages urban pedestrians' health. Furthermore, outdoor air pollutants can penetrate indoor spaces, leading to further health damage. Through better understanding

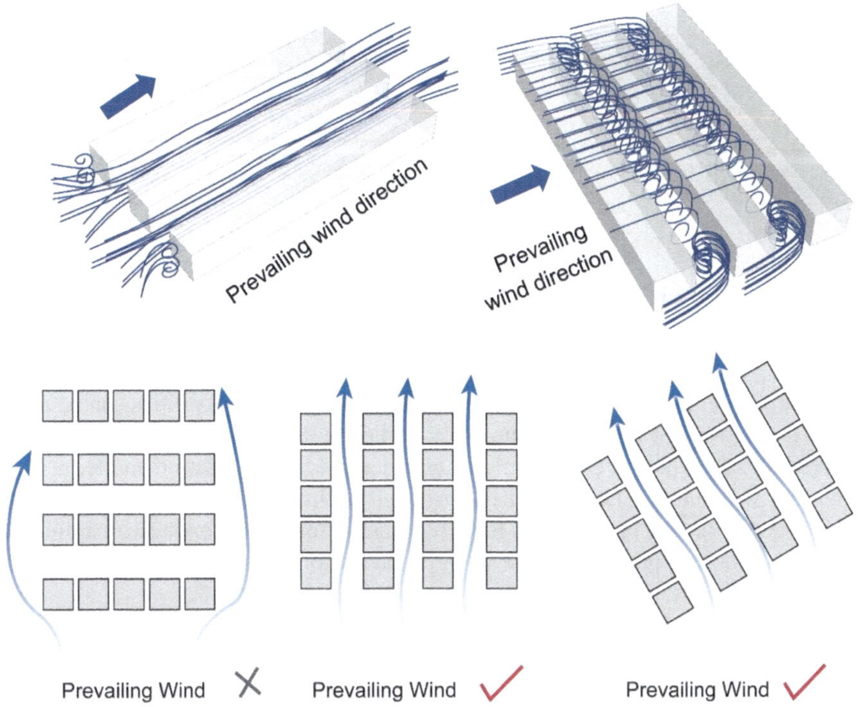


**Fig. 7.7** Schematic illustration of airflow pattern in streets parallel (left) and perpendicular (right) to the prevailing wind direction, adapted from Ng (2009)

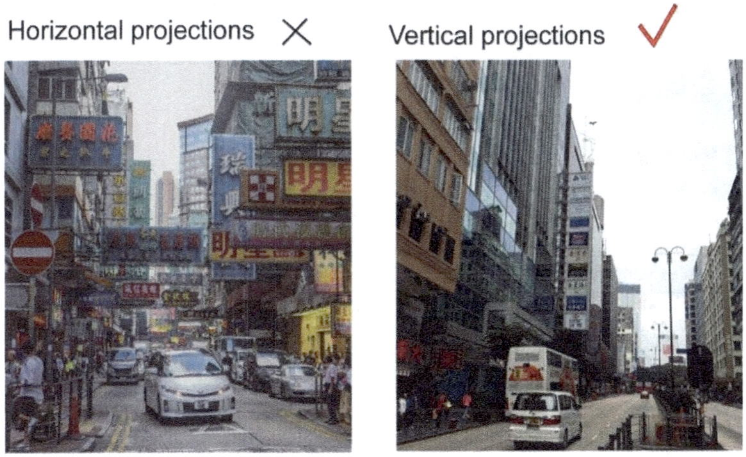


**Fig. 7.8** Horizontal and vertical projections in street canyons, adapted from Ng (2009)

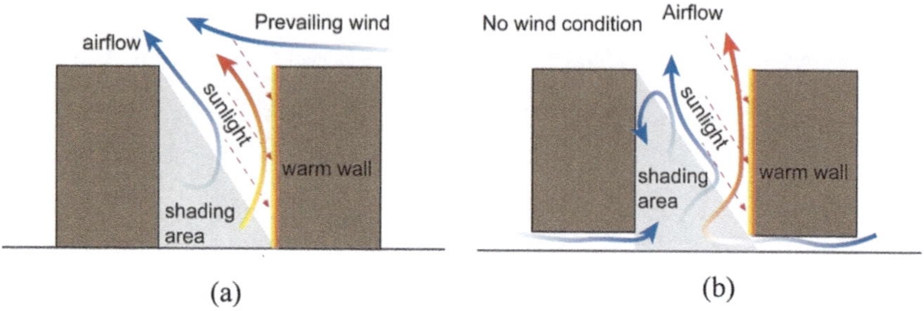


**Fig. 7.9** Taking advantage of thermal buoyancy to improve outdoor ventilation for **a** wind-driven conditions and **b** no wind conditions

of the urban ventilation mechanism, air pollution in high-density cities can be mitigated through urban design. Quantitative studies on urban ventilation have shown that proper urban design can effectively enhance outdoor ventilation and reduce air pollution without sacrificing land use efficiency. Urban morphology and thermal buoyancy are the two main factors for urban ventilation. Based on these quantitative studies, six design principles have been proposed to improve urban ventilation in high-density cities.

The design principles include "Lift-up" buildings, creating height variability, taking advantage of open space and central high-rise buildings, creating wind paths, avoiding projecting obstructions in the street canyon, and taking advantage of thermal buoyancy. By incorporating these climate-sensitive design strategies, high-density cities can improve their sustainability and resilience. In conclusion, improving urban ventilation in high-density cities is essential to mitigate air pollution and protect the health of urban residents. These design strategies offer a practical and effective solution to achieve this goal.

## References


Bennett DH, McKone TE, Evans JS, Nazaroff WW, Margni MD, Jolliet O et al (2002) Defining intake fraction. Environ Sci Technol 36(9). PMID: 12026996

Cai X (2012) Effects of differential wall heating in street canyons on dispersion and ventilation characteristics of a passive scalar. Atmos Environ 51:268–277. https://doi.org/10.1016/j.atmosenv.2012.01.010

Chan CK, Yao X (2008) Air pollution in mega cities in China. Atmos Environ 42(1):1–42. https://doi.org/10.1016/j.atmosenv.2007.09.003

Chen L, Hang J, Sandberg M, Claesson L, Di Sabatino S, Wigo H (2017) The impacts of building height variations and building packing densities on flow adjustment and city breathability in idealized urban models. Build Environ 118:344–361. https://doi.org/10.1016/j.buildenv.2017.03.042

Dai Y, Mak CM, Hang J, Zhang F, Ling H (2022) Scaled outdoor experimental analysis of ventilation and interunit dispersion with wind and buoyancy effects in street canyons. Energ Build 255. https://doi.org/10.1016/j.enbuild.2021.111688

Dai Y, Mak CM, Zhang Y, Cui D, Hang J (2020) Investigation of interunit dispersion in 2D street canyons: a scaled outdoor experiment. Build Environ 171. https://doi.org/10.1016/j.buildenv.2020.106673

Fan Y, Li Y, Hang J, Wang K, Yang X (2016) Natural convection flows along a 16-storey high-rise building. Build Environ 107:215–225. https://doi.org/10.1016/j.buildenv.2016.08.003

Gough HL, Barlow JF, Luo Z, King MF, Halios CH, Grimmond CSB (2020) Evaluating single-sided natural ventilation models against full-scale idealised measurements: impact of wind direction and turbulence. Build Environ 170. https://doi.org/10.1016/j.buildenv.2019.106556

Hang J, Chen L, Lin Y, Buccolieri R, Lin B (2018a) The impact of semi-open settings on ventilation in idealized building arrays. Urban Clim 25:196–217. https://doi.org/10.1016/j.uclim.2018.07.003

Hang J, Chen X, Chen G, Chen T, Lin Y, Luo Z et al (2020) The influence of aspect ratios and wall heating conditions on flow and passive pollutant exposure in 2D typical street canyons. Build Environ 168. https://doi.org/10.1016/j.buildenv.2019.106536

Hang J, Li Y (2011) Age of air and air exchange efficiency in high-rise urban areas and its link to pollutant dilution. Atmos Environ 45(31):5572–5585. https://doi.org/10.1016/j.atmosenv.2011.04.051

Hang J, Li Y, Buccolieri R, Sandberg M, Di Sabatino S (2012a) On the contribution of mean flow and turbulence to city breathability: the case of long streets with tall buildings. Sci Total Environ 416:362–373. https://doi.org/10.1016/j.scitotenv.2011.12.016

Hang J, Li Y, Sandberg M, Buccolieri R, Di Sabatino S (2012b) The influence of building height variability on pollutant dispersion and pedestrian ventilation in idealized high-rise urban areas. Build Environ 56:346–360. https://doi.org/10.1016/j.buildenv.2012.03.023

Hang J, Lin M, Wong DC, Wang X, Wang B, Buccolieri R (2016) On the influence of viaduct and ground heating on pollutant dispersion in 2D street canyons and toward single-sided ventilated buildings. Atmos Pollut Res 7(5):817–832. https://doi.org/10.1016/j.apr.2016.04.009

Hang J, Luo Z, Wang X, He L, Wang B, Zhu W (2017) The influence of street layouts and viaduct settings on daily carbon monoxide exposure and intake fraction in idealized urban canyons. Environ Pollut 220:72–86. https://doi.org/10.1016/j.envpol.2016.09.024

Hang J, Sandberg M, Li Y (2009a) Age of air and air exchange efficiency in idealized city models. Build Environ 44(8):1714–1723. https://doi.org/10.1016/j.buildenv.2008.11.013

Hang J, Sandberg M, Li Y (2009b) Effect of urban morphology on wind condition in idealized city models. Atmos Environ 43(4):869–878. https://doi.org/10.1016/j.atmosenv.2008.10.040

Hang J, Sandberg M, Li Y, Claesson L (2009c) Pollutant dispersion in idealized city models with different urban morphologies. Atmos Environ 43(38):6011–6025. https://doi.org/10.1016/j.atmosenv.2009.08.029

Hang J, Sandberg M, Li Y, Claesson L (2010) Flow mechanisms and flow capacity in idealized long-street city models. Build Environ 45(4):1042–1053. https://doi.org/10.1016/j.buildenv.2009.10.014

Hang J, Wang Q, Chen X, Sandberg M, Zhu W, Buccolieri R et al (2015) City breathability in medium density urban-like geometries evaluated through the pollutant transport rate and the net escape velocity. Build Environ 94(P1):166–182. https://doi.org/10.1016/j.buildenv.2015.08.002

Hang J, Xian Z, Wang D, Mak CM, Wang B, Fan Y (2018b) The impacts of viaduct settings and street aspect ratios on personal intake fraction in three-dimensional urban-like geometries. Build Environ 143:138–162. https://doi.org/10.1016/j.buildenv.2018.07.001

Hong Kong Traffic Department (2002) Hong Kong Annual Traffic Census. Traffic and Transport Survey Division Publication

Hu Y, Wu Y, Wang Q, Hang J, Li Q, Liang J et al (2022) Impact of indoor-outdoor temperature difference on building ventilation and pollutant dispersion within urban communities. Atmosphere 13(1). https://doi.org/10.3390/atmos13010028

Huang J, Jones P, Zhang A, Hou SS, Hang J, Spengler JD (2021) Outdoor airborne transmission of coronavirus among apartments in high-density cities. Front Built Environ 7. https://doi.org/10.3389/fbuil.2021.666923

Kammen DM, Sunter DA (2016) City-integrated renewable energy for urban sustainability. Science 352(6288):922–928. https://doi.org/10.1126/science.aad9302

Kazakos V, Luo Z, Ewart I (2020) Quantifying the health burden misclassification from the use of different PM2.5 exposure tier models: a case study of London. Int J Environ Res Public Health 17(3). https://doi.org/10.3390/ijerph17031099

Li Q, Liang J, Wang Q, Chen Y, Yang H, Ling H et al (2022) Numerical investigations of urban pollutant dispersion and building intake fraction with various 3D building configurations and tree plantings. Int J Environ Res Public Health 19(6). https://doi.org/10.3390/ijerph19063524

Li XX, Britter RE, Koh TY, Norford LK, Liu CH, Entekhabi D et al (2010) Large-Eddy simulation of flow and pollutant transport in urban street canyons with ground heating. Bound-Layer Meteorol 137(2):187–204. https://doi.org/10.1007/s10546-010-9534-8

Lin H, Qian ZM, Guo Y, Zheng Y, Ai S, Hang J et al (2018a) The attributable risk of chronic obstructive pulmonary disease due to ambient fine particulate pollution among older adults. Environ Int 113:143–148. https://doi.org/10.1016/j.envint.2018.01.029

Lin H, Tao J, Qian ZM, Ruan Z, Xu Y, Hang J et al (2018b) Shipping pollution emission associated with increased cardiovascular mortality: a time series study in Guangzhou, China. Environ Pollut 241:862–868. https://doi.org/10.1016/j.envpol.2018.06.027

Lin M, Hang J, Li Y, Luo Z, Sandberg M (2014) Quantitative ventilation assessments of idealized urban canopy layers with various urban layouts and the same building packing density. Build Environ 79:152–167. https://doi.org/10.1016/j.buildenv.2014.05.008

Lin Y, Chen G, Chen T, Luo Z, Yuan C, Gao P et al (2019) The influence of advertisement boards, street and source layouts on CO dispersion and building intake fraction in three-dimensional urban-like models. Build Environ 150:297–321. https://doi.org/10.1016/j.buildenv.2019.01.012

Liu J, Cui S, Chen G, Zhang Y, Wang X, Wang Q et al (2021a) The influence of solar natural heating and $NO_x$-$O_3$ photochemistry on flow and reactive pollutant exposure in 2D street canyons. Sci Total Environ 759. https://doi.org/10.1016/j.scitotenv.2020.143527

Liu S, Yang X, Yang H, Gao P, Hang J, Wang Q (2021b) Numerical investigation of solar impacts on canyon vortices and its dynamical generation mechanism. Urban Clim 39. https://doi.org/10.1016/j.uclim.2021.100978

Luo Z, Li Y, Lu WZ (2011) Effects of urban ventilation patterns on the carbon monoxide concentration in a high-rise mega city. Int J Vent 10(3):239–250. https://doi.org/10.1080/14733315.2011.11683952

Luo Z, Li Y, Nazaroff WW (2010) Intake fraction of nonreactive motor vehicle exhaust in Hong Kong. Atmos Environ 44(15):1913–1918. https://doi.org/10.1016/j.atmosenv.2010.02.016

Marshall JD, Riley WJ, McKone TE, Nazaroff WW (2003) Intake fraction of primary pollutants: motor vehicle emissions in the South Coast Air Basin. Atmos Environ 37(24):3455–3468. https://doi.org/10.1016/S1352-2310(03)00269-3

Mei SJ, Hu JT, Liu D, Zhao FY, Li Y, Wang HQ (2018) Thermal buoyancy driven canyon airflows inside the compact urban blocks saturated with very weak synoptic wind: plume merging mechanism. Build Environ 131:32–43. https://doi.org/10.1016/j.buildenv.2017.12.035

Mei SJ, Hu JT, Liu D, Zhao FY, Li Y, Wang Y et al (2017) Wind driven natural ventilation in the idealized building block arrays with multiple urban morphologies and unique package building density. Energ Build 155:324–338. https://doi.org/10.1016/j.enbuild.2017.09.019

Mei SJ, Liu CW, Liu D, Zhao FY, Wang HQ, Li XH (2016) Fluid mechanical dispersion of airborne pollutants inside urban street canyons subjecting to multi-component ventilation and unstable thermal stratifications. Sci Total Environ 565:1102–1115. https://doi.org/10.1016/j.scitotenv.2016.05.150

Mei SJ, Luo Z, Zhao FY, Wang HQ (2019) Street canyon ventilation and airborne pollutant dispersion: 2-D versus 3-D CFD simulations. Sustain Cities Soc 50. https://doi.org/10.1016/j.scs.2019.101700

Mei SJ, Yuan C (2021a) Analytical and numerical study on transient urban street air warming induced by anthropogenic heat emission. Energ Build 231. https://doi.org/10.1016/j.enbuild.2020.110613

Mei SJ, Yuan C (2021b) Three-dimensional simulation of building thermal plumes merging in calm conditions: turbulence model evaluation and turbulence structure analysis. Build Environ 203. https://doi.org/10.1016/j.buildenv.2021.108097

Mei SJ, Yuan C (2022) Urban buoyancy-driven air flow and modelling method: a critical review. Build Environ 210. https://doi.org/10.1016/j.buildenv.2021.108708

Ng E (2009) Policies and technical guidelines for urban planning of high-density cities—air ventilation assessment (AVA) of Hong Kong. Build Environ 44(7):1478–1488. https://doi.org/10.1016/j.buildenv.2008.06.013

Oliveira Panão MJN, Gonçalves HJP, Ferrão PMC (2009) Numerical analysis of the street canyon thermal conductance to improve urban design and climate. Build Environ 44(1):177–187. https://doi.org/10.1016/j.buildenv.2008.02.004

Pham MV, Plourde F, Kim SD (2005) Three-dimensional characterization of a pure thermal plume. J Heat Transf 127(6):624–636. https://doi.org/10.1115/1.1863275

Ruck B (1993) Wind-tunnel measurements of flow field characteristics around a heated model building. J Wind Eng Ind Aerodyn 50:139–151. https://doi.org/10.1016/0167-6105(93)90069-Z

Sha C, Wang X, Lin Y, Fan Y, Chen X, Hang J (2018) The impact of urban open space and 'lift-up' building design on building intake fraction and daily pollutant exposure in idealized urban models. Sci Total Environ 633:1314–1328. https://doi.org/10.1016/j.scitotenv.2018.03.194

UN-HABITAT (2010) The state of Asian cities 2010/11. http://www.unhabitat.org/pmss/listItemDetails.aspx?publicationID=3078

Wang Q, Hang J, Fan Y, Li Y (2020) Urban plume characteristics under various wind speed, heat flux, and stratification conditions. Atmos Environ 239. https://doi.org/10.1016/j.atmosenv.2020.117774

Wang Q, Sandberg M, Lin Y, Yin S, Hang J (2017) Impacts of urban layouts and open space on urban ventilation evaluated by concentration decay method. Atmosphere 8(9). https://doi.org/10.3390/atmos8090169

Wong CM, Ma S, Hedley AJ, Lam TH (2001) Effect of air pollution on daily mortality in Hong Kong. Environ Health Perspect 109(4):335–340. https://doi.org/10.2307/3454891

Xie X, Huang Z, Wang J, Xie Z (2005) The impact of solar radiation and street layout on pollutant dispersion in a street canyon. Build Environ 40(2):201–212. https://doi.org/10.1016/j.buildenv.2004.07.013

Yang X, Zhang Y, Hang J, Lin Y, Mattsson M, Sandberg M et al (2020) Integrated assessment of indoor and outdoor ventilation in street canyons with naturally-ventilated buildings by various ventilation indexes. Build Environ 169. https://doi.org/10.1016/j.buildenv.2019.106528

Yuan C, Ng E, Norford LK (2014) Improving air quality in high-density cities by understanding the relationship between air pollutant dispersion and urban morphologies. Build Environ 71:245–258. https://doi.org/10.1016/j.buildenv.2013.10.008

Yuan C, Zhu R, Tong S, Mei S, Zhu W (2022) Impact of anthropogenic heat from air-conditioning on air temperature of naturally ventilated apartments at high-density tropical cities. Energ Build 268. https://doi.org/10.1016/j.enbuild.2022.112171

Zhang K, Chen G, Wang X, Liu S, Mak CM, Fan Y et al (2019) Numerical evaluations of urban design technique to reduce vehicular personal intake fraction in deep street canyons. Sci Total Environ 653:968–994. https://doi.org/10.1016/j.scitotenv.2018.10.333

Zhou Y, Levy JI (2008) The impact of urban street canyons on population exposure to traffic-related primary pollutants. Atmos Environ 42(13):3087–3098. https://doi.org/10.1016/j.atmosenv.2007.12.037

**Zhiwen Luo** is Chair in Architectural and Urban Science at Cardiff University. His research focuses on designing smart, resilient, and healthy built environments at both building and urban scales with minimal energy expenditure, in response to climate change and public health emergencies such as COVID-19. He contributes this chapter with first author Dr. Shuo-Jun Mei, associate professor at Sun Yat-sen University, and third author Jian Hang, full professor at Sun Yat-sen University.

# Urban Ventilation

**Maria Kolokotroni**

**Abstract** This chapter focuses on ventilation to achieve satisfactory indoor air quality and provide thermal comfort in urban buildings. The impact of urban specific parameters of air and thermal (urban heat island) pollution on building ventilation is outlined. Urban external pollutants are discussed, and simple equations to calculate their concentration indoors are given. The urban heat island is explained and methods to account for this in the design of buildings is outlined. The impact of climate change on urban parameters is presented with suggestions on how this can be considered during the design through the use of appropriate weather files to include climate change and urban microclimate modifications. Case studies are presented with measured and simulated performance of urban ventilation to indicate suitable solutions.

**Keywords** Urban ventilation · Indoor air quality · Thermal comfort · Weather · Microclimate

## 1 Introduction

Ventilation of buildings has been a very important design consideration because it contributes to their function and performance in several ways: (a) maintain indoor air quality, usually considered as the primary function of ventilation, (b) avoid overheating when external conditions are favourable, usually in combination with other passive cooling strategies and (c) reduce energy use if natural ventilation is possible.

The COVID-19 pandemic has elevated the knowledge about the usefulness of ventilation amongst the occupants of the buildings. Therefore, while ventilation was only a background issue for occupants compared to, for example, insulation and renewables, it is now at the forefront of actions and interventions both for residential and commercial buildings. Natural ventilation through opening windows was encouraged during the pandemic. However, does natural ventilation provide healthy

M. Kolokotroni (✉)
Brunel University London, London, UK
e-mail: maria.kolokotroni@brunel.ac.uk

 
R. Yao (ed.), *Resilient Urban Environments*, Cities and Nature,
https://doi.org/10.1007/978-3-031-55482-7_8

air on all occasions? The simple answer is no as it depends on what the ambient conditions are and where the building is located. The obvious example is urban areas where the outside air is polluted.

In urban areas we have different categories of pollution: air pollution, thermal pollution and noise pollution. All three will impact on ventilation systems; high external air pollution will dictate some form of filtration is necessary, thermal pollution will not allow passive cooling strategies to work to their full potential and noise pollution will impose the necessity of noise attenuation which in most cases might exclude opening of windows. This chapter focuses on air and thermal pollution.

Section 2 presents the impact of the urban context in terms of air and thermal pollution on ventilation strategies; it describes the pollutants of concern and how they can be calculated and presents the effect of urban heat island on thermal comfort. Section 3 focuses on the prediction of ventilation rates considering the urban context and climate change through customised weather files and additional urban specific calculations. Section 4 presents three case studies of buildings in urban areas using suitable ventilation strategies where performance evaluation was carried out through on-site evaluation. Some concluding remarks are presented in Sect. 5.

## 2 Impact of the Urban Context

Urban areas create their own microclimate conditions which impact on ventilation's functions to maintain acceptable indoor air quality and reduce overheating.

### *2.1 Urban Air Pollution*

Acceptable indoor air quality in buildings is maintained by minimising pollutant sources, removing the pollutants at source and diluting the pollutants through ventilation. The last strategy assumes that air used for ventilation is "fresh", but how fresh is air in the cities? It is well known that air in the cities has higher pollution levels than air in the countryside due to city pollutant sources such as transport emissions, industrial processes and building exhaust pollution.

In many traditional calculations methods, $CO_2$ concentration is used as an indicator of Indoor Air Quality (IAQ) in buildings. The implication is that if $CO_2$ is within acceptable limits, then other pollutants will be also diluted to acceptable levels. What are the other pollutants?

Pollutants' concentration in buildings depends on two categories of pollutants: those generated inside the building by the occupants themselves, their activities (cooking, cleaning, bathing) and the building itself (materials and furniture). These pollutants are not considered in this chapter as we focus on pollutants coming into the building from outside. Outside pollutants in urban areas are: particulate matter (PM10, PM2.5), sulphur oxides ($SO_x$), nitrogen oxides ($NO_x$), volatile organic compounds (VOCs), carbon monoxide (CO) and ground-level ozone ($O_3$) formed as a chemical reaction between nitrogen oxides and VOC with sunlight.

The thresholds for some of the pollutants for acceptable IAQ are shown in Table 1 taken from various sources. $CO_2$ concentration shown in Table 1 is the concentration which has a direct impact on occupants. However, in ventilation applications the guide $CO_2$ concentration is much lower so that ventilation dilutes other pollutants to acceptable limits. The $CO_2$ concentration limit depends on the category of the space as shown in Table 2. It should also be noted that concentrations are quoted above external $CO_2$ levels to allow for the fact the external $CO_2$ is rising and so it does not impact calculations. So, for an internal space in an urban area where outside $CO_2$ concentration is 450 ppm, the internal $CO_2$ concentration for a space of Internal Environmental Quality category I should be $550 + 450 = 1000$ ppm.

There exist simple calculation methods to estimate the pollutant concentration in a space based on ventilation rate. Therefore, if we know how much ventilation is provided in a space and the external concentration of a pollutant and the internal sources (if they exist) we can estimate its concentration using equations Equations 1 and 2 which predict species concentration from known emission and ventilation rates (Persily and Polidoro 2019). The space-specific indoor pollutant concentration can then be calculated.

$$C_{(t)} = C_{(0)}e^{-\frac{q_V}{V_r}t} + \mathrm{Css}\left(1 - e^{-\frac{q_V}{V_r}t}\right) \tag{1}$$

**Table 1** Indoor air pollutants and threshold concentrations for acceptable IAQ

| Quantity | Threshold | | Refs. |
|---|---|---|---|
| | Limit | Duration | |
| $CO_2$ (PPM) | 5000 PPM[a] | 8 h mean | HSE UK Government (2019 |
| $CO_2$ (PPM) | See Table 2 | IAQ limit | |
| $CH_2O$ (mg/m$^3$) | 100 μg/m$^3$ | 30 min mean | WHO (2010) |
| $CH_2O$ (mg/m$^3$) | 2500 μg/m$^3$ | 8 h mean | HSE UK Government (2019) |
| VOC (PPM) | 300 μg/m$^3$ | 8 h mean | HMG, AD F (2021) |
| $PM_1$ (μg/m$^3$) | No stated limits | | |
| $PM_{2.5}$ (μg/m$^3$) | 25 μg/m$^3$ | 24 h mean | WHO (2010) |
| $PM_{10}$ (μg/m$^3$) | 50 μg/m$^3$ | 24 h mean | WHO (2010) |
| $O_3$ (μg/m$^3$) | 100 μg/m$^3$ | 8 h mean | WHO (2010) |
| $SO_2$ (μg/m$^{3)}$ | 20 μg/m$^3$ | 24 h mean | WHO (2010) |
| $SO_2$ (μg/m$^{3)}$ | 500 μg/m$^3$ | 10 min mean | WHO (2010) |
| CO | 100 | 15 min mean | WHO (2010) |
| CO | 10 | 8 h mean | WHO (2010) |
| $NO_2$ | 200 | 1 h mean | WHO (2010) |
| $NO_2$ | 40 | Annual mean | WHO (2010) |

**Table 2** $CO_2$ concentration limits for ventilation calculations (EN16798-1, 2019 Annex B)

| Indoor Environmental Quality (IEQ) Level of expectation | Category | Default design $CO_2$ concentration above outdoor concentration assuming a standard $CO_2$ emission of 20 L/(h per person) | |
|---|---|---|---|
| | | Default value | Default value for bedrooms |
| High | IEQ I | 550 | 550 |
| Medium | IEQ II | 800 | 800 |
| Moderate | IEQ III | 1350 | 1350 |
| Low | IEQ IV | 1350 | 1350 |

$$\text{Css} = C_{\text{out}} + \frac{G}{q_{\text{v}}} \tag{2}$$

where

$C_{(t)}$ the concentration in the room at time t in mg m$^{-3}$
$C_{(0)}$ the indoor concentration at time 0 in mg m$^{-3}$
$q_{\text{v}}$ the volume flow rate of supply air in m$^3$ s$^{-1}$
$G$ the mass flow rate of emission in the room in mg s$^{-1}$
$t$ the time in s
$C_{(\text{out})}$ the outdoor concentration
$V$ the volume of air in the room in m$^3$

Therefore if natural ventilation or mechanical ventilation without filtration is used some calculations based on outdoor pollution levels can be carried out to determine whether acceptable IAQ can be provided or whether some form of filtration is necessary.

## 2.2 *Urban Thermal Pollution*

The urban heat island (UHI) is a well-known and documented phenomenon. UHI describes the differences in air temperature between an urban area and surrounding countryside as shown in Fig. 1.

The urban thermal pollution causes are:

- Heat capacity and conductivity of materials: in cities we use high-density materials with high heat capacity and high thermal conductivity, in rural areas the ground is less dense, has a lower heat capacity and has an insulating layer above.
- Solar absorptivity (or reflectivity): Albedo (solar reflectivity) varies in both rural and urban areas and depends on the light spectrum characteristics of the surface of materials, this is usually controlled by the colour but selective pigments in the surface can alter it, leading to what we call cool materials that can be used to mitigate urban thermal pollution (CRRC 2022; ECRC, 2022).

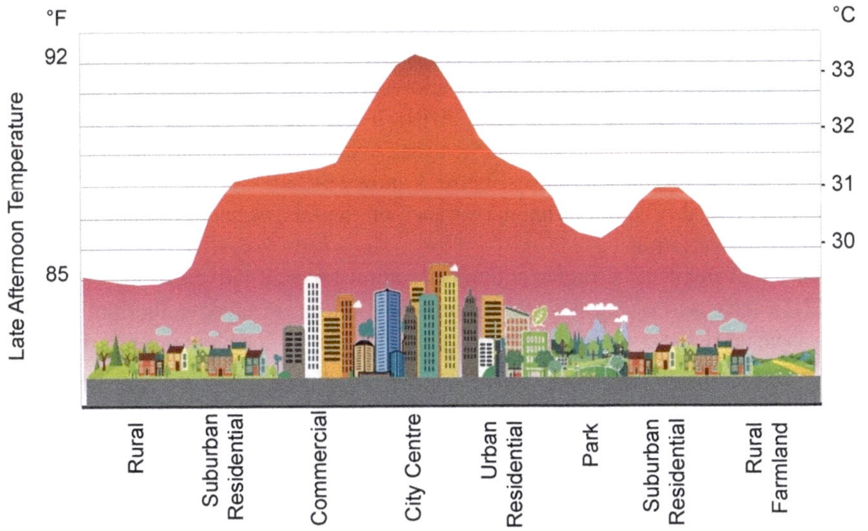


**Fig. 1** Urban heat island (drawn by author from various sources)

- Sky view factor: determines effectiveness of long-wave radiation for cooling; in deep urban canyons a small sky view factor will reduce this effectiveness for cooling to the sky.
- Wind speed: Average rural wind speeds are higher than urban ones because the ground surface is smoother. The "rougher" urban surfaces reduce wind speeds, but there are local variations especially within deep narrow street canyons where wind will create little disturbance at ground level.
- Energy consumption: releases heat; rural energy use is small compared to the energy received from the sun but energy use density in urban areas is much higher.
- Vegetation: To evaporate water requires energy - this helps keep plants and the air around them cool. Urban areas are "harder"; they have less vegetation, less evaporative cooling and less shading of the ground.

The above parameters contribute to higher ambient temperatures in urban areas and have been analysed in a number of review papers (e.g. Akbari and Kolokotsa 2016). In addition, the UHI has been documented through measurements and simulations in many cities. At global level, (Santamouris 2020) reports on recent progress of urban overheating and urban heat island. Air temperature experimental results from 400 cities show that the magnitude of the average temperature increase may exceed 4–5 °C, while at the peak may exceed 10 °C. The paper reviews UHI impact on energy demand by buildings and impacts on outdoor air quality and health. In terms of energy demand, the analysis show that the total increase in energy demand is close to 12% considering all locations and building types, and the actual magnitude is strongly correlated with the magnitude of UHI. In most cases, the reduction in heating energy demand is less than the increase in cooling energy demand resulting to a net increase. This net increase depends on ambient temperatures during the cooling season but is also related to building type and its energy efficiency strategies.

Two examples of UHI are included below; Fig. 2 shows the measured UHI in the city of Chongqing, China (Jiang et al. 2019). It can be seen that the UHI intensity is higher and the wind speed lower in the urban avenue throughout the year compared to rural and suburban locations. Figure 3 shows the measured UHI in the city of Lisbon, Portugal. The nocturnal appearance on the UHI is shown clearly as well as the seasonal variation.

The UHI intensity is important consideration for ventilative cooling (IEA EBC 2022), a term describing the provision of ventilation to improve thermal comfort. This is because the cooling effectiveness of outside air will be less in a city during the day. Also, UHI is mostly a nocturnal phenomenon, and this will impact on ventilative cooling strategies such as night cooling which will be less effective in a city even in high latitudes (Kolokotroni and Salvati 2021). Figure 4 shows the day and night

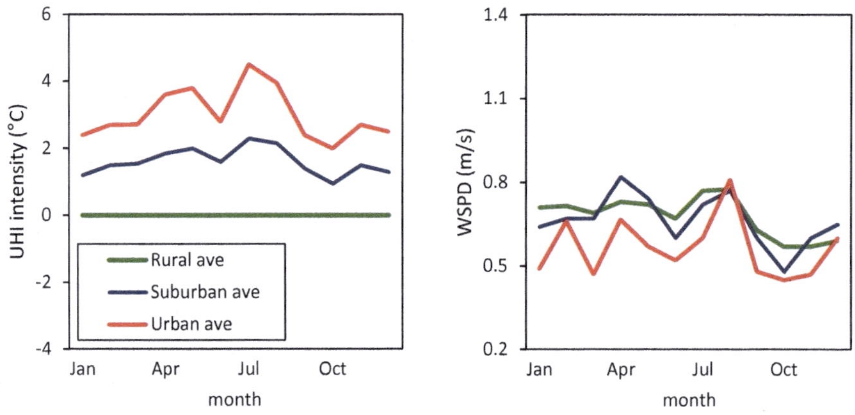


**Fig. 2** Measured UHI in Chongqing, China (adapted from Jiang et al. 2019).

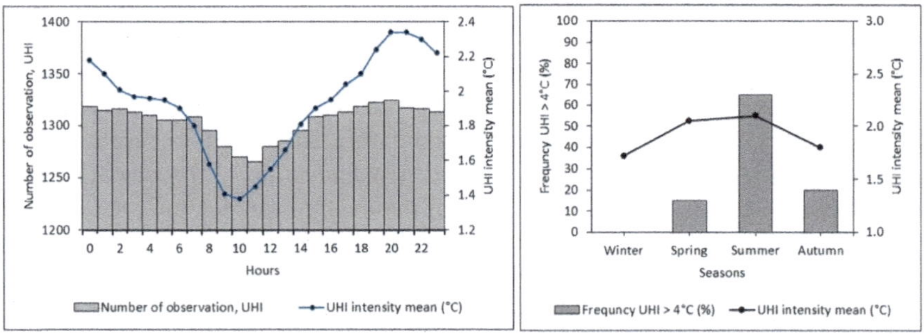


**Fig. 3** Measured UHI in Lisbon Portugal (adapted from Lopes et al. 2013)

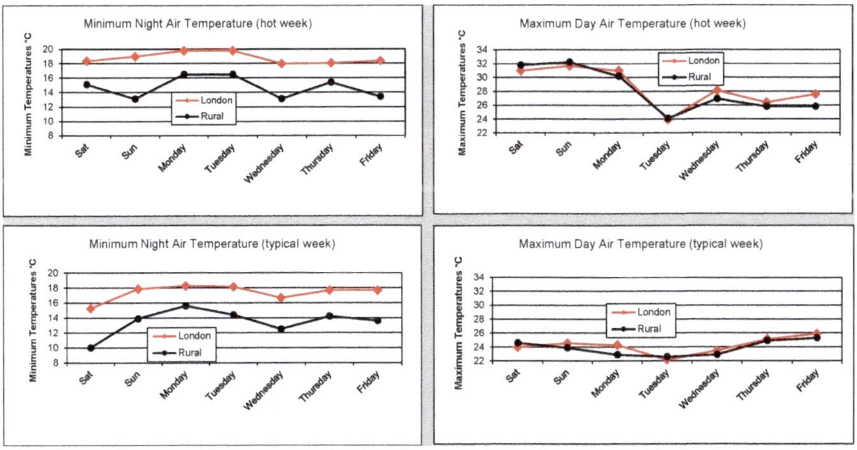


**Fig. 4** Air temperature in London and reference site during a typical and a hot week. (adapted from Kolokotroni et al. 2006)

temperature in a city and rural location in London, England. It shows that night UHI intensity has a similar magnitude on hot as well as typical summer weather.

Such differences in external conditions between city and countryside buildings should be considered during the design. The next section presents how this can be implemented.

## 3 Urban Weather Files for Building Simulation

For the prediction of ventilation in urban buildings the local climate needs to be considered; this can be done by using customised weather files. A method of how this can be achieved in a dense urban context is shown diagrammatically in Fig. 5 (Salvati et al. 2020). The methodology is based on the use of a suite of climate tools, including the Urban Weather Generator (Bueno et al. 2013; Mao et al. 2017) to calculate urban air and surface temperatures and empirical models to estimate wind speed in urban canyons. The calculation of the canyon wind speed was based on the URBVENT project approach (Ghiaus et al. 2004), applying different empirical models for different urban situations, depending on the geometry and orientation of the canyon and the wind speed and direction at the meteorological station. This modelling approach is suitable for building in dense urban areas, characterised by urban canyons with high height-to-width ratio. In less dense suburban areas, the impact of atmospheric and surface heat island is lower. However, the distribution of wind speed at the microscale may affect the thermal performance of naturally ventilated buildings also in suburban locations. The inclusion of local wind speed and the impact of urban context in external temperatures gives data for predicting ventilation flowrates in the specific urban context.

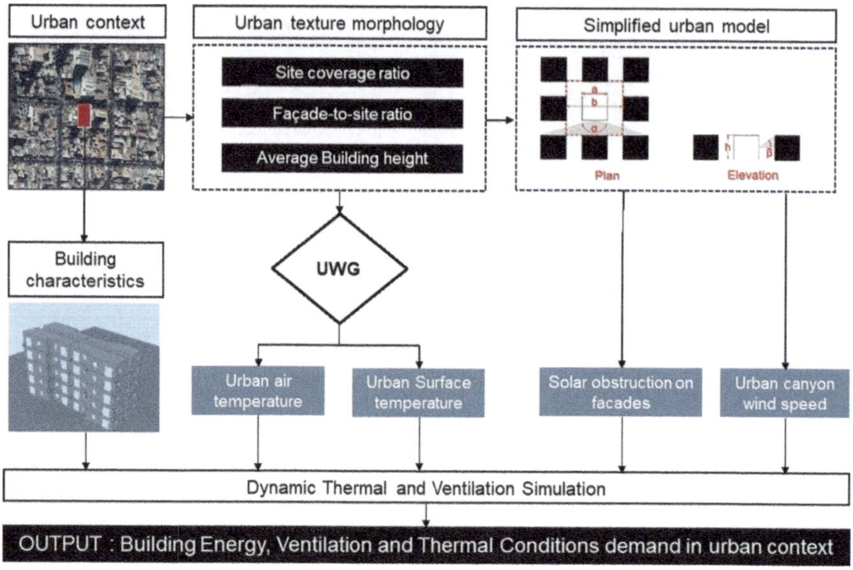


**Fig. 5** Urban considerations in building simulation (adapted from Salvati et al. 2020)

In addition, climate change impacts on UHI to exacerbate the phenomenon as well as on wind patterns with the consequence of influencing building ventilation. In terms of overheating a recent review (Tavakoli et al. 2022) on the resilience of buildings concluded that design could play a key role in reducing overheating and future designs should always be stress tested in future extreme conditions. The impact of the overlapping effect of global warming and urban microclimate on building thermal performance is therefore important for our understanding on how buildings will perform now and in the future.

A number of sources are available providing future weather files which take climate change into account. In many cases, future weather files are generated based on the Intergovernmental Panel on Climate Change (IPCC) scenarios which are regularly updated. The latest update was published in February 2022 (IPCC 2022) called the 6$^{th}$ assessment report, but some weather files available to designers have been generated using previous scenarios published in 2014 (IPCC 2014) and 2007 (IPCC 2007) called the 5th and 4th assessment, respectively. Future weather files can be obtained from a number of sources; in many cases regional professional bodies provide such files, for example, CIBSE in the UK (CIBSE 2014), based on scenarios by the UK Climate Impacts Programme (UKCIP) or provided by academic projects for example the weather generator (CC WorldWeatherGen 2022) or by specialised commercial programs, for example Meteonorm (Meteonorm 2022).

Methodologies which include both future weather and urban microclimate are less developed. Very few attempts have been reported to generate future weather data that also include urban effects, to investigate the thermal performance, overheating risk and resulting natural ventilation airflow rates of buildings under future urban climate

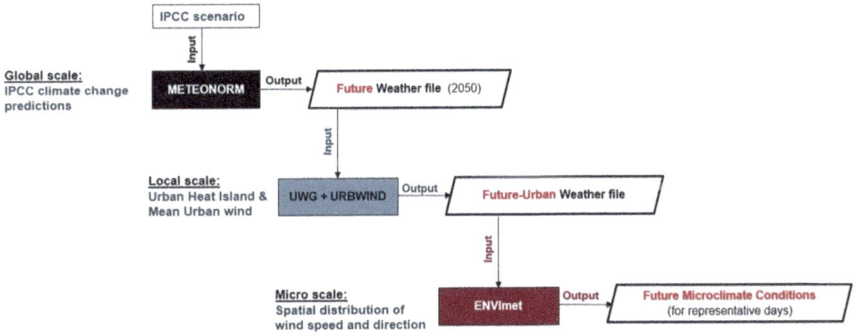


**Fig. 6** Method to generate future urban microclimate conditions for building simulation (*Source* Salvati and Kolokotroni 2022)

conditions. The methodology described above in this section based on the use of urban climate models (UWG), empirical wind models (URBVENT) can be used with future weather files to generate data about what happens around the building. The method is shown diagrammatically in Fig. 6.

In the study reported in (Salvati and Kolokotroni 2022), the software Meteonorm was used to generate future weather files based on the IPCC emission scenarios. The future weather files were used as input to the UWG (Bueno et al. 2013) and URBVENT (Ghiaus et al. 2004) models to generate future urban weather files, capturing the local modifications of air temperature and wind speed determined by an urban area at the local scale (Salvati et al. 2019, 2020). These simplified models do not account for the impact of the actual three-dimensional geometry of the building and its surrounding obstacles on the air flow around the building. However, important microclimate phenomena are also driven by the shape and arrangement of the buildings and other urban elements and their interaction with wind flow and solar radiation. These microscale phenomena have significant impact on the outdoor thermal comfort but also on the building energy and natural ventilation performance in urban context. Therefore, the computational fluid dynamics model ENVImet (ENVImet 2022) was used to predict these modifications in wind speed, direction and surface temperatures around the building; the outputs were used in a Dynamic Thermal Simulation model to predict airflow rates in the building.

An example of the difference in the prediction of ventilation flows for a case study in the temperate climate of London using the above methodology is shown in Fig. 7 (Salvati and Kolokotroni 2022).

Figure 7 shows that the reduced wind speed has a clear impact on the ventilation rates of the bedrooms, in particular those on the East facade (R21, R22, R23 and R24). Using the ENVImet wind data instead of the weather file data, the air change rate (ACH) of these rooms is reduced by about 35%, from an average of 3.1 to an average of 2 ACH. The other rooms are much less affected by reduced wind speed. In some cases, the ventilation rates increase using the ENVImet data instead of the pressure coefficients and weather files data. This happens for the bedroom R25,

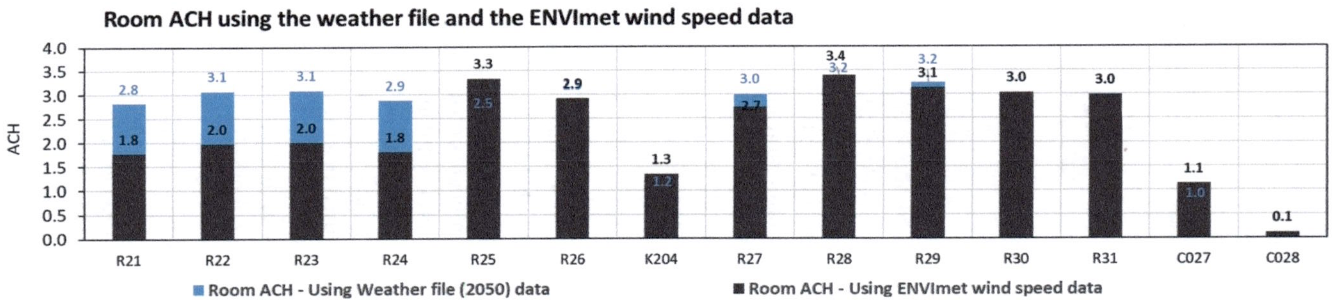


**Fig. 7** Prediction of air change rate in a case-study building in London (*Source* Salvati and Kolokotroni 2022)

which is located on a corner; this is probably due to the beneficial effect of a different wind speed and wind pressure on its window compared to the other windows on the same façade, contributing to create a pressure difference with the other spaces. The reduced ventilation rates of the east-façade rooms also have a significant impact on the indoor operative temperatures of the rooms, which increases by 1.0–1.1 °C on average. The maximum increase of the operative temperature due to the reduced ventilation reaches up to 3.7 °C.

This section described how urban weather files can be used to predict natural ventilation airflows and internal air temperatures for the calculation of overheating specifically for urban buildings and overheating. The predicted urban airflows can then be used also for the calculation of pollutant concentrations using the equations of Sect. 2.1.

## 4 Case Studies

A project run under the auspices of the International Energy Agency (IEA) Energy in Buildings and Communities (EBC) on Ventilative Cooling has investigated a number of built case-study buildings (O'Sullivan P and O'Donovan A, 2018). Some of the case-study buildings are set in an urban environment, and three were selected to be presented below covering a range of building types (residential, school and office), climates and locations. The case studies presented below are a summary of the materials included in the project (Annex 62) report (O'Sullivan P and O'Donovan A, 2018).

**The first case study** is a cluster of residential buildings in China located in the Hunan province. It was completed in 2007 and Fig. 8 presents a photo of the external of the buildings who are high rise residential as well as the floor plan of a typical apartment where measurements were carried out.

**Fig. 8** Typical apartment and external view of the MOMA residential buildings in China, Hunan (provided by Hunan University)

Because of the external pollution, the buildings are mechanically ventilated with a Dedicated Outdoor Air System (DOAS) for ventilation requirements, whereas a ceiling radiant cooling system provides thermal comfort. The air inlets are located on the floor and the air outlet in the upper level of the kitchen and the bathroom. There are windows in the buildings with the main function for daylighting and are normally closed and thus avoiding outdoor polluted air into the indoor space.

The system providing ventilation, heating and cooling is composed of two parts, (a) air supply and exhaust system and (b) cooling and heating water system. The fresh air supply and exhaust system consists of 8 units with heat recovery. The cold and hot water system for conditioning the fresh air consist of one set of ground source heat pump; in the summer cooling water comes from the geothermal wells (supply and returning water temperature is 32/27 °C) or the cooling tower (supply and returning water temperature is 7/12 °C). The ground source heat pump in winter takes heat from the ground source to produce 45/40 °C water supplying fresh air units to deal with outdoor fresh air. The systems are designed for the conditions shown in Table 3.

The system was simulated using a computational fluid dynamics (CFD) program, and measurements were carried out in a typical apartment to compare with the simulations. It was found that good thermal comfort and IAQ were provided to the apartments using the systems. In addition, a questionnaire was carried out with 100 respondents which indicated satisfaction with the systems and provided conditions.

**The second case study** is a high-rise office development in Osaka, Japan. The design aimed to introduce the external environment visible to the occupants and a natural ventilation strategy was included as part of this concept. In each floor, large natural ventilation openings were installed; in order to deal with the partitions between tenants, four voids at the corner and a void in the core region were designed to provide fresh air. At four floors in the upper storeys, the flow path connecting the first centre void is closed to prevent the backflow from the void into the office rooms, and a second centre void was provided instead. This was done to reduce the differences in natural ventilation rate amongst floors because this building is a multi-tenant office building. Figure 9 shows how the natural ventilator and the air handling unit (AHU) are controlled and the detail of the inlet diffuser for natural ventilation.

As a part of performance evaluation for the buildings, the local mean age of air was measured under natural ventilation operation by using $CO_2$ as a tracer gas. This verified that the age distribution is strongly dependant on the wind direction if the

**Table 3** External and internal conditions for the MOMA building

| Parameter | Value |
|---|---|
| $T_e$, Summer external temperature | 35.8 °C |
| $T_z$, Summer operative temperature | 25 °C |
| Overheating criteria | $T_z < 28$ °C for 99% $hr_{occ}$ |
| Minimum IAQ air supply rate | 10 $ls^{-1}$/person |
| Cooling air supply rate | 30 $ls^{-1}$/person |
| Noise level rating | 35 dB |

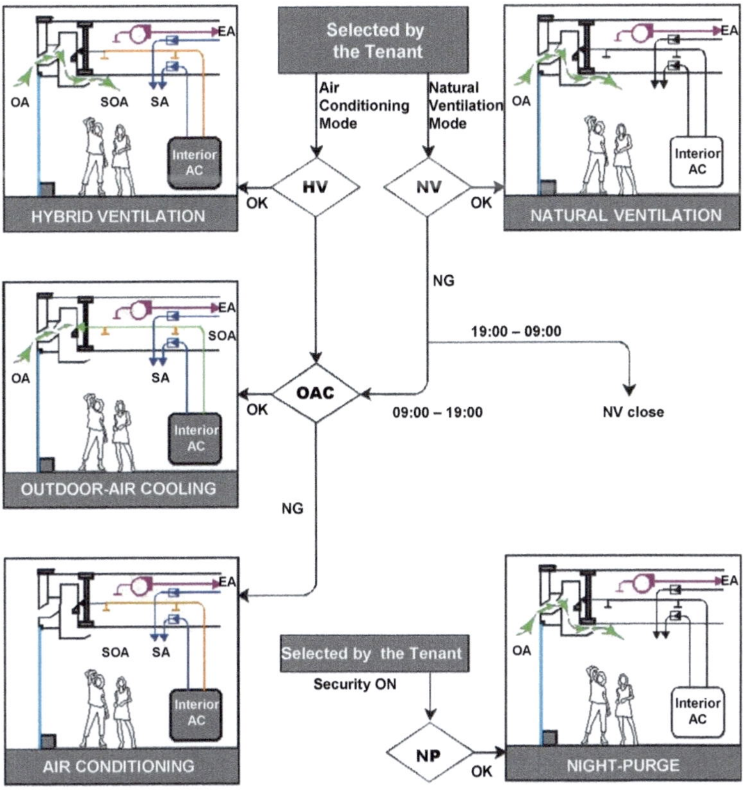


**Fig. 9** Ventilator and AHU control and the detail of the inlet diffuser (adapted from O'Sullivan and O'Donovan 2018).

horizontal cross-ventilation operation was applied to the building and that the natural ventilation system could work quite well by introducing buoyancy-induced natural ventilation using corner voids even when the external wind condition was calm.

Thermal performance was also measured; under the natural ventilation operation, indoor temperature is always higher than outdoor temperature, and sometimes considerably increases depending on natural ventilation rate. On the other hand, under the operation of hybrid ventilation, no significant temperature rise can be seen because mechanical air-conditioning is also operated in addition to the natural ventilation.

The internal $CO_2$ was measured and was shown that relatively low indoor $CO_2$ concentration of 500–600 ppm is observed when the natural ventilation is operated in both hybrid and natural ventilation modes, which verifies that low concentration is stably achieved by adopting natural ventilation system.

Since pollen could become a serious matter in spring in Japan, some tenants have declared not to operate natural ventilation to prevent it. Measuring the indoor pollen particle number; however, there observed no significant effect of natural ventilation.

This indicates that the impression of the occupants considerably affects the natural ventilation operation.

**The third case study** is a kindergarten building in Portugal. This is a small two-story building built in 2013 with a total area of 680 $m^2$ distributed in two floors with 3m floor to ceiling height. This school is located in the mild subtropical-Mediterranean climate of Lisbon, Portugal, characterised by mild winters (minimum temperature $\approx$ 4 °C) and dry summers with high levels of solar radiation (maximum temperature $\approx$ 37 °C). In spring and summer, there are many days with large temperature diurnal range (up to 18 °C), that can potentially make a night cooling approach very effective. Driven by the need to lower maintenance costs, the building is naturally ventilated, does not have a mechanical cooling or ventilation and the control system is manual. The walls have high thermal inertia. Heating is provided by passive convectors, fed by a heat pump. Natural ventilation air is introduced into the space through low-level grilles or by windows opening on the façade and is exhausted in the centre or back of the room, through one or two chimneys (depending on classroom dimensions).

The ventilation solution implemented in this kindergarten consists in a high-level openable window plus low-level grilles installed on the façade of each classroom that control the inflow air. During the colder months a hydraulic radiator that was installed in front of the grilles could be used to ensure users thermal comfort. The air will be exhausted in the back of the room, through one or two thermal chimneys (depending of classroom size). For optimal performance of the ventilative cooling systems designed, two operation modes were considered: winter and summer; Figure 10 shows the control modes.

The main thermal comfort problems occur in the summer, when it is necessary to increase ventilation (to maintain acceptable $CO_2$ concentration), but the outside air is warmer. Ideally, during this time all the openings should be closed to achieve comfortable interior air temperature but open to do not exceed 1625 ppm ($CO_2$ concentration). In these cases, the users will determine what comfort parameter is more relevant to his comfort and to define if the openings should be maintained closed or be opened.

The natural airflow rates were measured in controlled test conditions with constant sensible gains. The measurements were performed in two classroom with different chimney height and heat gains densities and showed good agreement with the simulations providing adequate air flow. Internal air temperature and $CO_2$ concentration measurements show that performance was as expected.

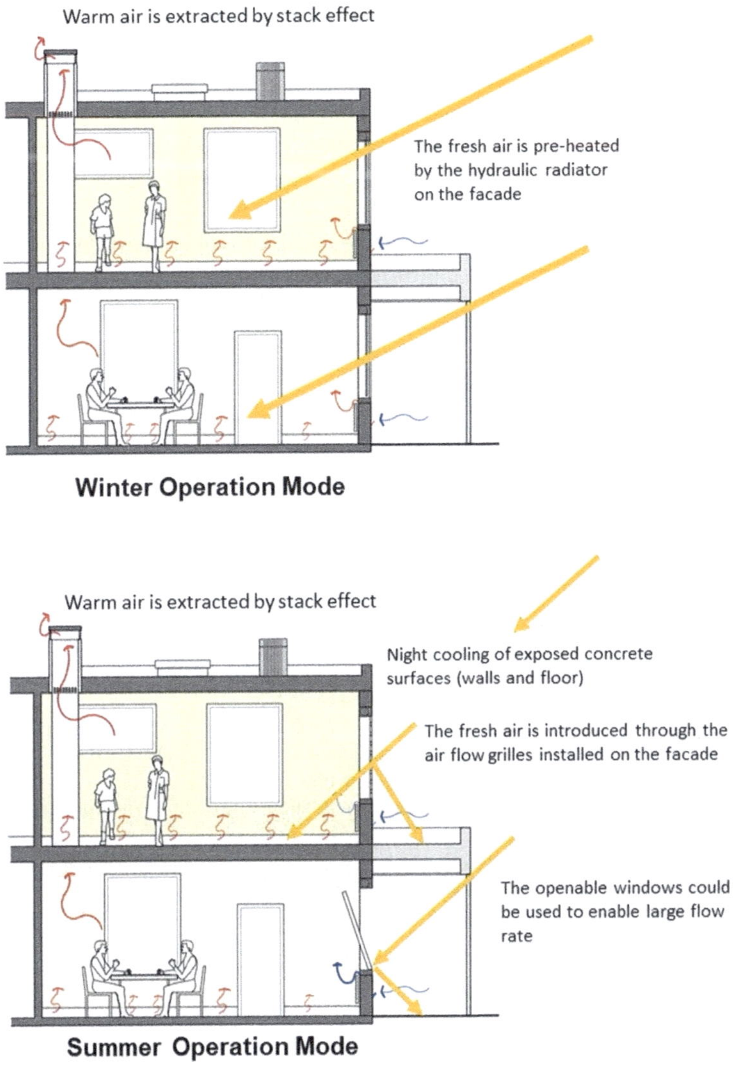


**Fig. 10** Winter and Summer operational modes (adapted from O'Sullivan and O'Donovan 2018).

## 5 Summary

Urban ventilation is impacted by air pollution in the cities and the urban heat island exacerbated by climate change. Therefore, tools are needed to help designers to consider specific urban characteristics when proposing ventilation strategies for urban buildings.

Such tools and approaches have been described in this chapter. First, equations are presented to facilitate calculation of pollutants' concentration inside the buildings depending on air flow rate provided and the external concentration of the pollutants. Then, a methodology was presented on how to incorporate urban heat island impact on thermal and ventilation simulations and how to calculate microclimatic variations around the building which will affect ventilation rates.

Resilience of proposed solutions has become important that the impact of climate change and how to test the solutions through future and urban weather files has been discussed.

The case studies show some examples of solutions which have been successful as on site measurements have demonstrated.

## References


Akbari H, Kolokotsa D (2016) Three decades of urban heat islands and mitigation technologies research. Energy Build 133:834–842

Bueno B, Norford L, Hidalgo J, Pigeon G (2013) The urban weather generator. J Build Perform Simul 6(4):269–281. https://doi.org/10.1080/19401493.2012.718797

CIBSE (2014) TM49 Design summer years for London, CIBSE

CC WorldWeatherGen (2022) Climate change world weather file generator for World-Wide Weather. https://energy.soton.ac.uk/ccworldweathergen/

CRRC (2022) https://coolroofs.org/. Assessed 23 Aug 2022

ESRC (2022) https://coolroofcouncil.eu/. Assessed 23 Aug 2022

ENVImet (2022) https://www.envi-met.com. Assessed 26 Sept 2022

Ghiaus C, Allard F, Santamouris M, Georgakis C, Roulet C-A, Germano M, Tillenkamp F, Heijmans N, Nicol F, Maldonado E, Almeida M, Guarracino G, Roche L (2004) URBVENT WP1 final report: soft computing of natural ventilation potential

HMG (2021) Approved Document F1: Means of ventilation. 2021 edition incorporating 2010 and 2013 amendments 2010

HSE UK Government. Workplace health, safety and welfare - L24 [Internet]. [cited 2019 Nov 22]. Available from: http://www.hse.gov.uk/pubns/books/l24.htm

IEA, ECB (2022) Annex 62 Ventilative Cooling. https://venticool.eu/information-on-annex-62/annex-62-home/. Accessed 23 Aug 2022

Jiang P, Liu X, Zhu H, Li Y (2019) Features of urban heat island in mountainous chongqing from a dense surface monitoring network. Atmosphere 10(2):67. https://doi.org/10.3390/atmos10020067

Kolokotroni M, Salvati A (2021) Comfort and energy implications of urban microclimate in high latitudes. In: Palme M, Salvati A (eds) Urban microclimate modelling for comfort and energy studies. Springer, Berlin. 978-3-030-65420-7

Kolokotroni M, Giannitsaris I, Watkins R (2006) The effect of the London Urban Heat Island on building summer cooling demand and night ventilation strategies. Solar Energy 80(4):383–392

Lopes A, Lopes A, Alves A, Alcoforado MJ, Machete R (2013) Lisbon Urban heat island updated: new highlights about the relationships between thermal patterns and wind regimes. In: Advances in meteorology, vol 2013 (2013), Article ID 487695, 11p. https://doi.org/10.1155/2013/487695

Mao J, Yang JH, Afshari A, Norford LK (2017) Global sensitivity analysis of an urban microclimate system under uncertainty: design and case study. Build Environ 124:153–170. https://doi.org/10.1016/j.buildenv.2017.08.011

Meteonorm—https://meteonorm.com/en

O'Sullivan P, O'Donovan A (2018) Ventilative cooling case studies, Report of IEA, ECB Annex 62. https://venticool.eu/information-on-annex-62/annex-62-publications/deliverables/. Assessed 2 Sept 2022)

Persily A, Polidoro BJ (2019) Residential application of an indoor carbon dioxide metric. Paper presented at the 8th TightVent—6th Venticool Conference, pp 15–16

Salvati A, Kolokotroni M (2022) Urban microclimate impact on ventilation and thermal performance of multi-family residential buildings: two case studies in different climates and urban settings. In: 42nd AIVC conference: ventilation challenges in a changing world, 5–6 Oct 2022, Conference, Rotterdam, The Netherlands

Salvati A, Palme M, Chiesa G, Kolokotroni M (2020). Built form, urban climate and building energy modelling: case-studies in Rome and Antofagasta. J Build Perform Simul. https://www.tandfonline.com/doi/full/10. 1080/19401493.2019.1707876

Santamouris M (2020) Recent progress on urban overheating and heat island research. Integrated assessment of the energy, environmental, vulnerability and health impact. Synergies with the global climate change. Energy Build 207:109482

Tavakoli E, O'Donovan A, Kolokotroni M, O'Sullivan PD (2022) Evaluating the indoor thermal resilience of non-residential passively cooled low energy buildings: a review, Building and Environment, Vo 222, 109376 https://doi.org/10.1016/j.buildenv.2022.109376

UN IPCC (2007) AR4 Climate Change 2007: Synthesis Report—IPCC

UN IPCC (2022), Climate Change 2022: Mitigation of Climate change, Working Group III contribution to the Sixth Assessment Report of the Intergovernmental Panel on Climate Change, Summary for Policymakers. https://www.ipcc.ch/report/ar6/wg3/downloads/report/IPCC_AR6_WGIII_SPM.pdf. Assessed 24 Aug 2022.

UNIPCC (2014) AR5 Synthesis Report: Climate Change 2014—IPCC

WHO (2010) World Health Organization. Regional Office for Europe. WHO guidelines for indoor air quality: selected pollutants. ISBN: 9789289002134

**Maria Kolokotroni** is Professor of Energy in the Built Environment at Brunel University London. She is a Chartered Engineer (CEng) and Fellow of CIBSE (Chartered Institute of Building Services Engineers) in the UK. She is a member of IEA (International Energy Agency) AIVC (Air Infiltration and Ventilation Centre) Board. She has co-ordinated and participated to a number of UK, European and International projects on Energy Efficient Technologies for Buildings, Nearly Zero Energy Building (NZEB) Retrofit, Low Energy Cooling and Ventilation, Urban Heat Island and Mitigation Strategies such as cool materials and microclimatic interventions as well as the impact of climate change on the energy performance of buildings and urban areas.

# Balancing Outdoor and Indoor Soundscapes in a Warming Climate

**Simone Torresin, Francesco Aletta, and Jian Kang**

**Abstract** The acoustic quality of living and working environments can affect health, well-being and quality of life of building occupants. Soundscape methodologies allow a comprehensive characterisation of acoustic perception by people in a specific context, thus guiding the design of built environments that are supportive, rather than not disturbing. However, the management of the indoor soundscape is strongly interrelated with that of the outdoor, urban soundscape, with building envelopes and ventilation openings being the main interface. A coherent integration between indoor and outdoor soundscapes will be increasingly urgent in the light of a warming climate, as positive urban soundscapes will be key for the successful implementation of passive cooling strategies in buildings. Building on the emerging concept of the indoor soundscape studies, this chapter will address the potential of a soundscape approach for the built environment to shape resilient cities, able to address the main global challenges posed by global warming.

**Keywords** Soundscape · Ventilation · Acoustics · Sustainability

## 1 Introduction

Environmental noise is one of the main public health threats in Europe, with 1 in 5 people living in areas where traffic noise levels are harmful to health (European Environment Agency 2020a). Indeed, among the most significant public health risks in Europe, noise comes second, following air pollution and followed by the impacts of climate change (e.g., heatwaves), causing 12,000 premature deaths and contributing

S. Torresin
Department of Civil Environmental and Mechanical Engineering, University of Trento, Via Mesiano 77, 38123 Trento, Italy

S. Torresin · F. Aletta · J. Kang (✉)
UCL Institute for Environmental Design and Engineering, University College London (UCL), Central House, The Bartlett14 Upper Woburn Place, London WC1H 0NN, UK
e-mail: j.kang@ucl.ac.uk

 
R. Yao (ed.), *Resilient Urban Environments*, Cities and Nature,
https://doi.org/10.1007/978-3-031-55482-7_9

to 48,000 new cases of ischaemic heart disease each year (European Environment Agency 2020b).

To deal with this concern, various authorities have provided guidance on the assessment and management of environmental noise, which so far has mainly been based on setting upper limits for sound pressure levels in outdoor environments (European Union 2002). Based on a series of systematic literature reviews commissioned by the WHO, noise exposure thresholds were set for various noise sources at which adverse health effects were documented (WHO 2018). For example, the target for traffic noise was to reduce sound levels below 53 dB(A) $L_{\mathrm{den}}$ and for aircraft noise below 45 dB(A) $L_{\mathrm{den}}$. Despite this regulatory framework, no urban area in Europe has reached such thresholds so far (European Environment Agency 2020a). The scientific community has questioned whether this will actually be reached in the future, as no improvement has been observed between the 2014 and 2020 reports. Moreover, the population exposed to environmental noise is expected to increase further due to future urban growth and increased demand for mobility (European Environment Agency 2020a). Looking at the exceptional drop in noise levels experienced during the lockdown of the COVID-19 pandemic (Aletta et al. 2020), it is clear that, realistically, we will not be able to meet the WHO exposure targets any time soon unless drastic infrastructural changes in society occur, such as a total shift towards quieter or non-motorised mobility. Meanwhile, it is necessary to implement actions to improve the quality of the urban sound environment.

The literature has indeed shown that the mere reduction in dB noise levels is not only often unpractical, but even not necessarily correlated with an improvement in acoustic perception (Cain et al. 2013; Kang et al. 2016) or reduced noise complaints (Tong et al. 2021). Acoustic perception is extremely complex and multifactorial, a function of several contextual, personal and situational factors and only partly explained by variations in sound levels. The soundscape framework embraces complexity, abandons the view of noise as waste and considers sound and its impacts holistically through a multidisciplinary approach. The concept of soundscapes has its roots in the context of urban planning and has recently been standardised by the International Organisation for Standardisation (ISO), where soundscape is defined as "acoustic environment as perceived, experienced or understood by people in context" (International Organization for Standardization 2014). The soundscape approach elevates the goal of acoustics from reducing the negative impacts of noise to using sound as a design resource in order to create environments that promote well-being and support the activities taking place in them.

In this context, two factors must be considered.

Firstly, industrialised society is basically indoors, spending about 90% of its time in buildings (Klepeis et al. 2001). Therefore, to achieve the desired effects on people's health and well-being, the goal must be the improvement of soundscapes within buildings. This has led in recent years to the adaptation of the framework on soundscape for its application in indoor environments, as will be discussed in Sect. 3 of this chapter. However, the indoor soundscape depends largely on the characteristics of the buildings, the sources within them, and the urban soundscape, which is mainly accessible through ventilation openings (e.g., windows). The interaction between indoor and outdoor soundscape will be the focus of Sect. 4.

Secondly, measures to mitigate air pollution and global warming often offer co-benefits in terms of reducing environmental noise, and environmental emissions. However, not all interventions are equally effective for the various stressors, and this calls for coordination between the communities of researchers, policymakers and stakeholders working to tackle noise, overheating, and air pollution.

One example is provided by space cooling and ventilation. Indoor cooling is currently among the major drivers of climate change and is responsible for nearly 5% of total global energy consumption and about 1 gigaton of $CO_2$ in 2020 (United Nations Environment Programme 2021). Extensive use of mechanical ventilation and cooling as a solution to heat effects would spike global energy consumption, peak electricity demand and greenhouse gas emissions, while generating new sources of noise inside and outside buildings. Bioclimatic architecture and passive building design strategies are among the principles suggested by the United Nations to reduce the environmental impact of indoor cooling, in order to avoid the use of active cooling as much as possible (United Nations Environment Programme 2021). Natural ventilative cooling (e.g., based on opening windows) can help avert the risk of overheating inside buildings with minimal or no use of active cooling. However, a prerequisite is a noise-free outdoor environment in which to open windows to prevent the occurrence of annoyance or other noise-related critical health outcomes identified by WHO (World Health Organization 2018). However, the availability of a quiet area or a positive urban soundscape can even be a co-benefit of natural ventilation, able to improve occupant comfort and well-being, as addressed in Sect. 5.

In the following, the potential of the soundscape approach applied in the built environment will be discussed, in light of the challenges provided by global warming, as an essential component of healthy environments and sustainable development.

## 2 Indoor Soundscaping: A Soundscape Approach for the Indoor Built Environment

The definition of "soundscape" provided by the ISO standard emphasises the "context" in which acoustic perception takes place, thus giving room for a full application of the soundscape framework in the context of the indoor built environment. Compared to outdoor settings, in the indoor built environments the building plays a fundamental role in defining the three elements that underlie the definition of soundscape:

(1) **the acoustic environment**: depending on its geometry and materials, buildings filter external sounds and shape the acoustic field (Kang 2007)
(2) **the person**: depending on its use, buildings contribute to determining the expectations and needs of the occupants

(3) **the context**: the building envelope delimits the space, separates the indoor environment from the outdoor one.

Considering the amount of time spent inside buildings, even more so after the spread of home-based work with the COVID-19 pandemic, the potential impact that an indoor soundscape approach can offer in terms of improving the living conditions of occupants inside their homes and workplaces is evident.

In order to be able to guide and evaluate design actions on the acoustic quality of cities and buildings, it is essential to be able to "measure" the soundscape. By measuring the acoustic environment and how it is perceived by the inhabitants, it is possible to derive numerical indicators and reference thresholds with which engineers, architects and urban planners can assess the acoustic quality of their design objects (e.g., a building or an urban space) right from the forecast phase (i.e., before the construction). These indicators must be able to represent the perception of the acoustic environment by end users. The actual acoustic quality can then be ascertained during the operational phase of the project through direct feedback from the users, from which indications on possible corrective actions can be derived.

At the current state of research, however, a predictive verification of the soundscape is not yet entirely possible. The ISO/TS 12,913–2 technical specification includes "people-based" measurement methods (e.g., sound walks, questionnaires, interviews) and "instrument-based" measurement methods (e.g., binaural measurements) (International Organization for Standardization 2018), which allow the assessment of perceptual descriptors of the soundscape (e.g., "pleasantness" or "annoyance") and the calculation of numerical indicators of the acoustic environment (e.g., acoustic and psychoacoustic parameters) (Aletta et al. 2016). As stated in the standard, there is still a fundamental gap between perceptual metrics and acoustic or psychoacoustic measurements and, currently, the characterisation of a soundscape is only possible through the listener's feedback. For this reason, the scientific community is devoting much effort to defining new indices capable of objectively describing soundscapes, supplementing or even replacing decibel-based metrics with new soundscape indices that are able to reflect human comfort levels (Kang et al. 2019). In anticipation of such developments, the predictive assessment of soundscapes can take place with the aid of virtual reality thanks to current auditory scene simulation techniques, as is already commonly done with the visual rendering of buildings and urban environments prior to their construction. Albeit with some limitations in terms of ecological validity, the combination of visual and auditory rendering (i.e., the "auralisation") of an environment prior to its realisation can allow potential users of the building to experience its sound and visual landscape in advance, before purchase.

As for the assessment of existing acoustic environments, this can be done through the data collection methodologies included in the ISO/TS 12,913–2 technical specification (International Organization for Standardization 2018). The document recommends binaural measurements for the characterisation of the "physical" acoustic environment. Binaural (i.e., "two-ear") listening allows humans to localise sound sources, to understand a target message by isolating it from the surrounding noise, and to form a spatial sense of their environment. Similarly, headphones and head and torso simulators for binaural recording employ two microphones to simulate listening through the two ears in a kind of artificial head.

The signal recorded from binaural measurements can be played back through headphones, for example, to conduct listening tests, thus reproducing the acoustic sensations of a listener located in the original environment more accurately than using a traditional single microphone with a sound level meter. Binaural recordings also make it possible to calculate psychoacoustics parameters (e.g., loudness, sharpness, roughness, fluctuation strength) that are representative of basic auditory sensations in humans and complement the information obtainable from sound level measurements.

The technical specification then reports guidelines for conducting soundwalks, questionnaires, and interviews, in order to gain information about the relationship between people and the acoustic environment, and their perceptual and behavioural response to the acoustic stimuli around them.

The emotional response to the acoustic environment can be assessed through the model of affective response to soundscape developed by (Axelsson et al. 2010). The model identifies the dimensions along which to evaluate people's perception. It was derived from listening tests in which participants were asked to rate fragments of urban sounds using 116 perceptual attribute rating scales (e.g., "cosy", "inhospitable"). Through Principal Component Analysis (PCA), the attributes were reduced to a limited number of principal dimensions that could describe most of the variance in the ratings of the audio signals, and which were interpreted according to the attributes with which they were most strongly associated. These dimensions, as adopted in the ISO technical specification, are pleasantness (i.e., how pleasant or unpleasant the soundscape is) and eventfulness (i.e., how rich in sound events the environment is). The model also included a second set of alternative, orthogonal dimensions, named calmness (i.e., how calm or chaotic the soundscape is) and excitement (i.e., how vibrant or monotonous the soundscape is), carried forward by a 45° rotation from the two main dimensions (see Fig. 1a). According to the model, soundscapes describable as "vibrant" are thus both "pleasant" and "eventful", "chaotic" soundscapes are "eventful" and "annoying", soundscapes describable as "monotonous" are both "annoying" and "uneventful", and "calm" soundscapes are "uneventful" and "pleasant".

The processing of (numerical) values obtained from the evaluation of the eight scales corresponding to the extremes of the four dimensions (i.e., pleasant, vibrant, eventful, chaotic, annoying, monotonous, uneventful, calm) makes it possible to plot the soundscapes in the perceptual plane described by the two dimensions pleasantness and eventfulness, depicted in Fig. 1a. This type of visualisation is particularly effective in that it makes it possible to set goals for design interventions and to assess their effectiveness in perceptual terms, by comparing the ante and post operam soundscapes in the perceptual space.

It should be noted that Axelsson's model was developed from listening tests in neutral laboratory environments, reproducing sounds recorded outdoors. Therefore, it is unclear whether the scales and dimensions reported in the ISO standards and technical specifications are equally valid for describing the affective response to

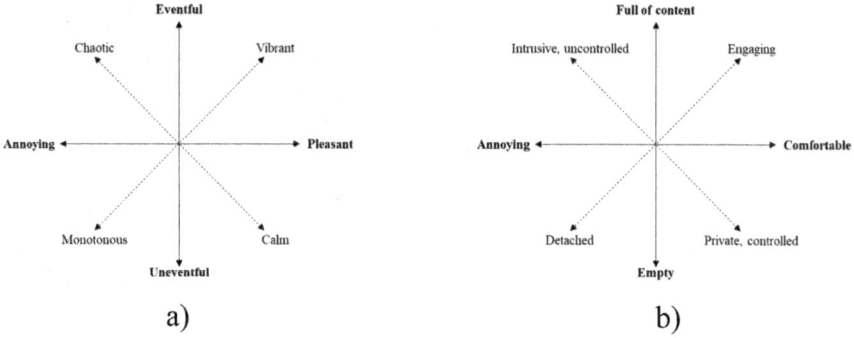


**Fig. 1** Models for the assessment of the affective response to acoustic environments in **a** outdoor settings (International Organization for Standardization 2019) and **b** indoor residential settings (adapted from Torresin et al. 2020a)

indoor soundscapes. Compared with outdoor contexts, the indoor soundscape is in fact characterised by:

- a combination of sounds generated from sources both external and internal to the building
- the presence of a reverberant sound field
- a greater variety of tasks performed by people (not just relaxing or transiting)
- a greater amount of time spent by people immersed in the acoustic environment
- a lower possibility of control over the acoustic environment (e.g., people are usually not free to move to a different and acoustically better place).

These peculiarities have led to question a direct application of the urban soundscape model to indoor environments, since some of the perceptual dimensions might be specific to outdoor contexts, and different ones might be relevant in the assessment of confined environments. Therefore, this prompted the implementation of laboratory listening tests to derive a model for assessing the affective response to indoor soundscapes. Tests were based on the experimental design of Axelsson's study, adapted to consider the peculiarities of the indoor soundscape in the case of a residential type environment (Torresin et al. 2020a).

Thirty-five participants individually took part in 20 consecutive listening sessions within a reconstructed living room at the acoustics laboratory at University College London. The 20 listening conditions were obtained from the combination of four types of sounds generated within the room (i.e., ventilation noise, music, television, and a condition in the absence of generated sounds) and five types of outdoor urban contexts (i.e., absence of sounds from outside, heavy road traffic, light road traffic, pedestrian area, and garden) filtered through a half-open window. During the experiment, the window was simulated by a screen that projected a view outward for the entire duration of the test. All acoustic stimuli were previously recorded indoors and, in the case of sounds from outside the building, in various half-open window environments over different urban settings in London. One-minute excerpts were then selected from the recordings and played back in the laboratory. The sound recording and playback techniques in the lab enabled the creation of highly immersive and realistic acoustic scenarios.

Participants were asked to sit in an armchair and relax, imagining that they were in their living room at home with a half-open window. Each session included a one-minute listening session, at the end of which participants had to describe the soundscape in which they were immersed through 97 rating scales, each defined by an attribute or set of attributes (e.g., "private", "familiar, intimate, personal") to which people associated a score between zero (i.e., the attribute does not correspond to the perception of the soundscape) and 100 (i.e., the attribute corresponds perfectly to the perception of the soundscape). The attributes were selected from the list of adjectives used in Axelsson's experiment and supplemented with additional attributes from focus groups and the literature in order to describe the peculiarities of perception of interior spaces.

Principal component analysis performed on the ratings of the 97 attributes returned three dimensions capable of explaining 58%, 25% and 7% of the total variance of the data, respectively. Their interpretation was based on the identification of the attributes most correlated with each of them and led to the designations of "comfort", "content", and "familiarity" labels. The first two dimensions, which together explain more than 80% of the total variance, describe the degree of comfort and content of sounds and events in the soundscape, similarly to the two dimensions identified by Axelsson (pleasantness and eventfulness). Analysis of the attributes and their positioning with respect to the two main dimensions (comfort and content) then led to the definition of secondary dimensions, rotated 45° in the same plane (see Fig. 1b), representing environments that are more or less attractive and engaging ("engaging" or "detached") and environments that are perceived as more or less private and under the control of the user ("private, controlled" or "intrusive, uncontrolled"). The latter dimension is particularly interesting for the type of environments under study, as it characterises a fundamental aspect of dwellings that we do not find in the study of outdoor urban environments: home as a private and intimate place where one can exercise control over the environment in which one lives.

The reference system allows measuring and representing the acoustic perception of people in residential settings and to determine which factors contribute most to it. Among other factors, of interest in the context of studying the relationship between acoustics and ventilation is understanding the interaction between indoor and outdoor sounds on the indoor soundscape, as addressed in the following section.

## 3 The Impact of Outdoor Soundscape on Indoor Soundscape

The outdoor acoustic environment can influence the soundscape inside buildings and potentially affect occupants' productivity, health, and well-being. Traditionally, the focus of research and policymakers has been on establishing dose–response relationships between disturbing noise sources and their negative impacts. However, in order to guide a management of sound environments aimed at promoting positive outcomes on occupants, it is interesting to delve into the set of negative and positive impacts that different urban sound environments can have on the indoor soundscapes.

The laboratory listening tests described in Sect. 3 made it possible to investigate the interaction effect between different types of urban sound environments and indoor sounds on the indoor soundscape and, in particular, its underlying dimensions (i.e., comfort and content).

Results show that in the simulated living room with a window ajar, acoustic comfort was mainly influenced by the type of sounds introduced by the window and coming from outside the building (Torresin et al. 2020a). In particular, soundscapes dominated by heavy traffic noise were found to be particularly annoying, those dominated by light traffic or anthropogenic noise (e.g., voices, footsteps) provided neutral comfort conditions, while natural sounds were pleasantly perceived. Interestingly, in the absence of indoor sound sources or in the presence of unpleasant noise sources in the room (e.g., mechanical ventilation noise), sounds from outside (e.g., people's voices or natural sounds) managed to improve the degree of comfort, despite the higher overall sound level.

As for the degree of room saturation or "content", the main effects were related to the type of external sounds and to the presence of music inside the room. In general, in the presence of heavy traffic noise, anthropogenic and natural sounds, the indoor soundscape was perceived as more saturated. The effect of indoor sound type on the acoustic saturation of the environment was only relevant in the absence of stimuli from outside, with higher content generally provided by music and television-generated sounds (Torresin et al. 2020a).

Attributes of the indoor soundscape model depicted in Fig. 1b were employed in an online questionnaire administered to adult participants (18–65 years), to assess indoor soundscape perception of people working from home during the COVID-19 lockdown as part of the project "Home as a place of rest and work: the ideal indoor soundscape during the Covid-19 pandemic and beyond" (Torresin et al. 2021). The survey targeted the city of London and Italy as study areas and was conducted in January 2021. Results were derived from 464 UK participants (181 males, 282 females, 1 other; mean age: 32.2 years; SD: 9.1 years) and 384 Italian participants (215 males, 166 females, 3 other; mean age: 29.6 years; SD: 8.9 years). Structural equation modelling (SEM) was used to investigate the patterns of effects of variables related to the perceived acoustic environment, building characteristics, situational and personal factors on the indoor soundscape, when evaluated with reference to different activities (i.e., working from home and relaxation) (Torresin et al. 2022a).

Further insights were gained from the thematic analysis of responses to open-ended questions investigating the ideal indoor soundscape for various activities, including working from home, relaxation, home training and sexual activities. The results from the UK sub-sample are reported in (Torresin et al. 2022c).

Findings showed that greater comfort was associated with working and relaxing in a room that overlooked a quiet urban area. The availability of accessible quiet areas and quiet sides in the home has indeed benefits in terms of reduced annoyance, increased health and quality of life (Kluizenaar et al. 2011; Van Renterghem and Botteldooren 2012; de Kluizenaar et al. 2013; Van Kamp et al. 2016). The increase in comfort emerged in both samples, Italian and UK ones, for both WFH and relaxation activities (Torresin 2022). Quietness was among the most desired elements in an ideal soundscape (see Table 1), especially in reference to relaxation (by 64.1% of respondents, N: 426), followed by work-from-home (WFH) activities (55.9%, N: 449), sexual activities (46.1%, N: 204) and home workout (30.7%, N: 296) (Torresin et al. 2022c). Similarly, a reduction in external noise in an ideal scenario was mainly desired for relaxation activities (16.4%), followed by WFH (12%), sexual activities (8.3%) and home workout (6.8%).

Concerning the contribution of specific sound sources to acoustic comfort, listening to natural sounds resulted in an improvement of comfort conditions while

**Table 1** Percentage of respondents who reported the willingness of presence, increase (+), or decrease (−) of a certain feature in the ideal soundscape for the different activities investigated

| | Relaxing (N: 426) (%) | WFH (N: 449) (%) | Sexual activities (N: 204) (%) | Home workout (N: 296) (%) |
|---|---|---|---|---|
| + Outdoor sounds (generic) | 1.6 | 2.9 | 0.0 | 2.0 |
| − Outdoor sounds (generic), + sound insulation from external noise | − 16.4 | − 12.0 | − 8.3 | − 6.8 |
| + Traffic | 2.8 | 5.3 | 5.4 | 2.7 |
| − Traffic | − 13.6 | − 22.9 | − 3.9 | − 6.1 |
| − Construction | − 6.1 | − 9.1 | − 1.0 | − 1.7 |
| + Human sounds from outside | 1.6 | 1.8 | 0.0 | 0.0 |
| − Human sounds from outside | − 3.8 | − 2.9 | − 0.5 | − 0.3 |
| + Natural sounds | 26.8 | 44.1 | 9.8 | 9.8 |
| − Natural sounds | − 3.5 | − 2.7 | − 2.0 | − 2.0 |
| Quietness | 64.1 | 55.9 | 46.1 | 30.7 |
| With some sound | 11.3 | 19.4 | 12.3 | 12.8 |

Percentages are calculated on the number of participants who addressed aspects relevant to the design of the indoor soundscape. Adapted from Torresin et al. (2022c)

WFH in London, whereas the direct effect on comfort was not significant either in Italy or during relaxation in London (Torresin 2022). However, natural sounds were frequently mentioned among the positively perceived sounds in relation to WFH but also to relaxation, both in London and Italy (Torresin 2022). When imagining an ideal soundscape, natural sounds were especially desired while WFH (26.8%), and to a lesser extent during relaxation (26.8%), physical training (9.8%) and sexual activities (9.8%). These results are in line with previous literature, which has repeatedly reported the positive perception of natural sounds, their effect of improving comfort and pleasantness (Axelsson et al. 2010; Torresin et al. 2020a), but also their ability to reduce stress, improve mood and cognitive performance (Kaplan 1995; Ratcliffe 2021).

Traffic was often cited as a source of annoyance in relation to both WFH and relaxation. However, comfort was not directly affected by the perceived predominance of traffic noise in the SEM results (Torresin et al. 2022a). This is likely due to the decrease in traffic flows and related noise levels in London (Aletta et al. 2020; Lee and Jeong 2021) and Italy (Bartalucci et al. 2020; Zambon et al. 2021), which might have led to a reduction in annoyance from external sources during the COVID-19 lockdown, compared to the pre-pandemic period. On the other hand, construction works were still ongoing and the noise generated could be more noticeable due to the reduced traffic flow (Lee and Jeong 2021). Construction works were indeed reported as one of the noise sources that negatively impacted work activities, especially in London (Torresin 2022). In an ideal soundscape, a reduction in traffic noise was advocated especially for WFH (22.9%), followed by relaxation (13.6%), sport (6.1%) and sexual activities (3.9%). However, background traffic noise is sometimes also desired, especially during sexual activities (5.4%) and while WFH (5.3%) (see Table 1). Construction noise, on the contrary, is never welcome, and its reduction, in an ideal soundscape, is particularly desired during WFH (9.1%) and relaxation activities (6.1%).

With regard to the sounds of people from outside (such as voices and footsteps), no link with acoustic comfort emerged in either the Italian or English sample (Torresin 2022). In an ideal soundscape, human sounds from outside are generally not mentioned and the maximum reduction is desired in relaxation activities by 3.8% of the respondents (see Table 1).

In general, external sound stimuli were reported to be disturbing and frustrating in relation to their sound level and temporal patterns, with loud and sudden noises being more disruptive. However, the impact was also highly dependent on the activity at hand. Sound stimuli were found to be more impactful for tasks requiring concentration and intelligibility (e.g., WFH, reading, meditation and yoga). External sound stimuli were sometimes deemed beneficial as they could provide a contact with the outside world and masking opportunities in presence of activities that generate noise that may cause disturbance to others (e.g., playing loud music) or embarrassment and privacy issues (e.g., during sexual activities).

As regards content scores, in presence of rooms overlooking quiet urban areas, the indoor soundscape was perceived as less saturated with sounds and resulted therefore emptier, i.e., with lower content scores.

The results gathered from the above two studies, based on laboratory research and an online questionnaire respectively, brought evidence on the interaction effects between outdoor and indoor soundscapes. How can this knowledge be used to optimise ventilation, occupant well-being, and take into account sustainability aspects in light of climate change?

## 4 Ventilation: An Interface Between Indoor and Outdoor Soundscapes

Although working in rooms exposed to noisy urban settings is not ideal, results from the online survey during the COVID-19 pandemic showed that people tended to keep their windows open, even in winter (Torresin et al. 2022b). Many factors drive the opening of the window (e.g., feeling the sense of breeze) and among them is sound contact with the outside environment. Participants, while expressing a request for "quietness", wouldn't like an entirely "silent" environment and emphasised the importance and significance associated with positive sound stimuli. This is consistent with what has been reported in the soundscape literature regarding the definition of quietness and quality of the indoor soundscape, which would not only or necessarily be related to a reduction in dB sound level but would rather be a complex construct. Below certain sound level thresholds, function for instance of the activity at hand, control opportunities and exposure time (Torresin et al. 2020b), positively perceived sounds and multisensory safety indicators would be fundamental for the quality of the indoor soundscape (Andringa and Lanser 2013; Kang et al. 2016; Aletta and Kang 2019; Torresin et al. 2020b).

Building façades and in particular façade ventilation devices (in the simplest case, windows) can therefore be a pathway for incoming noise but also for welcome sounds, depending on the external urban context. If quietness also depends on the quality of sounds available, it follows that sound insulation, particularly of façades, should ideally be "selective". This means that building façades should ideally be able to selectively filter out noise and let sounds in according to people's perception and be tunable to changing activities and needs. Compared to noise control approaches, which have historically focused on the conflict between ventilation and acoustic comfort, the soundscape approach paves the way for an evaluation of the potential of natural ventilation to improve the soundscape inside buildings, as a potential co-benefit of natural ventilation (Torresin et al. 2019).

This thus leads to rethinking the role of window opening and natural ventilation devices as tools capable of blocking, transmitting, or adjusting external sounds to provide a connection with the outside and a sense of place, release (pleasant) sounds, or mask (annoying) noise (Torresin et al. 2019). This could be possible in the future thanks to some of the solutions (or a combination of them) that are currently being researched, such as window automations informed by the internal and external soundscape (Torresin et al. 2019; Fusaro and Kang 2021), active noise control technologies applied to façade openings (Lam et al. 2021) and metamaterial design (Fusaro et al. 2022), couple with more traditional passive solutions (Kang and Brocklesby 2005).

It is therefore evident how the balance between indoor and outdoor soundscape based on people's perception can be an important driver of technological innovation that can promote on the one hand the health and well-being of the occupants and on the other hand the use of ventilation solutions and overheating mitigation strategies that minimise energy consumption and emissions into the environment.

## 5 Design Suggestions

The indoor soundscape approach is able to turn the traditional conflicting relationship between acoustics and ventilation into an opportunity both for people and the environment. Ventilation mechanisms can be rethought as a vehicle for positive sounds, and this can be an additional trigger for prioritising the care of the urban soundscape. It is therefore crucial that the soundscape approach goes along with the cost–benefit assessments of the technological solutions being considered: if applied to the case of ventilation, for example, in favourable urban soundscape contexts, it is clear that the soundscape approach is able to value acoustic comfort as a possible co-benefit of natural ventilation. However, in acoustically polluted urban contexts, the use of natural ventilation may not be feasible or may require design solutions that are only feasible in the early stages of planning. This leads to two main recommendations.

First, it is important to address issues related to the perceived acoustic environment in the early stages of urban planning or building design (and retrofit). Pending soundscape prediction indices, reference could be made to environmental, building, and room acoustics parameters, supplemented by narratives from professionals who can identify and describe critical noise sources, quiet areas to be protected, soundscape areas of interest, and related design interventions. Interventions could include, among others, the creation of green areas, noise barriers, water features, or possible interventions for soundscape mitigation and/or enhancement, taking cues from (and populating) online catalogues that collect examples of "soundscape interventions" (Moshona et al. 2022). The narratives could then be supplemented with visualisations/auralisations in virtual and augmented reality, thus enabling a full perceptual evaluation from the design stages.

Second, it is important to address problems related to acoustics, air quality, and overheating in a coordinated manner. This is because sources of possible pollution are often common, but not all solutions are necessarily effective for all environmental stressors. For this reason, there is a growing call for multidomain research (Schweiker et al. 2020), and there are early examples of coordination at the policy level, who have led to joint guidelines across domains. Examples are the The Acoustics, Ventilation and Overheating Guide by the UK Association of acoustics and noise consultants (Association of Noise Consultants 2020), the Approved Doc O—Overheating in Building regulation in England, which includes aspects related to the acoustic environment upstream of the choice of overheating mitigation strategy (Department for

Levelling Up Housing and Communities 2022) or the Revised planning guidance in relation to air quality, noise and soundscape by the Welsh government (Planning Policy Branch Welsh Government 2022).

## 6 Conclusions

The soundscape framework, when applied to the indoor built environment, can help improve occupants' health and well-being by valuing acoustic perception within the factors to be considered in the design of buildings and their technological solutions. The example of soundscape application to the field of building ventilation also demonstrated how essential a soundscape perspective can be for the sustainable development of communities (Kang et al. 2023). The consideration of soundscape aspects from the earliest stages of urban planning and building design would in fact favour conditions that allow for the adoption of passive ventilation (and cooling) strategies on the one hand, and occupants' health and well-being on the other, based on the use of sound as a design resource.

Overall, the chapter highlighted the need for diverse scientific communities and practitioners from a wide range of disciplines to come together and collaborate across discipline boundaries for sustainable development, including soundscape quality among the crucial factors for the environment and humans in a warming climate.

## References

Aletta F, Kang J (2019) Promoting healthy and supportive acoustic environments: Going beyond the quietness. Int J Environ Res Public Health 16

Aletta F, Kang J, Axelsson Ö (2016) Soundscape descriptors and a conceptual framework for developing predictive soundscape models. Landsc Urban Plan 149:65–74. https://doi.org/10.1016/j.landurbplan.2016.02.001

Aletta F, Oberman T, Mitchell A et al (2020) Assessing the changing urban sound environment during the COVID-19 lockdown period using short-term acoustic measurements. Noise Mapp 7:1–12. https://doi.org/10.1515/noise-2020-0011

Andringa TC, Lanser JJL (2013) How pleasant sounds promote and annoying sounds impede health: a cognitive approach. Int J Environ Res Public Health 10:1439–1461. https://doi.org/10.3390/ijerph10041439

Association of Noise Consultants (2020) Acoustics ventilation and overheating—residential design guide

Axelsson Ö, Nilsson ME, Berglund B (2010) A principal components model of soundscape perception. J Acoust Soc Am 128:2836–2846. https://doi.org/10.1121/1.3493436

Bartalucci C, Borchi F, Carfagni M (2020) Noise monitoring in Monza (Italy) during COVID-19 pandemic by means of the smart network of sensors developed in the LIFE MONZA project. Noise Mapp 7:199–211. https://doi.org/10.1515/noise-2020-0017

Cain R, Jennings P, Poxon J (2013) The development and application of the emotional dimensions of a soundscape. Appl Acoust 74:232–239. https://doi.org/10.1016/j.apacoust.2011.11.006

de Kluizenaar Y, Janssen SA, Vos H et al (2013) Road traffic noise and annoyance: a quantification of the effect of quiet side exposure at dwellings. Int J Environ Res Public Health 10:22582270. https://doi.org/10.3390/ijerph10062258
Department for Levelling Up Housing and Communities (2022) Overheating: approved document O. https://www.gov.uk/government/publications/overheating-approved-document-o
European Environment Agency (2020a) Environmental noise in Europe—2020
European Environment Agency (2020b) EEA report: healthy environment, healthy lives: how the environment influences health and well-being in Europe. Pubblications Office of the European Union, Luxembourg
European Union (2002) EU DIRECTIVE 2002/49/EC of 25 June 2002, relating to the assessment and management of environmental noise. Off J Eur Union 12–25
Fusaro G, Kang J (2021) Participatory approach to draw ergonomic criteria for window design. Int J Ind Ergon 82:113–114. https://doi.org/10.1016/j.ergon.2004.08.001
Fusaro G, Kang J, Asdrubali F, Chang W (2022) Assessment of acoustic metawindow unit through psychoacoustic analysis and human perception. Appl Acoust 196:108885. https://doi.org/10.1016/j.apacoust.2022.108885
International Organization for Standardization (2014) ISO 12913-1:2014—Acoustics—Soundscape Part 1: definition and conceptual framework
International Organization for Standardization (2018) ISO TS 12913–2:2018—Acoustics—Soundscape part 2: data collection and reporting requirements
International Organization for Standardization (2019) ISO TS 12913–3:2019—Acoustics—Soundscape part 3: data analysis
Kang J (2007) Urban sound environment. CRC Press, London
Kang J, Aletta F, Gjestland TT et al (2016) Ten questions on the soundscapes of the built environment. Build Environ 108:284–294. https://doi.org/10.1016/j.buildenv.2016.08.011
Kang J, Aletta F, Oberman T et al (2019) Towards soundscape indices. In: International Congress on Acoustics—ICA, pp 2488–2495
Kang J, Aletta F, Oberman T et al (2023) Supportive soundscapes are crucial for sustainable environments. Sci Total Environ 855:158868. https://doi.org/10.1016/j.scitotenv.2022.158868
Kang J, Brocklesby MW (2005) Feasibility of applying micro-perforated absorbers in acoustic window systems. Appl Acoust 66:669–689. https://doi.org/10.1016/j.apacoust.2004.06.011
Kaplan S (1995) The restorative benefits of nature: toward an integrative framework. J Environ Psychol 15:169–182. https://doi.org/10.1016/0272-4944(95)90001-2
Klepeis N, Nelson WC, Ott WR et al (2001) The National Human Activity Pattern Survey (NHAPS): a resource for assessing exposure to environmental pollutants. J Expo Sci Environ Epidemiol 11:231–252
de Kluizenaar Y, Salomons EM, Janssen SA et al (2011) Urban road traffic noise and annoyance: the effect of a quiet façade. J Acoust Soc Am 130:1936–1942
Lam B, Gan WS, Shi DY et al (2021) Ten questions concerning active noise control in the built environment. Build Environ 200:107928. https://doi.org/10.1016/j.buildenv.2021.107928
Lee PJ, Jeong JH (2021) Attitudes towards outdoor and neighbour noise during the COVID-19 lockdown: a case study in London. Sustain Cities Soc 67:102768. https://doi.org/10.1016/j.scs.2021.102768
Moshona CC, Aletta F, Henze H et al (2022) What is a soundscape intervention? Exploring definitions and identification criteria and a platform to gather real-world examples. In: Proceedings of the Internoise 2022: 51st International Congress and Exposition on Noise Control Engineering, pp 1–9
Planning Policy Branch Welsh Government (2022) Revised planning guidance in relation to air quality, noise and soundscape. https://www.gov.wales/revised-planning-guidance-relation-air-quality-noise-and-soundscape
Ratcliffe E (2021) Sound and soundscape in restorative natural environments: a narrative literature review. Front Psychol. https://doi.org/10.3389/fpsyg.2021.570563

Schweiker M, Ampatzi E, Andargie MS et al (2020) Review of multi-domain approaches to indoor environmental perception and behaviour. Build Environ 106804. https://doi.org/10.1016/j.buildenv.2020.106804
Tong H, Aletta F, Mitchell A, Oberman T (2021) Increases in noise complaints during the COVID-19 lockdown in Spring 2020 : a case study in Greater London, UK. Sci Total Environ
Torresin S (2022) Indoor soundscape modelling: rethinking acoustic comfort in naturally ventilated residential buildings. University of Trento
Torresin S, Albatici R, Aletta F et al (2019) Acoustic design criteria in naturally ventilated residential buildings: new research perspectives by applying the indoor soundscape approach. Appl Sci 9:1–25. https://doi.org/10.3390/app9245401
Torresin S, Albatici R, Aletta F et al (2020a) Indoor soundscape assessment: a principal components model of acoustic perception in residential buildings. Build Environ 182:107152. https://doi.org/10.1016/j.buildenv.2020.107152
Torresin S, Aletta F, Babich F, Bourdeau E et al (2020b) Acoustics for supportive and healthy buildings: emerging themes on indoor sustainability. Sustainability 6054. https://doi.org/10.3390/su12156054
Torresin S, Albatici R, Aletta F et al (2021) Indoor soundscapes at home during the COVID-19 lockdown in London—Part I: associations between the perception of the acoustic environment, occupantś activity and well-being. Appl Acoust 183:108305. https://doi.org/10.1016/j.apacoust.2021.108305
Torresin S, Albatici R, Aletta F et al (2022a) Indoor soundscapes at home during the COVID-19 lockdown in London – Part II: A structural equation model for comfort, content, and well-being. Appl Acoust 185:108379. https://doi.org/10.1016/j.apacoust.2021.108379
Torresin S, Albatici R, Aletta F, et al (2022b) Associations between indoor soundscapes, building services and window-opening behaviour during the COVID-19 lockdown. Build Serv Eng Res Technol 43. https://doi.org/10.1177/01436244211054443
Torresin S, Ratcliffe E, Aletta F, et al (2022c) The actual and ideal indoor soundscape for work, relaxation, physical and sexual activity at home: a case study during the COVID-19 lockdown in London. Front Psychol. https://doi.org/10.3389/fpsyg.2022.1038303
United Nations Environment Programme (2021) 2021 Global status report for buildings and construction: towards a zero-emission, efficient and resilient buildings and construction sector. Nairobi
Van Kamp I, Klaeboe R, Brown AL, Lercher P (2016) Soundscapes, human restoration and quality of life. In: Kang J, Schulte-Fortkamp B (eds) Soundscape and the built environment. CRC Press, Boca Raton, pp 43–68
Van Renterghem T, Botteldooren D (2012) Focused study on the quiet side effect in dwellings highly exposed to road traffic noise. Int J Environ Res Public Health 9:4292–4310. https://doi.org/10.3390/ijerph9124292
WHO (2018) Environmental noise guidelines for the European Region
Zambon G, Confalonieri C, Angelini F, Benocci R (2021) Effects of COVID-19 outbreak on the sound environment of the city of Milan, Italy. Noise Mapp 8:116–128. https://doi.org/10.1515/noise-2021-0009

**Jian Kang** is Professor of Acoustics and Soundscape, and Head of the University College London's Acoustics Group. He is recipient of the prestigious Advanced ERC Grant Award, currently working internationally on developing Soundscape Indices. He is Fellow of Royal Academy of Engineering, International Member (Academician) of the Chinese Academy of Engineering, and Member of Academia Europaea -The Academy of Europe. He contributes chapter 9 with first author **Simone Torresin**, Assistant Professor at the University of Trento, and second author **Francesco Aletta**, an architect and urban sound planner by training, Lecturer (Teaching) in Building Physics and Soundscape at University College London (UK), together.

# Biotic Homogenization and Rewilding of Urban Greening

Cheng Jin, Siwei Hu, Liangjun Da, Yuandong Hu, and Yongchuan Yang

**Abstract** Cities have become the main abodes for humans, and landscape plant species with their notable influence on the quality of life, are important components of the urban ecosystem. Human preferences for the plant species of urban greening have greatly facilitated the spread of non-native species, resulting in the homogenization of landscape plant communities across spatial scales. More concretely, homogenization makes a limited cohort of familiar species with widespread popularity, while these plants usually exhibit traits such as easy propagation, ornamental characteristics, and have a tropical origin. This has contributed to accommodating many climatically mismatched plants in cities. Therefore, the need to explore the climatic suitability and sustainability of landscape plants is especially relevant due to globalization and climate change. Landscape plant selections should not only pay attention to the existing landscape needs but also consider the sustainability of landscape plant species to climate change, especially for long lifespan woody plants. In the face of these serious problems in urban greening, the research found that the development of near-to-nature forests is an effective way to solve these problems, and the excellence of near-to-nature forests has been proven in Shanghai. The application of spontaneous urban plants (SUPs) is another effective way, and the biodiversity-friendliness of SUPs has also been proven in 59 major cities in China. To sum up, nature-based solutions with rewilding and providing room for natural processes are important to improve urban biodiversity and develop green spaces that are more natural and provide people with more genuine nature experiences.

**Keywords** Urban greening · Biotic homogenization · Climate change · Spontaneous urban plants · Rewilding

---

C. Jin · S. Hu · Y. Yang (✉)
Key Laboratory of the Three Gorges Reservoir Region's Eco-Environment, Ministry of Education, Chongqing University, Chongqing 400045, China
e-mail: ycyang@cqu.edu.cn

L. Da
Institute of Science and Engineering of Ecology in Arid and Semi-arid Areas, Xi'an University of Architecture and Technology, Xi'an 710055, China

Y. Hu
College of Landscape Architecture, Northeast Forestry University, Harbin 150006, China



R. Yao (ed.), *Resilient Urban Environments*, Cities and Nature,
https://doi.org/10.1007/978-3-031-55482-7_10

# 1 Biotic Homogenization of Urban Greening

**Abstract** Human preferences for the species in urban greening have greatly facilitated the spread of non-native species, resulting in the homogenization of urban greening across spatial scales. We introduced the biotic homogenization of urban greening through a study of the 11 major cities along the Yangtze River in China. The results showed that China's urban communities are becoming homogenized, as the urban communities of different cities are highly similar to each other despite their geographical separation. The spread of non-native species by anthropogenic activities plays a substantial role in biotic homogenization, since 27% of the landscape plant species we found in the 11 cities were cultivars and introduced species, and 25% were used outside of their native distribution ranges in China. This may be explained by the market and urban planners who tend to favor greening plants that are highly profitable and have esthetic ornamental traits, rather than spending time introducing and acclimatizing the native species in each city, despite the level of biodiversity in China is extraordinarily high.

Urbanization has profoundly affected global biodiversity (Lambin et al. 2001; McKinney 2006; Grimm et al. 2008; McDonald et al. 2008). As critical outcomes of urbanization, habitat alteration and human-assisted dispersal of species have led to the reduction or extirpation of regional and endemic native species, while expanding the distribution of cosmopolitan, non-native species (or introduced, alien, exotic, and non-indigenous species). This generally indicates the increasing similarity of species composition between different areas, known as "biotic homogenization" (McKinney and Lockwood 1999).

This chapter introduces the biotic homogenization of urban greening through a study of the 11 major cities along the Yangtze River in China (Fig. 1). In this study, the question asked was whether landscape plant communities (an abbreviation of "urban communities") show high similarity across major cities in the Yangtze River basin region. Then, the study investigated the similarities between urban and natural communities within each city and addressed the distance decay of similarity relationships for all pairs of cities. The purpose was to see if divergence was occurring between the urban and natural communities at the city and geographic scale. Lastly, the study looked into species' origins and ornamental traits to understand the preferences for landscape plants in the Yangtze River basin region, and to infer how the human selection of plants would determine the compositional characteristics of the current urban communities.

The study relied on the Jaccard similarity index ($J$) to measure the biotic homogenization of urban greening in different locations (McKinney 2004). $J$ was defined as: $J = c/(a + b + c)$, where $a$ is the number of species unique to the first site, $b$ is the number of species unique to the second site, and $c$ is the number of species in common at both compared sites. $J$ ranges between 0 and 1, where a result of 0 means that the two sites have no species in common, and a result of 1 means that the species compositions of the two sites are identical.

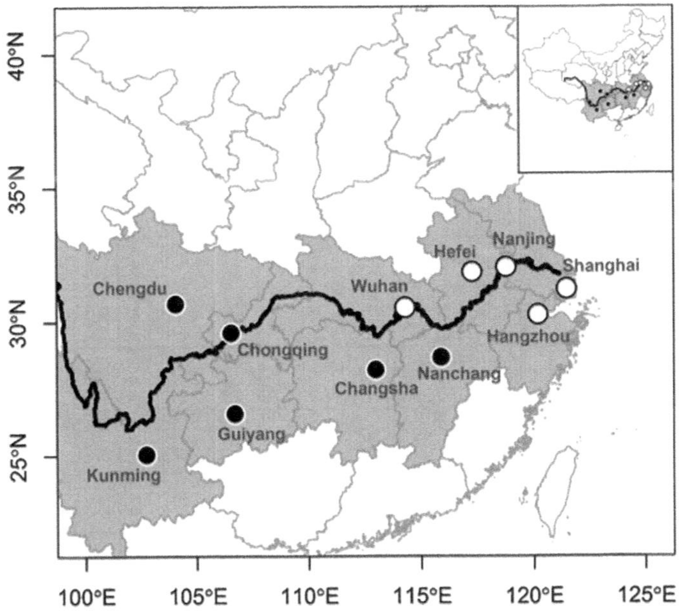


**Fig. 1** Distribution of the 11 major cities along the Yangtze River in China. The solid black line represents the Yangtze River; solid circles indicate the six cities in the middle subtropics, and hollow circles indicate the five cities in the northern subtropics; gray areas show the provinces to which each city belongs

## *1.1 Biotic Homogenization of Urban Communities*

The study found the urban communities showed high similarities among the 11 major cities (Fig. 2a). Furthermore, within-city similarities between the urban and natural communities were quite low for the 11 cities (Fig. 2b). Further, the distance decay of similarity relationships addressed varied between urban and natural communities: the Jaccard similarity index ($J$) calculated for natural communities in pairs of cities rapidly decreased to less than 0.1 as the geographical distance between cities increased; while, J calculated for urban communities in pairs of cities always remained higher than 0.3, indicating a high proportion of shared species with increasing geographical separation (Fig. 3). These facts indicate that China's urban communities are becoming homogenized, as the urban communities of different cities are highly similar to each other despite their geographical separation. What's more, these artificial communities have also diverged greatly from the natural communities at both the city and geographic scale.

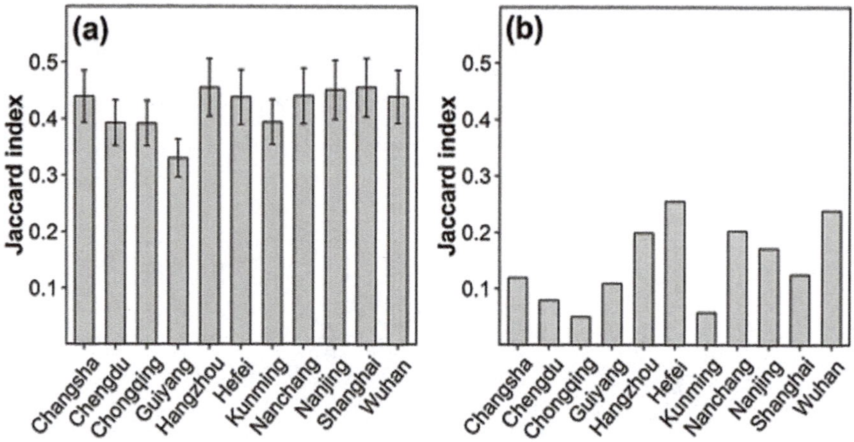


**Fig. 2** Jaccard similarity index summarized by city: **a** Jaccard similarity index calculated for urban communities in pairs of cities. Error bars show the standard errors, and **b** Jaccard similarity index calculated between urban and natural communities within each city. Cities are sorted alphabetically

**Fig. 3** Jaccard similarity index values expressing similarity in the composition of urban and natural communities plotted against geographical distance. Yellow circles indicate pairs of urban communities, and blue circles indicate pairs of natural communities. Lines suggest significant trends fit by the generalized additive models. The gray-shaded area represents the 95% confidence interval

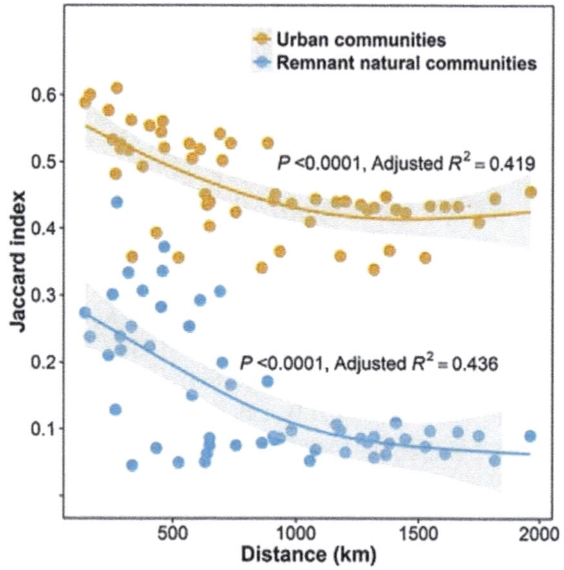


## 1.2 *Causes of the Homogenization*

The spread of non-native species by anthropogenic activities plays a substantial role in biotic homogenization (McKinney 2006; La Sorte et al. 2014). China is one of the world's "megabiodiversity countries" (McNeely et al. 1990), and the Yangtze River basin region, the focus of this study, provides a wealth of biological resources.

Nevertheless, among the landscape plant species that were collected, 9.89% were cultivars and 17.58% were introduced species. In some European countries, such as Great Britain, Ireland, and Spain, the proportions of non-native species account for 41.1%, 23.9%, and 24.9% of the current floras, respectively (Winter et al. 2009). However, the total number of species in these three countries is only about 1000–3000, differing from China (Preston et al. 2002; Winter et al. 2009). Despite the extraordinarily high level of biodiversity in China (> 30,000 vascular species), 27% of the landscape plant species were cultivars and introduced species.

The northern and middle subtropical areas that were studied differ in terms of climatic conditions and floristic compositions (Mackinnon et al. 1996; López-Pujol et al. 2006; Tang et al. 2006). In our study, the similarity in the natural communities in the middle subtropics was less than that of the northern subtropics, possibly because of a higher level of biodiversity in the middle subtropics, compared to the northern subtropics (Fig. 4). This regional difference in the composition of natural flora was shown earlier by the rapid distance decay of similarity relationships between natural communities with increasing geographical separation. Human selection of landscape plant species has altered this relationship: as the geographical distance between pairs of cities increased, the similarities between urban communities stabilized at a high level, because of the high proportion of human-selected, common landscape plant species across the 11 major cities (Fig. 3). In this study, the study found that 25% of the greening species were used outside of their native distribution ranges in China (23.08% was originally distributed in the middle subtropics and 2.20% was distributed in the northern subtropics, though they were used in all 11 cities). As China is now at a stage of rapid economic growth and urban expansion, to meet the requirements of the current market economy, existing examples of urban greening may be referred and implemented, instead of introducing and acclimatizing local, indigenous species (Niemelä 1999; Breuste 2004; Da and Song 2008; Song et al. 2009).

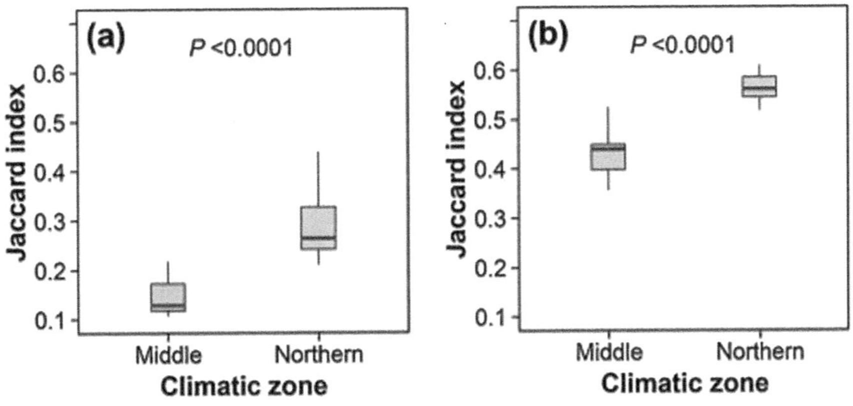


**Fig. 4** Jaccard similarity index by climatic zones for: **a** remnants of natural communities and **b** urban communities. Significance test was performed using the Mann–Whitney U test

Meanwhile, a total of 91 woody species were recorded that were used in urban greening in all 11 cities. Of the 91 species, some had ornamental traits in common, like colorful leaves (13.18%) or flowers (48.35%), while some species were fast-growing or resistant to atmospheric pollution. This is evidence that commercial nurseries in China tend to favor highly profitable landscape plants, with esthetic ornamental traits (e.g., colorful leaves and flowers) and favorable growth characteristics (e.g., fast-growing and resistance to atmospheric pollution). The preferences for a limited cohort of familiar plants have tended to restrict species pool for urban greening, and the options left for urban planners were limited (Da et al. 2009; Song et al. 2009, 2010).

The selection of landscape plant species in China, which is inseparable from the unique domestic market and economic conditions, could be the basis for the homogenization. Generally, the planning of urban green spaces and selection of appropriate landscape plant species is given top priority, rather than prioritizing the commercial benefits and human demands and then choosing from the existing options (Miller 1997; Jim 1999; Niemelä 1999; Nagendra and Gopal 2010). Furthermore, to establish communities that are suitable for specific local conditions and to minimize species expiration, planners need to consider ways for increasing the proportion of native species (Gamborg and Larsen 2003; Da and Song 2008; Nagendra and Gopal 2010; Bürgi 2015; Guo et al. 2015). Unfortunately, only a few cities in China have plans in place to construct "near-natural" urban green spaces with native, indigenous species (e.g., Da and Song 2008; Da et al. 2009; Guo et al. 2015). This eventually led to the current situation within the 11 major Chinese cities where the urban communities have diverged greatly from the remnant natural communities (Fig. 2b). Given the current trend of homogenization, urban planning strategies need to be carefully designed with measures that emphasize the importance of native species in the future.

## 2 Diverse Origins and Climatic Congruence of Landscape Plants

**Abstract** It is important to explore the climatic suitability and sustainability of landscape plants under globalization and climate change, as biotic homogenization of urban greening makes many climatically mismatched plants accumulated in cities. We took 36 major cities in China as an example to exhibit the climatic congruence of landscape plants, based on the geographical and climatic origins of landscape plants. The results showed that most alien species of landscape plants in China were derived from tropical regions in the Americas, Asia, and Australia, and domestic species mainly originated from temperate southern China. Arid cities with harsher environmental stresses accumulated more climatically mismatched plants. Under climate change, the climatic sustainability of landscape plants was projected to vary among climate zones, being influenced by a rich presence of temperate species. Overall, our findings emphasized that landscape plant selections should not only pay attention to the existing landscape needs, but also consider the climatic sustainability of landscape plant species to climate change, especially for long lifespan woody plants.

Biotic homogenization makes a limited cohort of familiar plants with widespread popularity, while these plants tend to have traits such as easy propagation, ornamental characteristics, and tropical origin. This has contributed to accommodating many climatic mismatched plants in cities. Therefore, it is necessary to explore the climatic suitability and sustainability (climatic sustainability means the sustainability of climatic suitability for landscape plants, under climate change) of landscape plants due to globalization and climate change (Ordóñez and Duinker 2015; Burley et al. 2019), especially for woody landscape plants, given the long lifespan of most trees (Burley et al. 2019; Jin et al. 2020).

The geographic and climatic origins of landscape plant species provide the basis for exploring the climatic suitability and sustainability of landscape plants. The geographical origin and accumulation of plants have attracted increasing attention, but the climatic origin and/or accumulation of plants are often ignored. The inadequate knowledge base has hindered the exploration of the climatic suitability and sustainability of landscape plants.

The study took 36 major cities in China as an example to exhibit the climatic congruence of landscape plants, based on the geographical and climatic origins of landscape plants. First, the geographical and climatic origins of landscape plants in major Chinese cities were identified. Second, the climatic suitability and sustainability of landscape plants were ascertained, based on the species' geographical and climatic origins.

The climatic mismatched ratio (CMR) was used to measure the climatic suitability of landscape plants. CMR characterized the degree of climate mismatch between the climatic origins of landscape plant species (the species that are currently planted at each city) and the city's present climate condition. The study used the equation $\mathrm{CMR} = B/C$ to calculate the CMR of each city, where B is the number of the species with climatic origins mismatching with the city's climate (species' climatic origins excluded the city's climate), and $C$ is the total number of landscape plant species of the city. The value of CMR ranges between 0 and 1, where 0 denotes no mismatch and 1 indicates that all species are mismatched. In addition, the study calculated the future CMR based on the future climate condition of each city and compared them with the current CMR to explore the climatic sustainability of landscape plants for each city. Scenario RCP8.5 was used to predict the future climate condition of each city (2071–2100) according to Beck et al. (2018). When the magnitude between current and future CMR shifted by over 10%, it was interpreted as a change between the current and future ones.

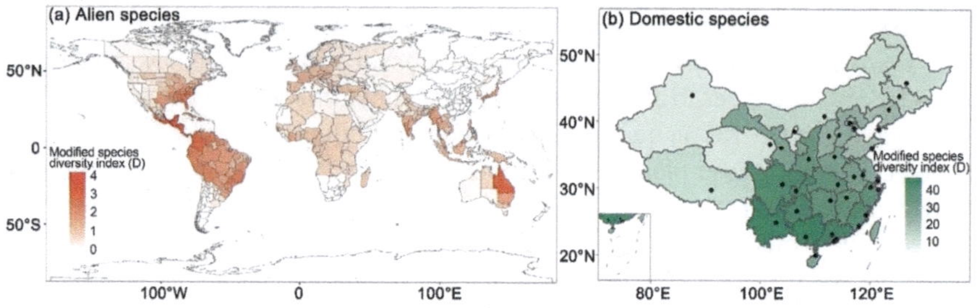


**Fig. 5** Geographical origins of **a** alien species and **b** domestic species of landscape plants in China's 36 major cities. The color saturation in the map legends corresponds to the magnitude of the modified species diversity index (*D*). Black dots in (**b**) were locations of the 36 cities

## 2.1 Geographical and Climatic Origins of Landscape Plant Species in China's Major Cities

The landscape plant species in China's major cities mainly originated from climate zones with precipitation and temperature regimes conducive to vigorous and diverse plants. For instance, most alien species were derived from tropical regions in the Americas, Asia, and Australia (Figs. 5a, 6a), and domestic species mainly originated from temperate southern China (Figs. 5b, 6b). As a possible explanation, the mainly low-latitude donor regions with suitable precipitation and temperature conditions for plants, are rich in biological resources (Cadena et al. 2012; Chejanovski and Wiens 2014). The inherent biodiversity endowment could provide plant explorers and introducers with a large pool of options for landscaping purposes (Hooper et al. 2005; Svitok et al. 2016).

Furthermore, human landscape plant preferences offered another reason, especially the landscape fashion for tropical scenery, that resulted in the extensive spread of species from the tropical zone (Faggi and Ignatieva 2009; Rozzi 2013; Yan and Yang 2017). Nevertheless, due to the confined tropical area of China, fewer domestic species could originate from the intra-country tropical climate zone (Fig. 6b). In the map of the Köppen–Geiger climate classification (Beck et al. 2018), the tropics of China included only the Hainan Island, the southernmost part of Yunnan Province, and Taiwan Island. Hence, China has limited tropical plant resources (Connor and McCoy 1979).

## 2.2 Climatic Suitability and Sustainability of Landscape Plants in Different Cities

The CMRs of landscape plants in arid cities were higher than those in other cities (Fig. 7). This finding might result from the limited availability of indigenous biological resources in arid climate zones, especially species that were suitable for urban

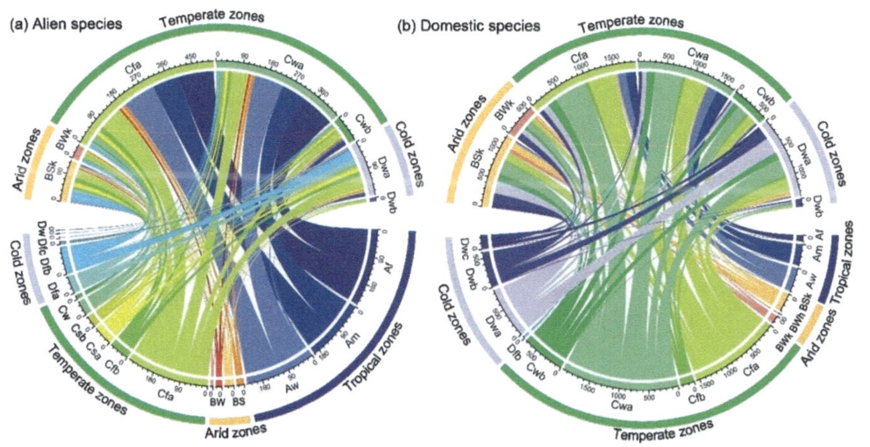


**Fig. 6** Multiple species introduction pathways between the donor and recipient climate zones for **a** alien species and **b** domestic species. The upper half of each chord diagram represents the climate zones of China's 36 major cities, which are the recipients of species, and the lower half represents the climate zones of donor regions. The abbreviations and color scheme of climate zones follow the Köppen–Geiger climate classification

areas (Cadena et al. 2012; Chejanovski and Wiens 2014; Avolio et al. 2015a). Such inherent deprivation restricted the adoption of native species in these regions for urban landscaping (Hooper et al. 2005; Svitok et al. 2016). With a confined pool of pre-urbanization flora, urbanization could raise the species richness of the recipient cities by the introduction of exotic horticultural species to satisfy the innate desire for diversity in urban greening (Avolio et al., 2015a, b). With that in mind, the small market of the limited number of major cities in the arid climate zones could be homogenized or transformed to align with the dominant market and the landscape fads in the temperate and tropical cities (Da et al. 2009; Qian et al. 2016; Avolio et al. 2018).

Cities in different climate zones demonstrated different trends in CMRs under global climate change (Table 1). This may be explained by the rich temperate species in all climatic cities (Fig. 6). For instance, five cold cities were projected to become temperate in the future (2071–2100), so their CMRs might decrease significantly in the future (Table 1). Nevertheless, two temperate cities were projected to become tropical, so their CMRs might rise (Table 1 and Fig. 7b). In addition, the landscape plant species compositions of arid cities have been projected to be very unsuitable, since the high CMRs at present or in the future (Fig. 7).

Under climate change, the climatic sustainability of landscape plants was projected to vary among cities, calling for differential responses and treatments. Further, in applying improvement measures to urban greening, the trade-off between the current needs and the future sustainability of landscape plants under climate change should be carefully handled (Ahern 2013; Westerink et al. 2013; Garcia-Garcia et al. 2016). In addition, for landscape plants in arid cities, considering that

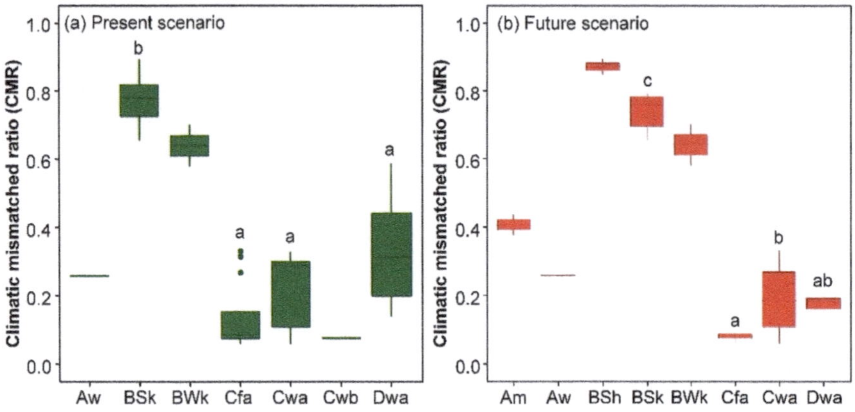


**Fig. 7** Climatic mismatched ratios (CMRs) of landscape plants in different cities based on their climate classification: **a** present scenario and **b** future scenario in the future (2071–2100). Rank sum test and multiple comparison analysis (Dunn's test) were used to compare the CMR (only climate types contained in at least three cities were compared). The different letters above the boxes indicate a significant difference ($P < 0.05$) and the same letters above the boxes indicate a non-significant difference

**Table 1** Climatic mismatched ratios (CMRs) of landscape plants in the seven cities [Köppen climatic primary classification of these cities will change in the future (2071–2100)] at present and in the future (2071–2100)

| City | Present climate | Projected climate[a] | Present CMR | Projected CMR[a] | Trend |
|---|---|---|---|---|---|
| Shenzhen | Dry-winter humid subtropical (Cwa[b]) | Tropical monsoon (Am) | 0.30 | 0.44 | ↑ |
| Macao | Dry-winter humid subtropical (Cwa) | Tropical monsoon (Am) | 0.29 | 0.38 | ↑ |
| Dalian | Hot summer continental (Dwa) | Dry-winter humid subtropical (Cwa) | 0.30 | 0.23 | ↓ |
| Beijing | Hot summer continental (Dwa) | Dry-winter humid subtropical (Cwa) | 0.33 | 0.23 | ↓ |
| Qingdao | Hot summer continental (Dwa) | Dry-winter humid subtropical (Cwa) | 0.48 | 0.17 | ↓ |
| Tianjin | Hot summer continental (Dwa) | Dry-winter humid subtropical (Cwa) | 0.43 | 0.19 | ↓ |
| Zhengzhou | Hot summer continental (Dwa) | Dry-winter humid subtropical (Cwa) | 0.58 | 0.09 | ↓ |

Arrows indicate the change trends of the climatic discordance index of each city between now and the future. The red arrow means rising, the blue arrow means descending

[a] Projected climate classification or CMR (the ratio of species which mismatch with city climate) of the 7 cities in the future (2071–2100)

[b] The abbreviations follow the Köppen–Geiger climate classification

their current species compositions have notable climatically mismatched components that had to be maintained at a high economic cost (Garcia-Garcia et al. 2016), revamping their landscape plants to increase the proportion of arid species that demand less management was recommended.

# 3 Development of Near-Natural Forests and Rewilding of Urban Greening

**Abstract** In recent years, the development of urban forests has been characterized by the blind introduction of exotic species and the uniform use of species with high ornamental value but low climatic suitability, resulting in the loss of local biodiversity and inter-regional biological homogenization. By summarizing the practice and effectiveness of near-natural forestry in Shanghai over the past 20 years, this chapter demonstrates that the development of near-natural forests is an effective way to mitigate urban biodiversity loss and regional biohomogenization. At the same time, in order to achieve urban rewilding, this chapter also shows an overview of the composition and diversity of spontaneous urban plants (SUPs) in 59 major cities in China. This part identified 1211 SUP species through an extensive search of existing field research studies and fieldwork. The species composition pattern of SUPs. The abundance of SUPs and the high proportion of native species suggested that limited urban habitats can still contribute to the enrichment and accumulation of urban biodiversity, which represents the biodiversity-friendliness of this taxon, i.e., creating space for SUPs or using SUPs in urban planning and management is another effective way to react the urban greening challenge.

In recent years, the development of urban forests has been characterized by the blind introduction of exotic species and the uniform use of species with high ornamental value but low climatic suitability, resulting in the loss of local biodiversity and inter-regional biological homogenization. As an important way to conserve urban biodiversity, the study has found that the development of near-natural forests is an effective way to solve these problems, and the first near-natural forests have been built in China for more than 20 years, and their superiority has been proven in Shanghai. At the same time, to rewild cities and bring them closer to nature, the use of spontaneous urban plants (SUPs) in urban greening is another effective way. The biodiversity-friendliness nature of SUPs has been proven in 59 major cities in China. Therefore, using the properties of SUPs, low-maintenance and sustainable urban green landscapes can be established in cities.

## 3.1 Effective Way to Solve Urban Greening Problems

The "near-natural forests" is derived from Akira Miyawaki's concept of "environmental protection forest" (Miyawaki 1999, 2004). The theoretical basis for the development of "near-natural forests" is the ecological theory of community succession and potential natural vegetation, and the selection of suitable native species and the use of container seedlings to "simulate nature", through the combination of early artificial planting and later natural vegetation growth, to build. The aim is to create a "less artificially managed" forest with a complete community structure, high species diversity and a stable state of development, following the natural succession pattern at a later stage (Da and Xu 2003). It can be further defined as an artificial forest that is close in species composition and community structure to the top regional communities, based on natural growth at a later stage and using a "nature simulation" approach (Cheng et al. 2004; Da et al. 2004).

Potential natural vegetation is the type of natural vegetation that can be developed under the influence of environmental conditions such as climate, soil, and topography after the disturbing effects of human activities have ceased, and can generally be inferred from the natural conditions of the area and the existing knowledge of vegetation, as well as from the type of top communities. Ecological greening in urban areas and the creation of "near-natural forests" means determining the target forest type, selecting the plant species, and configuring the tree species accordingly, based on the potential natural vegetation of the area (Da and Xu 2003). In the area of subtropical broadleaved evergreen forest, it takes several hundred years of practice to evolve from bare ground to a top community under natural conditions, but if the top community is planted directly with seedlings of the dominant species in the tree layer and shrub layer with human involvement and appropriate modification of soil and other environmental conditions, it can be reduced to one-tenth of the original time.

### 3.1.1 Near-Natural Forestry Practices in Shanghai

The practice of near-natural forestry in China began in 1997 with the reforestation of barren hills around the Great Wall, and was later applied to the revegetation of tailings sites in Maanshan and the development of green belts on both sides of highways in Qingdao, Xuzhou, and Ningbo (Cheng et al. 2004; Da and Xu 2003). The first "near-natural forest" in Shanghai was constructed in the Pudong New Area, with a 3000 $m^2$ sample plot in June 2000 and a 2000 $m^2$ sample plot in April 2001, after which plant growth and community structure changes were tracked. It has now been 19 years since the plants have grown well and are largely established. Near-natural forest sample plots were subsequently constructed in Songjiang District and Minhang District in 2007 and 2009. The near-natural forest in Songjiang District is located from Qianbujing to Tangkou in Songjiang District, with a total length of 12.6 km and a total area of 207.6 $hm^2$ at the end of 2009. The dominant species in

**Fig. 8** Seed collection before the development of the near-natural forest in Shanghai. The collection site (left, Dajinshan Island) and the seeds used for seedlings (right)

**Fig. 9** Predevelopment seedlings in Shanghai near-natural forest

the top stages of the succession of local potential natural vegetation were selected as establishment species for the development of the near-natural forest in Shanghai, including *Machilus thunbergii*, *Quercus glauca*, *Quercus myrsinifolia*, *Castanopsis sclerophylla*, *Neolitsea sericea*, etc. Seeds of all these species were collected from Da Jinshan Island in the southeast of Shanghai (Fig. 8).

Before planting, the study centralized the collected seeds for the nursery and prepared the land for the selected planting area (Fig. 9). Then, to combine with science and nature education, the study organized local elementary school students to plant trees to complete the plant establishment of the near-natural forest (Fig. 10).

The re-survey was conducted from 2017 to 2018 to assess the ecological benefits of the restoration of the near-natural forests, all three of which have been restored for at least a decade.

### 3.1.2 Effectiveness of Near-Natural Forest Restoration in Shanghai

A total of 67 species of woody plants (plants $\geq$ 1.5 m in height) belonging to 54 genera and 33 families were recorded from three near-natural forests in Shanghai. Of

**Fig. 10** Land preparation for the near-natural forest in Pudong District, Shanghai (left); Organize tree-planting activities for elementary school students and complete near-natural forest planting (right)

these, 44 species in 35 genera of 23 families of trees and 23 species in 21 genera of 16 families of shrubs were recorded. Among the sample sites, there are 47 species of 23 families and 29 genera in the near-natural forests of Songjiang District, 7 families, and 11 genera and 13 species in the near-natural forests of Minhang District, and 16 families and 23 genera and 25 species in the near-natural forests of Pudong District.

The main species composition of the three near-natural forest plots was still dominated by the early-growing species, with a small number of sedentary species. And there were some differences in the dominant species of the upper wood layer (height greater than 1.5 m). The plant density and basal area of the tree layer showed an increasing trend in the three plots. The plant density of the Minhang plot was significantly higher than that of the Songjiang and Pudong plots. The highest growth rate of the basal area was Minhang and Songjiang's was the lowest. With the community recovering, the species composition of the three plots became more and more similar over recovery time, among which the community composition of Pudong and Minhang tended to be the same basically after 10 years of recovery because their initial planting patterns were relatively similar.

The diameter-level distribution of woody plants in three near-natural forest plots generally showed as intermittent development. The number of individuals with a small diameter class was dominant. In addition, with the increase in diameter class, the number of individuals decreased. Target restoration species planted in near-natural forests, such as Cyclobalanopsis glauca and Cyclobalanopsis myrsinifolia, have produced a large number of naturally regenerated seedlings after 10 years of restoration, and both the population structures were L-shaped.

In summary, although with different development patterns, the three near-natural forest plots gradually became more consistent with the community restoration; and the evergreen broad-leaved trees such as Cyclobalanopsis glauca, Cyclobalanopsis myrsinifolia and Machilus thunbergii are gradually occupied the dominant position in the community, and can regenerate well. The target restoration community of near-natural forests has initially formed (Fig. 11).

**Fig. 11** Shanghai Pudong district near-natural forest from development to restoration

## 3.2 *Spontaneous Plants Can Be Fully Utilized in Urban Rewilding*

In the last decade, spontaneous urban plants (SUPs), consisting of plants not intentionally propagated and introduced by humans, have gradually been perceived as an essential component of urban vegetation (Cervelli et al. 2013). The emergence of SUPs does not require economic investment. They are more authentic to the local character and perform valuable ecosystem services to benefit city inhabitants (Bonthoux et al. 2019a), such as regulating microclimate, supplying suitable habitats or refuges for insects and other arthropods (Robinson and Lundholm 2012), providing plant resources for pollinators (Larson et al. 2014; Yang et al. 2019), and hosting a diverse range of native species (Huang et al. 2019). The use of SUPs to create low-maintenance and sustainable landscapes may be more in line with the characteristics of native natural ecosystems. They can contribute to the "naturalness" of urban landscapes while enhancing biodiversity (Berthon et al. 2021; Li et al. 2019). Therefore, understanding and comparing the composition and diversity patterns characteristics of SUPs in different regions and at different scales will facilitate the proper management and allocation of such plants to urban landscapes in the future. They can further promote urban rewilding and biodiversity conservation, which could contribute to establishing eco-friendly cities (Omar et al. 2018; Qian et al. 2020).

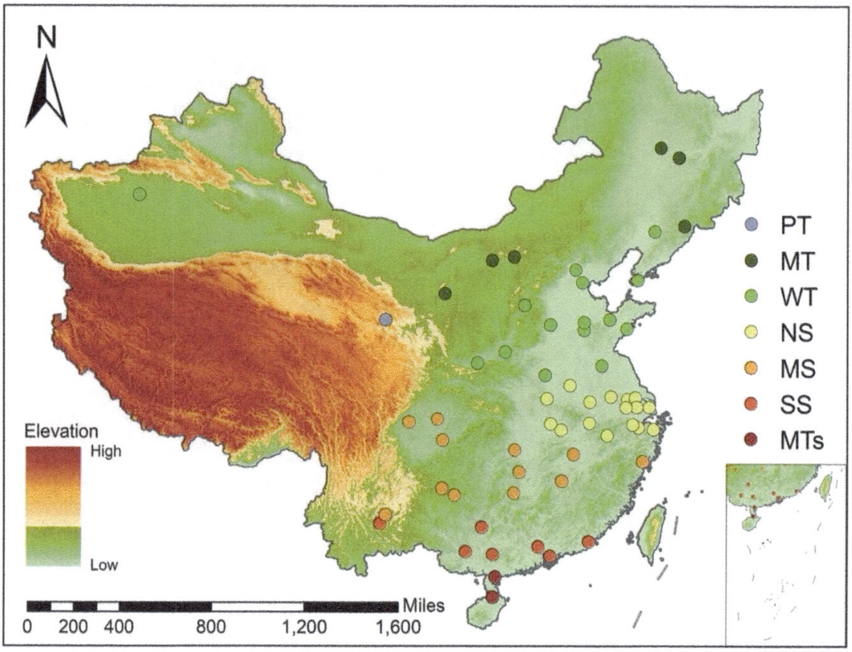


**Fig. 12** Locations of the 59 major Chinese cities included in the study. The climatic zones are denoted in different colors: MTs, marginal tropical; SS, southern subtropical; MS, mid-subtropical; NS, northern subtropical; WT, warm temperate; MT, mid-temperate; PT, plateau temperate

The chapter collected records of SUP occurrence in the urban areas of 59 major cities in China and analyzed the composition profile and diversity patterns of SUPs at different geographical scales, while making referable recommendations for urban ecological management (Fig. 12).

### 3.2.1 Overall Spontaneous Urban Plants Profiles in China's Major Cities

The chapter represents a meaningful attempt to characterize SUPs at the regional scale in China. Our curated SUP species list signaled that the 59 study cities accommodated over 1200 species, of which 81.33% are native species (Fig. 13). These rich SUP species endowments comprised about six percent of all China's herbaceous species (Flora of China Editorial Committee 2013). The survival of SUPs in cities is threatened by the double jeopardy of limited amenable space and relentless artificial disturbance and removal (Bonthoux et al. 2019b). In addition, most of the SUPs were native species. Therefore, in terms of overall species richness in 59 study cities and the high proportion of native species, the SUP has played a positive role in urban

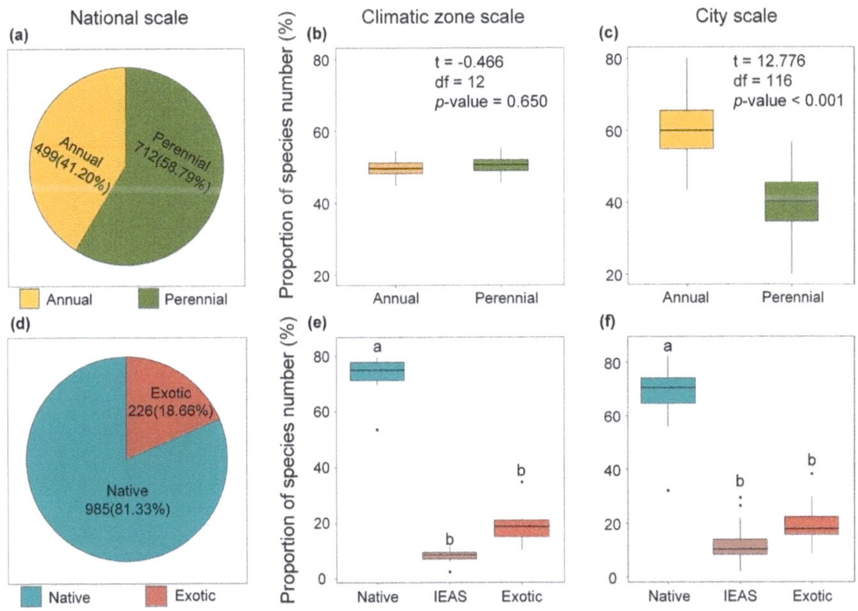


**Fig. 13** Compositional overview of SUPs at the national scale, climatic zone scale, and city scale. **a**, **d** show the proportion of different life forms and different biogeographic origin species in the 1211 SUPs; **b**, **c** show the compositional differences between annual and perennial species at the climatic zone and city scales; **e**, **f** show the compositional differences between native species, intra-country established alien species and exotic species at the climatic zone and city scales

biodiversity conservation despite its limited aggregate habitat space. The occasional species has promoted inter-urban heterogeneity and contributed to biodiversity.

The study found that Asteraceae and Poaceae were the most common families associated with SUPs at the city, climatic zone, and national scales (Fig. 14). Previous urban vegetation studies have also noted the dominance of Asteraceae and Poaceae species at the local scale (some for a single city or location) (Cervelli et al. 2013; Omar et al. 2018). The consistent dominance of these two families from local to regional scales can be explained by their status as major angiosperm families among herbaceous plants. Most constituent species have exceptionally broad ecological amplitudes to thrive in the diverse and heterogeneous habitats in the urban environment, permitting some members to achieve a cosmopolitan distribution (Katz et al. 2013; Zhuang et al. 2021). The dominant families also include Fabaceae, Lamiaceae, and Cyperaceae. Asteraceae and Lamiaceae have high ornamental and ecological value, with beautiful flowers and food sources for various insects such as bees and butterflies (Robinson and Lundholm 2012; Yang et al. 2019). In recent years, Poaceae and Cyperaceae plants have been increasingly adopted in landscaping urban green spaces (Asselin and Payette 2005). In short, SUPs can be used for unique "wild enjoyment" landscape configurations and aid the urban rewilding process.

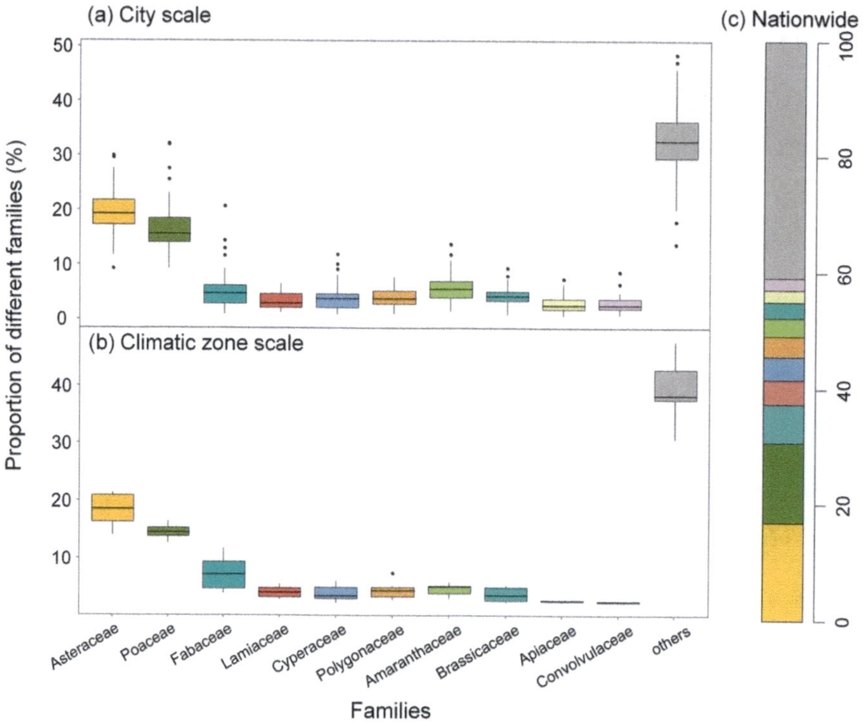


**Fig. 14** Most common SUP families and their proportions at the city scale (**a**), climatic zone scale (**b**), and national scale (**c**)

### 3.2.2 More Endemic Perennial Species Versus More Widespread Annual Species

Plant life-form synthesizes various morphological traits related to the access to resources that are heterogeneously distributed, both above ground (i.e., leaf longevity, plant height, or plant regrowth system) and below ground (i.e., rooting depth or root density) (Raunkiaer 1934). In Raunkiaer's life-form theory, recurring extreme environments (drought or low temperature) would favor species with their buds protected either by the seed coat (as in all annual species) or by their position progressively near, at, or even below the soil surface (appears as some dwarf plants) (Raunkiaer 1934; Schulze 1982). For spontaneous plants living in complex urban environments, the adaptive strategies of annuals may offer survival advantages. To be precise, they stabilize the population by adaptive molding into smaller individuals, shorter life cycles, and earlier flowering and fruiting, thereby reducing the survival risks (Bitomský et al. 2018). Perennials generally take longer to complete their life cycle and often suffer from premature termination of their lives before flowering and fruiting due to human mowing and weeding (Chollet et al. 2018). Even so, perennials actively

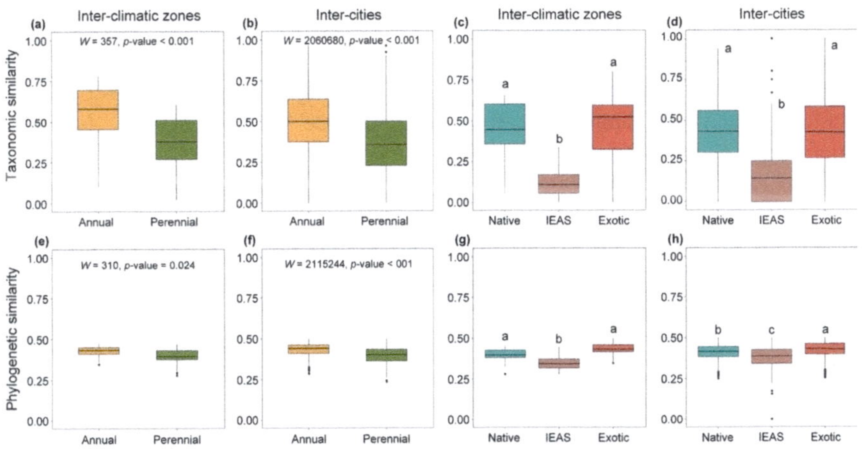


**Fig. 15** Compositional similarity profiles of SUPs with different life forms and different biogeographical origins: **a–d** for taxonomic similarity, and **e–h** for phylogenetic similarity; (**a**, **e**, **c**, **g**) for inter-climatic zones, and (**b**, **f**, **d**, **h**) for inter-cities

adapt to the urban environment through asexual reproductive organs such as underground bulbs and rhizomes, as expressed by Cyperus rotundus and Oxalis corniculata (Table 1). These characteristics related to life forms explain the higher abundance of annuals at the city scale. Habitats with less natural or anthropogenic stresses may accommodate perennials.

However, in expanding the geographical range, the proportion of perennial species has gradually become prominent (Fig. 13). This phenomenon can be well explained by the fact that annuals were significantly more similar in composition than perennials (Fig. 15). In short, although annual plants are more adapted to the urban environment, different local areas at the regional scale may share a relatively small number of annual species. In contrast, perennial species are more endemic and geographically specific. The findings could be applied to enhance biodiversity and achieve urban rewilding. Urban ecological managers could embrace sustainable solutions by creating and maintaining more perennial-friendly habitats while focusing on ecological and ornamental SUP species.

### 3.2.3 Suggestions for Urban Ecological Management

Based on the above discussion, the study proposed two recommendations to provide sustainable solutions for urban ecological planning and management of spontaneous nature in cities.

(1) To allow more time and space for SUPs to grow. Adopting moderate rather than aggressive or manicured management of urban green spaces, roads, and other habitats was recommended. Such a subdued approach will facilitate the colonization and long-term establishment of perennial species to enhance the heterogeneity and

regional characteristics of SUPs (Lerman et al. 2018). (2) To apply SUPs to urban rewilding. Careful observation of the occurrence of SUPs in the environment. The work can focus on native and occasional species, and select species with splendid ecological and ornamental traits to facilitate unique "wild enjoyment" landscape configurations.

## 4 Summary

In general, this chapter provides an in-depth discussion of biotic homogenization and rewilding of urban greening through three major aspects, including the current status of homogenization of plants in cities and its related factors, diverse origins and climatic congruence of landscape plants, and rewilding of urban greening through the case of near-natural forest development and the utilize of SUPs. However, there is still a long way to go to change the situation of urban biotic homogenization and rewilding of urban greening, which requires the synergistic development of basic research on urban biodiversity and practical application of urban rewilding.

## References


Ahern J (2013) Urban landscape sustainability and resilience: the promise and challenges of integrating ecology with urban planning and design. Landsc Ecol 28:1203–1212. https://doi.org/10.1007/s10980-012-9799-z

Asselin H, Payette S (2005) Late Holocene opening of the forest tundra landscape in northern Québec, Canada. Global Ecol Biogeogr 14:307–313. https://doi.org/10.1111/j.1466-822X.2005.00157.x

Avolio M, Pataki D, Gillespie T et al (2015a) Tree diversity in southern California's urban forest: the interacting roles of social and environmental variables. Front Ecol Evol 3:15. https://doi.org/10.3389/fevo.2015.00073

Avolio M, Pataki D, Pincetl S et al (2015b) Understanding preferences for tree attributes: the relative effects of socio-economic and local environmental factors. Urban Ecosyst 18:73–86. https://doi.org/10.1007/s11252-014-0388-6

Avolio M, Pataki D, Trammell T et al (2018) Biodiverse cities: the nursery industry, homeowners, and neigh-borhood differences drive urban tree composition. Ecol Monogr 88:259–276. https://doi.org/10.1002/ecm.1290

Beck H, Zimmermann N, McVicar T et al (2018) Present and future Koppen-Geiger climate classification maps at 1-km resolution. Sci Data 5:180214. https://doi.org/10.1038/sdata.2018.214

Berthon K, Thomas F, Bekessy S (2021) The role of 'nativeness' in urban greening to support animal biodiversity. Landsc Urban Plan 205:103959. https://doi.org/10.1016/j.landurbplan.2020.103959

Bitomský M, Mládek J, Cimalová Š (2018) Light limitation shapes the community seed mass of annual but not of perennial weeds. Perspect Plant Ecol Evol System 31:1–6. https://doi.org/10.1016/j.ppees.2017.11.005

Bonthoux S, Chollet S, Balat I, et al. (2019a) Improving nature experience in cities: what are people's preferences for vegetated streets? J Environ Manage 230:335–344. https://doi.org/10.1016/j.jenvman.2018.09.056
Bonthoux S, Voisin L, Bouché-Pillon S et al (2019b) More than weeds: spontaneous vegetation in streets as a neglected element of urban biodiversity. Landsc Urban Plan 185:163–172. https://doi.org/10.1016/j.landurbplan.2019.02.009
Breuste J (2004) Decision making, planning and design for the conservation of indigenous vegetation within urban development. Landsc Urban Plan 68:439–452. https://doi.org/10.1016/S0169-2046(03)00150-6
Bürgi M (2015) Close-to-nature forestry. In: Kirby K, Watkins C (eds) Europe's changing woods and forests: from wildwood to managed landscapes. CABI Publishing, London
Burley H, Beaumont L, Ossola A et al (2019) Substantial declines in urban tree habitat predicted under climate change. Sci Total Environ 685:451-462. https://doi.org/10.1016/j.scitotenv.2019.05.287
Cadena C, Kozak K, Gomez J et al (2012) Latitude, elevational climatic zonation and speciation in New World vertebrates. Proc R Soc B 279:194–201. https://doi.org/10.1098/rspb.2011.0720
Cervelli E, Lundholm J, Du X (2013) Spontaneous urban vegetation and habitat heterogeneity in Xi'an, China. Landsc Urban Plan 120:25–33. https://doi.org/10.1016/j.landurbplan.2013.08.001
Chejanovski Z, Wiens J (2014) Climatic niche breadth and species richness in temperate treefrogs. J Biogeogr 41:1936–1946. https://doi.org/10.1111/jbi.12345
Cheng W, Da L, Chen K et al (2004) Application of "Miyawaki's ecological method to reforestation" in the construction of outerringroad greenbelt in Shanghai. J Chin Urban Forest 5:21–23
Chollet S, Brabant C, Tessier S et al (2018) From urban lawns to urban meadows: reduction of mowing frequency increases plant taxonomic, functional and phylogenetic diversity. Landsc Urban Plan 180:121–124. https://doi.org/10.1016/j.landurbplan.2018.08.009
Connor E, McCoy E (1979) The statistics and biology of the species-area relationship. Am Natural 113:791–833. https://doi.org/10.2307/2460305
Da L, Song Y (2008) The construction of near-natural forests in the urban areas of Shanghai. In: Carreiro M, Song Y, Wu J (eds) Ecology, planning, and management of urban forests: international perspectives. Springer, New York
Da L, Li Y, Zhang J et al (2009) Consideration on the construction of urban land-mark plant community in Shanghai. Urban Probl 8:37–39
Da L, Xu D (2003) An attempt to build a "close-to-nature forest" in Shanghai. J Chin Urban Forest 1:17–20
Da L, Yang Y, Song Y (2004) Population structure and regeneration types of dominant species in an Evergreen Broadleaved Forest in Tiantong National Forest Park, Zhejiang Province, Eastern China. Chin J Plant Ecol 3:376–384
Faggi A, Ignatieva M (2009) Urban green spaces in Buenos Aires and Christchurch. ICE Proc Municipal Eng 162: 241–250. https://doi.org/10.1680/muen.2009.162.4.241
Flora of China Editorial Committee (2013) Flora of China. http://www.iplant.cn/frps
Gamborg C, Larsen J (2003) 'Back to nature'—a sustainable future for forestry? Forest Ecol Manage 179:559–571. https://doi.org/10.1016/S0378-1127(02)00553-4
Garcia-Garcia M, Sanchez-Medina A, Alfonso-Corzo E et al (2016) An index to identify suitable species in urban green areas. Urban Forest Urban Greening 16:43–49. https://doi.org/10.1016/j.ufug.2016.01.006
Grimm N, Faeth S, Golubiewski N et al (2008) Global change and the ecology of cities. Science 319:756–760. https://doi.org/10.1126/science.1150195
Guo X, Li W, Da L (2015) Near-natural silviculture: sustainable approach for urban renaturalization? Assessment based on 10 years recovering dynamics and eco-benefits in Shanghai. J urban Plan Develop 141:A5015001. https://doi.org/10.1061/(ASCE)UP.1943-5444.0000276
Hooper D, Chapin F, Ewel J et al (2005) Effects of biodiversity on ecosystem functioning: a consensus of current knowledge. Ecol Monogr 75:3–35. https://doi.org/10.1890/04-0922

Huang L, Qian S, Li T et al (2019) Masonry walls as sieve of urban plant assemblages and refugia of native species in Chongqing, China. Landsc Urban Plan 191:103620. https://doi.org/10.1016/j.landurbplan.2019.103620
Jim C (1999) A planning strategy to augment the diversity and biomass of roadside trees in urban Hong Kong. Landsc Urban Plan 44:13–32. https://doi.org/10.1016/S0169-2046(98)00113-3
Jin C, Zheng M, Huang L et al (2020) Co-existence between humans and nature: Heritage trees in China's Yangtze River region. Urban Forest Urban Greening 54:126748. https://doi.org/10.1016/j.ufug.2020.126748
Katz O, Lev-Yadun S, Bar P (2013) Plasticity and variability in the patterns of phytolith formation in Asteraceae species along a large rainfall gradient in Israel. Flora Morphol Funct Ecol Plants 208:438–444. https://doi.org/10.1016/j.flora.2013.07.005
La Sorte F, Aronson M, Williams N et al (2014) Beta diversity of urban floras among European and non-European cities. Global Ecol Biogeogr 23:769–779. https://doi.org/10.1111/geb.12159
Lambin E, Turner B, Geist H et al. (2001) The causes of land-use and land-cover change: moving beyond the myths. Global Environ Change 11:261–269. https://doi.org/10.1016/S0959-3780(01)00007-3
Larson J, Kesheimer A, Potter D (2014) Pollinator assemblages on dandelions and white clover in urban and suburban lawns. J Insect Conserv 18:863–873. https://doi.org/10.1007/s10841-014-9694-9
Lerman S, Contosta A, Milam J, et al. (2018) To mow or to mow less: lawn mowing frequency affects bee abundance and diversity in suburban yards. Biol Conserv 221:160–174. https://doi.org/10.1016/j.biocon.2018.01.025
Li X, Fan S, Kühn N et al (2019) Residents' ecological and aesthetical perceptions toward spontaneous vegetation in urban parks in China. Urban Forest Urban Greening 44:126397. https://doi.org/10.1016/j.ufug.2019.126397
López-Pujol J, Zhang F, Ge S (2006) Plant biodiversity in China: richly varied endangered, and in need of conservation. Biodivers Conserv 15:3983–4026. https://doi.org/10.1007/s10531-005-3015-2
Mackinnon J, Sha M, Cheung C et al (1996) A Biodiversity review of China. WWF International, Hong Kong
McDonald R, Kareiva P, Forman R (2008) The implications of current and future urbanization for global protected areas and biodiversity conservation. Biol Conserv 141:1695–1703. https://doi.org/10.1016/j.biocon.2008.04.025
McKinney M (2004) Do exotics homogenize or differentiate communities? Roles of sampling and exotic species richness. Biol Invas 6:495-504. https://doi.org/10.1023/B:BINV.0000041562.31023.42
McKinney M (2006) Urbanization as a major cause of biotic homogenization. Biol Conserv 127:247-260. https://doi.org/10.1016/j.biocon.2005.09.005
McKinney M, Lockwood J (1999) Biotic homogenization: a few winners replacing many losers in the next mass extinction. Trends Ecol Evol 14:450–453. https://doi.org/10.1016/S0169-5347(99)01679-1
McNeely J, Miller K, Reid W et al (1990) Conserving the World's Biological Diversity. IUCN, World Resources Institute, Conservation International, WWF-US and the World Bank, Washington, DC.
Miller R (1997) Urban forestry: planning and managing urban greenspaces. Prentice Hall, New Jersey
Miyawaki A (1999) Creative ecology restoration of native forests by native trees. Plant Biotechnol 16:15–25
Miyawaki A (2004) Restoration of living environment based on vegetation ecology: theory and practice. Ecol Res 19:83–90. https://doi.org/10.1111/j.1440-1703.2003.00606.x
Nagendra H, Gopal D (2010) Street trees in Bangalore: density diversity, composition and distribution. Urban Forest Urban Greening 9:129–137. https://doi.org/10.1016/j.ufug.2009.12.005

Niemelä J (1999) Ecology and urban planning. Biodiv Conserv 8:119–131. https://doi.org/10.1023/A:1008817325994

Omar M, Al Sayed N, Barré K et al (2018) Drivers of the distribution of spontaneous plant communities and species within urban tree bases. Urban Forest Urban Greening 35:174–191. https://doi.org/10.1016/j.ufug.2018.08.018

Preston C, Pearman D, Dines T (2002) New atlas of the British and Irish flora. In: An Atlas of the vascular plants of Britain, Ireland, the Isle of Man and the Channel Islands. Oxford University Press, New York

Ordóñez C, Duinker P (2015) Climate change vulnerability assessment of the urban forest in three Canadian cities. Clim Change 131:531–543. https://doi.org/10.1007/s10584-015-1394-2

Qian S, Qin D, Wu X et al (2020) Urban growth and topographical factors shape patterns of spontaneous plant community diversity in a mountainous city in southwest China. Urban Forest Urban Greening 55:126814. https://doi.org/10.1016/j.ufug.2020.126814

Qian S, Qi M, Huang L et al (2016) Biotic homogenization of China's urban greening: a meta-analysis on woody species. Urban Forest Urban Greening 18:25–33. https://doi.org/10.1016/j.ufug.2016.05.002

Raunkiaer C (1934) The life forms of plants and statistical plant geography. Clarendon Press, Oxford

Robinson S, Lundholm J (2012) Ecosystem services provided by urban spontaneous vegetation. Urban Ecosyst 15:545–557. https://doi.org/10.1007/s11252-012-0225-8

Rozzi R (2013) Biocultural ethics: from biocultural homogenization toward biocultural conservation. In: Rozzi R, Pickett S, Palmer C et al (eds) Linking ecology and ethics for a changing world: values, philosophy, and action. Springer, Dordrecht, pp 9–32. https://doi.org/10.1007/978-94-007-7470-4_2

Schulze E (1982) Plant life forms and their carbon, water and nutrient relations. In: Lange O, Nobel P, Osmond C et al (eds) Physiological plant ecology II: Water relations and carbon assimilation. Springer, Berlin, pp 615–676. https://doi.org/10.1007/978-3-642-68150-9_19

Song K, Qin J, Gao K et al (2009) Homogenization of plant diversity in Shanghai residential areas. Ying Yong Sheng Tai Xue Bao 20:1603–1607 (in Chinese with English abstract)

Song Y, Shi Z, Yu S et al (2010) Influence of urbanization on the characters of plants composition in urban greenspace in Shenzhen, China 19:615–620 (in Chinese with English abstract)

Svitok M, Hrivnak R, Kochjarova J et al (2016) Environmental thresholds and predictors of macrophyte species richness in aquatic habitats in central Europe. Folia Geobotanica 51:227–238. https://doi.org/10.1007/s12224-015-9211-2

Tang Z, Wang Z, Zheng C et al (2006) Biodiversity in China's mountains. Front Ecol Environ 4:347–352. https://doi.org/10.1890/1540-9295(2006)004[0347:BICM]2.0.CO;2

Westerink J, Haase D, Bauer A et al (2013) Dealing with sustainability trade-offs of the compact city in peri-urban planning across European City regions. Eur Plan Stud 21:473–497. https://doi.org/10.1080/09654313.2012.722927

Winter M, Schweiger O, Klotz S et al (2009) Plant extinctions and introductions lead to phylogenetic and taxonomic homogenization of the European flora. Proc Natl Acd Sci USA 106:21721–21725. https://doi.org/10.1073/pnas.0907088106

Yan P, Yang J (2017) Species diversity of urban forests in China. Urban Forest Urban Greening 28:160–166. https://doi.org/10.1016/j.ufug.2017.09.005

Yang F, Ignatieva M, Wissman J et al (2019) Relationships between multi-scale factors, plant and pollinator diversity, and composition of park lawns and other herbaceous vegetation in a fast growing megacity of China. Landsc Urban Plan 185:117–126. https://doi.org/10.1016/j.landurbplan.2019.02.003

Zhuang H, Wang C, Wang Y et al (2021) Native useful vascular plants of China: A checklist and use patterns. Plant Divers 43:134–141. https://doi.org/10.1016/j.pld.2020.09.003

**Yongchuan Yang** coming from Key Laboratory of the Three Gorges Reservoir Region's Eco-Environment, Ministry of Education, Chongqing University, Chongqing 400045, China. He contributes the chapter 10 with Cheng Jin (Chongqing University), Siwei Hu (Chongqing University), Liangjun Da (Xi'an University of Architecture and Technology), and Yuandong Hu (Northeast Forestry University)

# Urban Green Infrastructure

**Prashant Kumar, Sisay Debele, Arvind Tiwari, K. V. Abhijith, Jeetendra Sahani, and Soheila Khalili**

**Abstract** Urban greening is on rise worldwide due to its wide-ranging direct and indirect benefits. These direct benefits can span from working as a passive barrier between vehicles and pedestrians in near-road environments, capturing particulate matter from the atmospheric environments and mitigating the impact of urban heat island at city scale, acting as nature-based solutions to abate the impact of natural hydro-meteorological hazards such as flooding and heatwaves, to indirectly foster biodiversity. A careful planning of green infrastructure (GI) in urban areas can thus offer various ecosystem services if implemented in a holistic manner by considering any downsides such as biogenic volatile organic compound emissions, pollen, or the implementation of the wrong types at the wrong place. This chapter discusses the diverse application of GI in urban environments from street-scale implementation such as in roadside environments to the city scale for exploiting their benefits at city scale. The chapter covers their importance, specific considerations needed for their implementation, and benefits brought by them whilst considering any unintended downsides. Whilst GI cannot be the sole answer to air pollution and climate change mitigation issues in cities, they are an important, cost-effective part of a holistic solution, which embraces urban design, reduces exposure to emissions, natural hazards (e.g. floods, droughts, landslides, etc.), and contributes to improved health and well-being.

**Keywords** Urban greening · Nature-based solutions · Urban resilience · Ecosystem services · Air quality and biodiversity

P. Kumar (✉) · S. Debele · A. Tiwari · K. V. Abhijith · J. Sahani · S. Khalili
Global Centre for Clean Air Research (GCARE), School of Sustainability, Civil and Environmental Engineering, Faculty of Engineering and Physical Sciences, University of Surrey, Guildford GU2 7XH Surrey,, UK
e-mail: P.Kumar@surrey.ac.uk


R. Yao (ed.), *Resilient Urban Environments*, Cities and Nature,
https://doi.org/10.1007/978-3-031-55482-7_11

# 1 Introduction

Urban spaces are made up of complex built-up structures that are putting undue stress on their natural ecosystem. The proportion of the population living in urban areas grew from 29% in 1950 to 50% in 2010 (Kim and Miller 2019). In Africa only, the current 1.1 billion population is expected to double mostly in cities by 2050 (Gulati and Scholtz 2020), and 69% of the global population, or 6.3 billion people, are expected to live in urban areas (Kim and Miller 2019). Hence, there is a competing interest between the space for daily utilities for living and the empty green spaces for leisure and supporting biodiversity in urban areas. The problems of air quality degradation, increasingly frequent heatwaves, droughts, and floods as well as health-related effects of climate change are worsening the situation in urban settings. Climate change in itself is one of the most critical challenges to urban societies today. Urbanisation and climate change affect people's health, well-being, and the functioning of ecosystems and their services in various ways, e.g. by triggering and increasing impacts of several natural hazards (Debele et al. 2019; Sahani et al. 2019; Kumar et al. 2020; Kumar et al., 2024). It amplifies the already existing pressure of anthropogenic activities, such as air pollution, particularly in cities that are built along waterways or near coasts (Kumar 2021). As a result, cities all over the world are confronting intertwined environmental, social, and economic problems and are intended to become resilient to climate change and promote well-being of all their citizens. Adjusting natural and human systems in urban areas is therefore important to mitigate the impact of pollution and climate change and support resilient urban environments.

Green infrastructure (GI) is sustainably managed or restored ecosystems in order to tackle urban and societal challenges—such as air pollution, natural hazard, climate change, food security, water security, and human health and well-being. GI is principally established by well-functioning biophysical systems to which some protection, sustainable management, restoration, and implementation may apply. The term's definition is determined by the context in which it is used. It can refer to trees and vegetation that provide ecological benefits in cities, as well as engineered structures like sustainable urban drainage systems (Debele et al. 2019). The GI refers to street trees, hedges, bushes, green walls, green roofs, green spaces (parks), and many others in urban planning (Debele et al. 2019; Sahani et al. 2019; Kumar et al. 2020). GI is considered to be an interconnected green-space network including natural areas and features, public and private conservation lands, and other protected open spaces that are planned and managed for their natural resources and values and for the associated benefits to the societies (Di Marino et al. 2019). Despite its local variations, urban GI is characterised by multi-functionality and connectivity, enabling management and restoration efforts to combine ecological, social and economic functions of green spaces. In this regard, Hermoso et al. (2020) pointed out that the opportunities for designing the GI as a multi-functional network that plays a key role in the life of the city are huge.

GI is viewed as a potential method of reducing pollution impacts. The interaction between green infrastructure design (for example, species selection and spatial positioning) and air pollutants can have a negative and positive impact on personal exposure and thus human health (Abhijith et al. 2017). The effects are determined by the surrounding built environment as well as the type, location, and configuration of GI. Strategic GI planting near air pollution sources can provide the greatest benefits because high concentrations of air pollutants lead to higher deposition. However, the GI potential in reducing air pollutant concentrations is rarely studied in terms of city-scale strategic GI development, whereas substantial studies evaluated air pollution abatement potential of GI in near-road environments.

Due to the harmful impacts of urban heat island (UHI) affecting human health and well-being, employing GIs has become an effective strategy to mitigate the UHI and improve humans' thermal comfort. Human thermal comfort is related to meteorological parameters and personal factors. GIs alter the micro-climate through shading and evapotranspiration, resulting in air temperature, surface temperature, and mean radiant temperature (MRT) reductions. Therefore, GI improves humans' thermal comfort, especially in dense urban areas with low albedo surfaces and intense UHI effect. Based on the significant effects of GIs, evaluating different GI strategies in order to adopt the most effective ones has attracted the research community's attention in recent years.

Healthy ecosystems play a significant role in protecting communities from natural hazards at different scales. GI is fundamental part of nature-based solutions (NBSs) which can provide multiple services to the urban population. GI is increasingly being acknowledged and used as an effective and cost-efficient way of managing natural hazards such as floods, droughts, landslides, storm surges, helping people adapt to climate change and building community resilience and sustainability. They can address all three parts of the natural hazard risk component by: (1) preventing or mitigating the onset and propagation probability of natural hazards, (2) reducing people's exposure to hazards, and (3) limiting people's vulnerability to natural hazards. For example, urban GI helps keep cities cool during heatwaves, reduce urban flooding and droughts, and improve human health and well-being as well as provide habitats for different species. Besides reducing exposure to natural hazards (e.g. urban flooding and droughts and excessive heat), urban GI such as parks, playgrounds, and residential greenery can promote mental and physical health and reduce morbidity and mortality in urban residents by providing psychological relaxation and stress alleviation, stimulating social cohesion, and supporting physical activity.

Despite the known multi-dimensional benefits of green infrastructure (GI) (Fig. 11.1), challenges of their implementation in terms of both quantity and appropriateness exist for urban greening. This is further exacerbated by the lack of comprehensive guidance for urban planners and policymakers regarding their deployment (Kumar et al. 2019). Policymakers and city planners seeking to build urban resilience to climate change in cities and other urban dynamics will need to consider both layers of insurance in order to utilise the powerful role of GI in creating a sustainable, healthy, and liveable urban environment. This chapter summarises the design

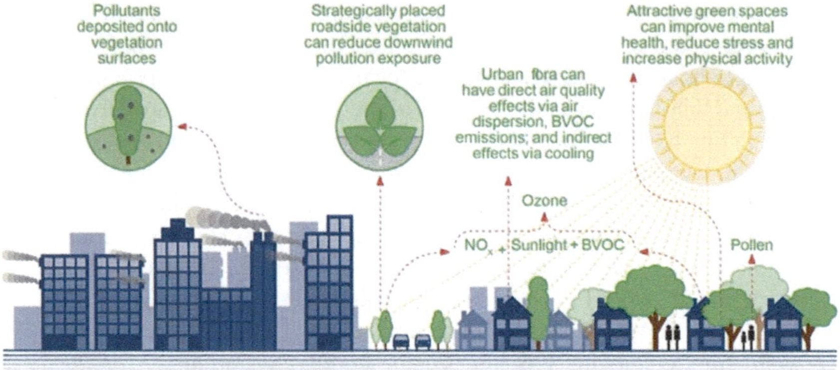


**Fig. 11.1** Multidimensional benefits of green infrastructure deployment at different spatial scales in urban environments. *Source* Authors own work from Kumar et al. (2019)

principles and the role of GI for the abatement of air pollution (Sect. 2), urban heat (Sect. 3), and natural hazards (Sect. 4).

## 2 GI for Passive Air Pollution Abatement

### *2.1 Near-Road Environment*

Green infrastructure or urban vegetation in near-road environments includes trees, hedges and bushes, and green walls and green roofs. Near-road environments can be classified into two built-up geometries—street canyons and open-road conditions—based on the relative positions of buildings, roads, and surrounding landscape. Street canyons in urban areas are roads with continuous buildings along both sides. GI in street canyons is often planted as part of urban landscaping plans of the local government, whereas open road consists of streets with open areas on either side with generally detached or discontinued buildings allowing lesser hindrance to the natural winds compared to street canyons. The GI in open road can be well planned and maintained or naturally occurring vegetation. GI in both urban and near-road environments helps in reducing air pollution exposure through dispersion and deposition of pollutants. Air pollutants such as particulate matter and gas molecules get deposited or absorbed to the GI surfaces when they are in close proximity. The deposition results in reduction in pollutant concentration by temporary removal from air to the plant surfaces. These deposited particles may be resuspended by wind or washed down by rain. The total quality of deposited air pollutants depends on various factors such as leaf area index (LAI), deposition velocity, pollutant concentration, time, micro- and macro-morphology of leaves, and meteorological conditions. Air pollutant reduction through dispersion involves physical transportation and dilution of the pollutants away from the receptors. The overall reduction in pollutant concentration is dominated by the dispersion (Jeanjean et al. 2016). In the near-road environment, GI serves as a porous physical barrier between the traffic and pedestrians. This passive air pollution reduction method alters pollutant dispersion patterns by deflecting streamlines, introducing turbulence, thereby enhancing dilution, and resulting in a longer pathway between source and receptors. GI-induced pollutant dispersion is influenced by topography of the built environment, characteristics of the GI, and local meteorological conditions.

In general, GI reduces air pollution exposure in open-road conditions. Downwind air pollution concentration is reduced up to 50% when GI has sufficient thickness, height, density, and no gaps in the vegetation barrier (Abhijith and Kumar 2020, 2019; Ottosen and Kumar 2020). Highest reduction concentration was observed for ultrafine particles, black carbon, and fine particulate matter ($PM_1$) in the presence of GI clearly indicating its effectiveness in mitigating traffic emissions (Abhijith and Kumar 2019). Gaps in the GI barriers and absence of canopy close to ground, such as in case of trees, result in increase in downwind pollutant concentration. Green screens can be used to reduce exposure when availability of space for planning is limited such as school fences and it can lower pollutant concentration above 30% (Tomson et al. 2021). GI in open road enhances mixing of pollution from the traffic and forces the airflow to loft over the barrier resulting in dilution and deposition on leaves and other parts of GI (Fig. 11.2). Combinations of trees and hedges recorded highest reductions in pollutant concentrations. This can be attributed to their fully covered canopy from close to ground to top meeting requirements for downwind pollution reduction. Lower LAI (less dense) and large gaps in the vegetation barrier can lead to unintended increase in pollution concentration (Abhijith et al. 2017; Baldauf 2017). Planting denser GI barriers along the boundaries of schools and parks and providing adequate maintenance can improve air quality in downwind direction.

Presence of GI increases the complexity of airflow and pollutant dispersion in street canyons which depends on aspect ratio, meteorological conditions, and traffic characteristics. Figures 11.3 and 11.4 provide an overview of variations in airflow in presence of GI in street canyons. Trees with denser and high-level canopies increase the air pollution concentration and exposure in regular ($0.5 < H/W < 2$) and deep ($H/W > 2$) street canyons (Abhijith et al. 2017; Barwise and Kumar 2020). Trees having lighter crowns with large spacing between them can be planted in shallow street canyons. However, hedges, green walls, and green screens can reduce air pollution exposure in shallow ($H/W < 0.5$) or regular ($0.5 < H/W < 2$) street canyons. Optimum dimensions of the hedges vary with the aspect ratio of the street canyon and the wind direction. Due to complex air pollution dispersion characteristics of street canyon, site-specific GI design considerations are required for desired exposure reduction (Abhijith et al. 2017; Tomson et al. 2021; Barwise and Kumar 2020). Few GI design guidelines (Baldauf 2017 and Kumar et al. 2019) summarised existing research findings and provided practical implementation recommendations for both street canyon and open road conditions. These guidelines help citizens, local authorities, and other interested groups to select suitable type of GI, species, and its physical

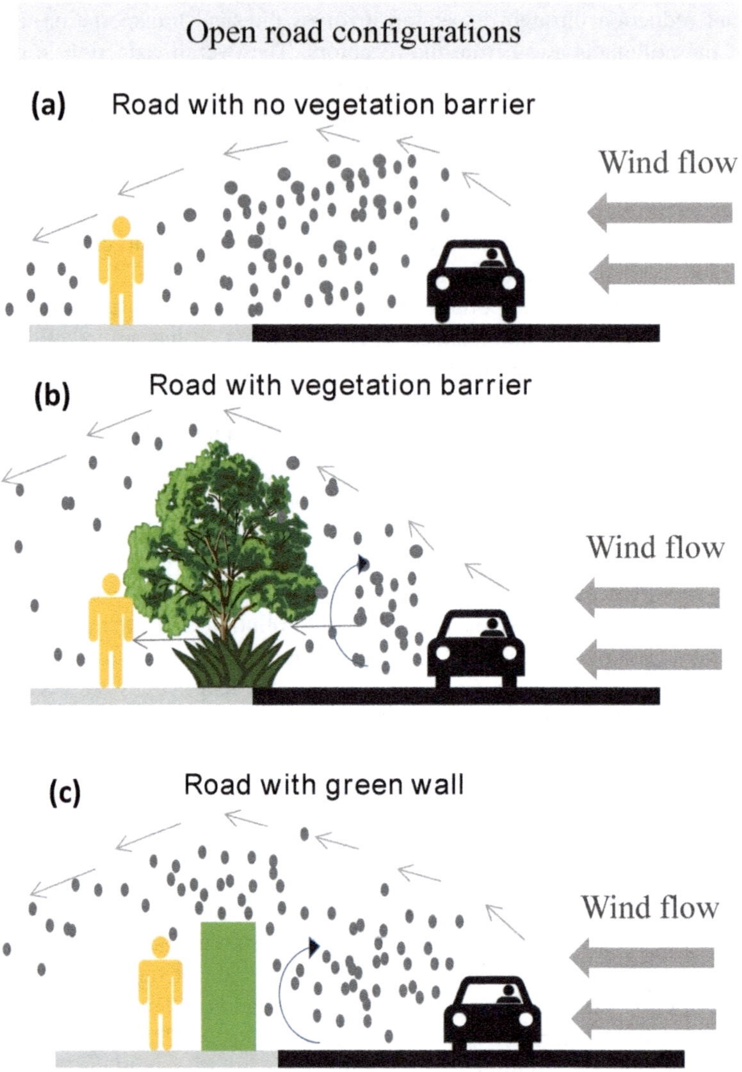


**Fig. 11.2** Dispersion patterns of road pollutants under open-road configurations **a** without vegetation barrier, **b** with vegetation, and **c** with green wall. *Source* Authors own work from Abhijith et al. (2017)

dimensions. Moreover, publicly available GI tools such as GI4RAQ tool (Levine et al. 2021) and HedgeDATE enable designing site-specific vegetation barriers in near-road conditions (Barwise et al. 2021).

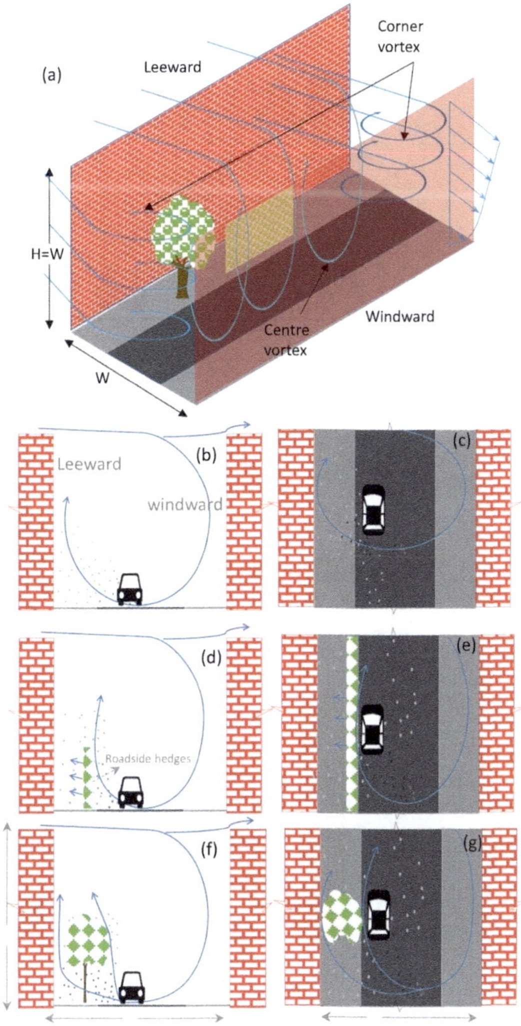


**Fig. 11.3** Schematic of the changes in the formation of vortex (shown by blue lines) in urban street canyon during the interaction of the flow with the leeward GI, showing vortex formation **a** in street canyon (width/height = 1) under perpendicular wind flow (Gromke 2011) as well as cross-section and top view of **b**, **c** GI-free street canyon, **c**, **d** street canyon with hedges, **e**, **f** street canyon with tree. *Source* Authors own work from Tiwari et al. (2019)

## 2.2 City-Scale Environment

The GI at the city scale is recognised as a passive method to improve the air quality near/far from the source. The following specific conditions make GI considerations distinct in urban planning: (i) their aerodynamic effects cause the air pollutant concentration redistributions in the atmosphere (Hefny et al. 2015); (ii) their surface area offers deposition of air pollutants (Nowak et al. 2018); and (iii) their surface roughness affecting enhanced atmospheric turbulence and mixing (Barnes et al. 2014). As detailed in Sect. 2.1, the roadside GI such as hedges may act as a filtering barrier between air pollution sources and receptors that can reduce personal exposures for

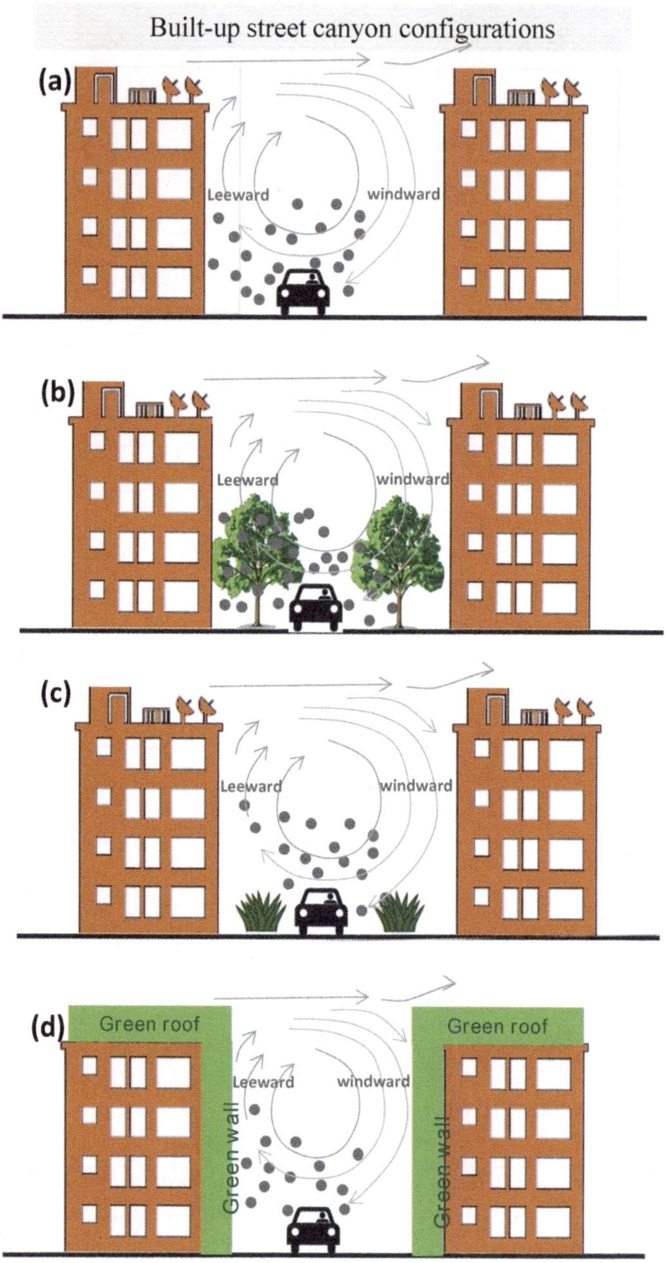


**Fig. 11.4** Description of flow and pollutant dispersion patterns in a street canyon with and without different types of vegetation: **a** vegetation free street canyon, **b** street canyon with trees, **c** street canyon with hedges, and **d** street canyon with green roof and green wall. *Source* Authors own work from Abhijith et al. (2017)

nearby populations. However, city-scale GI such as urban forests, parks, gardens, and hedges collectively increases atmospheric dilution as well as act as a sink (in terms of deposition) for atmospheric pollutants. Previous studies mentioned that aerodynamic effects have a greater impact on local air pollution levels than deposition near-road conditions (Vos et al. 2013), but both aerodynamic and deposition effects are important at the city scale (Tong et al. 2016). In addition, some species of GI emit a considerable amount of biogenic volatile organic compounds (BVOCs). The importance of BVOC emission into the urban atmosphere is related to their reactivity with some of the compounds released from anthropogenic sources, especially nitrogen oxides (NOx). Ozone, peroxyacyl nitrates, aldehydes and ketones, hydrogen peroxide, secondary organic aerosol, and particulate matter can be formed by the photochemically driven reaction between NOx and BVOCs (Calfapietra et al. 2013). Furthermore, air pollutant dispersion at the city scale is dominated by characteristics including source-induced turbulence, pollutant chemistry, local meteorology, source geometry and the surrounding buildings, terrain, flow alteration, and many more aspects (Tiwari and Kumar 2020).

The presence of GI at the city scale makes air pollutant dispersion more complex by adding GI-related characteristics to air pollutant dispersion (Tiwari et al. 2019). These characteristics include BVOC-pollutant chemistry, enhanced atmospheric turbulence, pollutant deposition, and the creation of heating and cooling shadows. GI surfaces such as leaves, stems, barks, fruits and heads serve as effective deposition sites for different air pollutants in different ways. For example, GI typically helps in the biological degradation of gaseous air pollutants via leaf stomata, whilst deposited PM is either washed by rain or dropped with the leaf to the ground depending on prevalent environmental conditions and GI characteristics (Gourdji 2018). In addition, climate and the specific composition of urban GI, in particular, influence the absorption and emissions of gases by vegetation influencing the composition of the atmosphere and air quality in urban environments. BVOC and ozone formation are enhanced in Mediterranean areas, where summers are typically characterised as high temperatures and scarce precipitation, and their interactions require greater attention. Overall, there are many pieces of evidence that reported the GI benefits to mitigate air pollution at the urban scale. For example, Baró (2015) estimated higher than 6% average percent air quality improvements for $PM_{10}$ in urban areas with a hypothetical 100% tree cover such as contiguous forest stands. Yin et al. (2011) showed that GI such as urban parks in Shanghai, China, could remove pollution at ground level by a maximum of 35% for total suspended particles and 21% for $NO_2$. In addition, Yang et al. (2008) estimated that 19.8 ha of green roofs removed 1675 kg of air pollutants in total in a year for Chicago city, with $O_3$ making up 52%, $NO_2$ (27%), $PM_{10}$ (14%), and $SO_2$ (7%) of the total.

An Air Pollution Health Risk Assessment (APHRA) is a mathematical approach that estimates the expected health impact of short- or long-term exposure to air pollutants in different variables such as age group, environmental conditions, and socioeconomic status (WHO 2016). An APHRA framework presented in Fig. 11.5 shows the link of health outcomes with short- or long-term exposure alteration due to the presence of GI. Several epidemiological studies have reported a wide range of health impacts, such as respiratory symptoms, hospital admissions, and/or premature deaths, associated with excess exposure to air pollution. These health outcomes could be quantified by the number of premature deaths, years of life lost (YLLs), disability-adjusted life years (DALYs), or change in life expectancy (WHO, 2016). For instance, Pope et al. (2009) reported a life expectancy increase by 0.64 years per 10 $\mu g/m^3$ decrease in $PM_{2.5}$ for 51 US cities. Knowledge of linkages between public health outcomes and GI-induced air pollution reduction is important to assess public health benefits (Fig. 11.5). Therefore, the linkages between GI and physical health through improved air quality have been explored in many scientific articles. However, the linkages between urban GI planning and other benefits including human mental health (such as mental health, stress reduction, and greater social cohesion) are rarely studied from a holistic point of view. Furthermore, Kumar et al. (2019) evaluated the nexus between air pollution, GI, and human health via summarising the limited evidence to establish links between pollution mitigation through GI and quantification of their health benefits. They highlighted the challenges (such as aerodynamic dispersion and deposition effects of GI) to quantifying health benefits that can be achieved from varying GI designs, owing to the limited approaches to simulate the effect of GI at the city scale.

By simulating a 4 $km^2$ area in Leicester city centre in the UK using a wind-tunnel validated CFD model, Jeanjean and coworkers have investigated the efficiency of trees to disperse traffic emissions (Jeanjean et al. 2016). They claimed that due to increased pollutant dilution brought on by higher air turbulence levels brought on by trees, the concentrations of traffic-generated air pollutants at pedestrian height decreased by 7–9%. However, due to the substantial amount of resources needed to construct large-scale wind-tunnel or CFD models, doing such research at bigger scales (such as city and regional scales) is extremely difficult. As a result, there are not many of these investigations in the literature that have studied the impact of GI on urban air quality using CFD tool. Tiwary et al. (2009) proposed an integrated modelling approach using ADMS-Urban to assess the effect of GI on $PM_{10}$ concentration levels' reduction and health benefits. Tallis et al. (2011) compared the Tiwary et al. (2009) approach with the UFORE (now known as "i-Tree" developed by United States Forest Service) for estimating $PM_{10}$ deposited amount and highlighted that deposition values used by Tiwary et al. (2009) are higher than i-Tree due to a number of presumptions, including (i) the average $PM_{10}$ concentration takes into account particle resuspension rate, (ii) the use of species-specific deposition velocity ($V_d$) values to estimate deposition, and (iii) the assumption that surface resistance remains constant throughout the season. To increase the model's accuracy, i-Tree needs a lot of data, including the types of species, land-use criteria, total tree height, and crown size, which must be gathered by knowledgeable researchers (Pace et al. 2018). Additionally, the i-Tree model assumes that the concentration of air pollutants is uniform throughout the domain, excluding the effect of GI dispersion (Cabaraban et al. 2013).

To overcome the highlighted limitation, Tiwari and Kumar (2020) have developed an integrated GI assessment approach that can quantify the aerodynamic dispersion and deposition effects of GI on pollutant concentration at an urban scale based on the type of GI and its location. They demonstrated the modelling approach for developing

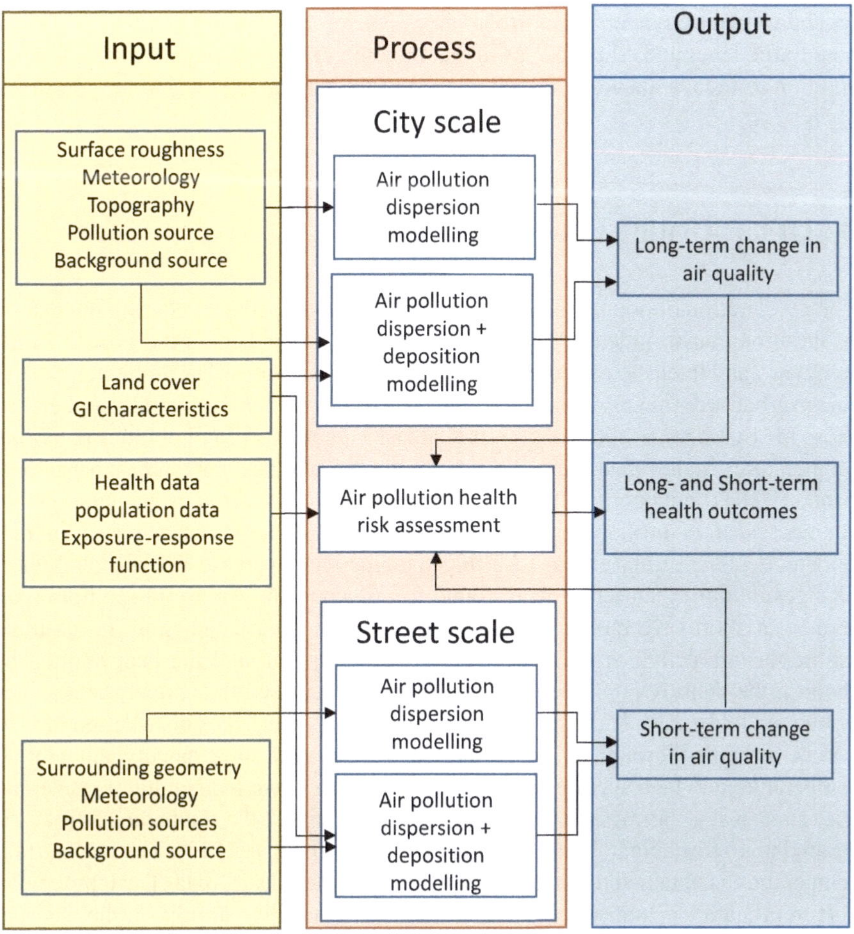


**Fig. 11.5** Air Pollution Health Risk Assessment (HPRA) framework for linking between GI, air pollution, and public health. *Source* Authors own work from Tiwari et al. (2019)

GI-influenced air quality map and estimating the annual averaged reduction in air pollutant concentrations under different GI scenarios to show the potential of GI planning in controlling the concentration of different traffic-related air pollutants ($NO_x$, $PM_{10}$, $PM_{2.5}$, ToNC), taking a typical UK town Guildford as a case study. The annual pollutant deposition was found to vary from 0.27 to 2.77 $t{\cdot}yr^{-1}{\cdot}km^{-2}$ for $NO_x$, 0.46 to 1.03 t $yr^{-1}$ $km^{-2}$ for $PM_{10}$, 0.08 to 0.23 t $yr^{-1}$ $km^{-2}$ for $PM_{2.5}$, and 28 to 4800 $\times$ $10^{19}$ particles for ToN, depending on the percentage share of GI type and traffic emissions. They concluded that planting coniferous trees can promote enhanced turbulence flow and offer more surface for deposition compared with other GI species (deciduous trees and grassland). Since BVOC emission is species-specific, the contribution to the photochemical reactivity in an urban environment is very much related to plant biodiversity in urban areas. Therefore, planting coniferous species near traffic lanes and other species in the remaining places could be a more effective solution to reduce annual pollutant concentration and ensure biodiversity in urban areas.

## 3 GI for Abating Urban Heat Stress

The acceleration of population growth and urban construction is altering the quality of the urban environment. A primary concern of researchers is the growth in the frequency and intensity of heatwaves and urban temperatures. The ambient temperature in urban areas or any densely populated built environment is generally higher than those measured at the surrounding rural areas. This phenomenon of cities recording temperatures higher than their non-urbanised surroundings is known as urban heat island (UHI) (Santamouris 2014). The anthropogenic alteration of the natural environment (such as development of buildings and impervious surfaces) results in trapping more radiation and energy resulting in temperature increase. The morphological changes due to urbanisation were found to influence an increase in multiple scatterings of shortwave radiation and trapping the long-wave radiation. In addition, anthropogenic activities such as heating, transportation, and the heat emitted by them in the densely populated urban areas have increased the temperatures, when compared to the rural background areas (Zhou et al. 2017). The distribution of UHI can be estimated through two different temperature measures, namely, ambient air temperature and land surface temperature (LST). Ambient temperature is generally measured using the meteorological monitoring stations, the temperature from the ground up to tree height, the "canopy layer" (Schwarz et al. 2012). The land surface temperature is obtained through satellite and airborne sensors data. The temperature rise in urban areas associated with the UHI has a negative impact on the environment, human health, and well-being, and above all, it increases mortality rates. The main challenge is how the harmful effects of these changes that affect human thermal comfort can be reduced.

The relationship between urbanisation and human thermal comfort is complex since there are a number of factors that need to be taken into consideration. Firstly, the changes in surface geometry alter the sky view factor (SVF) and the vertical surface area, which leads to changes in energy, water exchange (Susca et al. 2011), and airflow (Yang et al. 2011). Secondly, the huge thermal mass of paved surfaces (buildings, roads, and other urban infrastructures) results in an increase in thermal admittance and urban warming intensification (Akbari et al. 2001). Thirdly, anthropogenic heat emissions, including human metabolic heat and fuel combustion heat release, and low ventilation affect the surface and air temperature (Yang et al. 2011). Eventually, the insufficiency of natural elements and low evapotranspiration in urban areas exacerbate the negative impacts on inhabitants' thermal comfort.

GI regulates the micro-climate through shading and evapotranspiration, which results in an improvement in thermal comfort conditions. Evapotranspiration refers to transpiration from plants and evaporation from water bodies and soils. The absorbed solar energy is converted into the latent heat of evaporation. Thus, the temperature of the surrounding area is cooled. Through the phenomenon of evapotranspiration, GI reduces the mean radiant temperature, improves environmental conditions, and controls the reflection of radiant light (Salata et al. 2017). GI also reduces surface temperature by providing shading on the surfaces and exchanging the airflow near the surface through the convective flow, resulting in reduced temperature and improved ventilation of the environment (Morakinyo et al. 2017).

GI affects human thermal comfort in terms of physical, physiological, behavioural, and psychological aspects. Several studies aimed to evaluate the effects of different GI strategies on thermal comfort indices using on-site measurement, questionnaire surveys, and simulation. Imran et al. (2019) evaluated the effectiveness of urban vegetation patches such as mixed forest, a combination of mixed forest and grasslands, and a combination of mixed shrublands and grasslands in reducing urban heat island effects in the city of Melbourne during one of the most severe heatwave events using WRF model coupled with the Single-Layer Urban Canopy Model. The findings of the study were that by increasing fractions from 20 to 50%, mixed forest reduces near-surface (2 m) urban heat island by 0.6–3.4 °C, mixed forest and grasslands by 0.4–3.0 °C, and mixed shrublands and grasslands by 0.6–3.7 °C during the night. It is well known that thermal comfort can be improved by GI, including tree canopies, green open spaces, vertical greenery, and green roofs (Fig. 11.6). The results of previous studies emphasise that green infrastructure should be implemented on different scale levels, as the accumulated effect of all green elements and green spaces contributes to improved thermal comfort conditions in urban environments. The research community has paid great attention to taking full advantage of GI and achieving the best possible thermal comfort in urban design. However, with the rapid population growth and climate change crisis, it remains a challenge. Meanwhile, further studies should be expanded to enrich our knowledge about the effects of different GI types on thermal comfort conditions in each geographical area due to discrepancies between each climate type thermal conditions. Given the benefits of GI for the environment and public communities, there is an intense interest in understanding how different GI types affect thermal comfort conditions.

## *3.1 Green Open Spaces*

In terms of psychological aspects, green open spaces provide a positive effect on thermal comfort sensation. To examine the effects of greenery in a medium-scale park, Karimi et al. (2020) carried out field studies using simulation tools and on-site monitoring in a cold semi-arid climate type. The results of analysing different scenarios revealed that during the warm range (between 10:00 and 17:00), the average Physiological Equivalent Temperature (PET) thermal comfort index declined by 0.6–2.7 °C, whilst during nighttime PET increased by 0.5 °C causing a slight change from coolness to comfort conditions. Fabbri et al. (2020) evaluated the role of vegetation

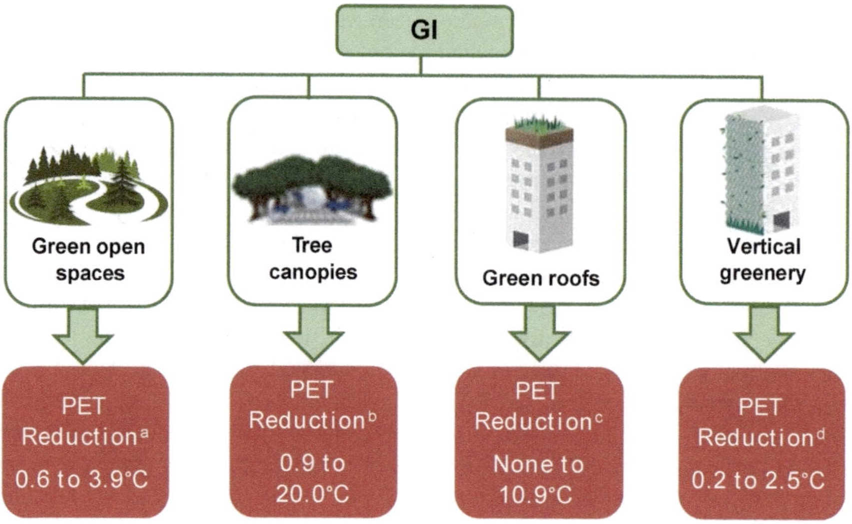


**Fig. 11.6** Summary of the effects of different GI types on PET thermal comfort index based on previous studies. *Source* Authors own work

in archaeological areas and how it affects the tourists' thermal comfort conditions by using PET analysis. They found out that there is a correlation between the number of trees and the amount of green area, and the PET values registered to result in an improvement in the scenarios with more trees and green areas.

Aram et al. (2020a) investigated the cooling effect of a large urban park (125 ha) on thermal comfort outside the park located in the centre of Madrid, Spain (hot-summer Mediterranean climate). The average PET value during the data collection days at the point 855 m away from the park was 41.3 °C, which was 3.9 °C higher than the locations 600 and 445 m away from the park (37.4 °C). It was also found that the PET at the 130 m and 280 m distance of the park was respectively 9.3% and 5.4% less than the PET in the heat island domain (Aram et al. 2020b). The examination of cognitive maps drawn by citizens showed that out of 145 respondents, 68.3% marked the park as the place they experience the greatest thermal comfort level. The results confirmed that with the decrease in temperature and consequently the decrease in PET in the areas around the park, people living and working in these areas exhibit higher levels of perceived thermal comfort (Aram et al. 2020a).

A study on the effects of the shade provided by park trees on thermal comfort conditions in Birjand, Iran (hot-arid climate), indicated that the annual percentage of subjects who voted in the thermal comfort range was 9% higher in the park shaded location compared with the tree-free location (Khalili et al. 2022).

Numerous studies confirmed the positive effect of green open spaces on humans' thermal comfort conditions. There are differences between the effects of green open spaces on thermal comfort conditions of previous studies, highlighting the significant influence of the crown height, trunk height, upwind vegetation cover, and other physical characteristics.

## *3.2 Tree Canopies*

Tree canopy affects the mean radiant temperature, significantly influencing the subjective perception of the thermal environment. According to previous surveys in the tree-lined street canyon, the MRT was found about 30 °C lower (Holst and Mayer 2011). In a study conducted by Park et al. (2012), it was found that the MRT was 24 °C lower inside the sidewalk tree canopy. Lee et al. (2013) determined that the reduction in PET reached 15.7 °C by the shading effect of tree canopies in Freiburg, Germany, during typical summer days. The impact of trees on reducing UHI effects occurs not only at tree heights but also at higher urban levels, taking advantage of energy efficiency in high-rise buildings and dense urban areas (Wang and Akbari 2016). By using a numerical model, shading trees were added to a street canyon on a summer day in the tropical city of Sao Paulo, Brazil (Spangenberg et al. 2008). The results of this simulation determined that incorporating street trees in the urban canyon causes a reduction of up to 1.1 °C in the air temperature whilst contributing to a general decline of MRT at pedestrian height (up to 24.0 °C). Employing PET as a reliable index for assessing thermal comfort conditions showed that street tree canyons improve pedestrians' thermal comfort considerably by reducing PET up to 12.0 °C.

An experimental study conducted in Guangzhou on the influence of tree plantings on pedestrian thermal comfort in street canyons with various aspect ratios revealed that big crown trees reduce daytime PET (4.0 °C) and index of thermal stress (ITS) (285 W) to improve thermal comfort. The results of this study indicated that at night, trees brought a rise in PET (0.9–1.5 °C), improving the thermal comfort of the human body (Chen et al. 2021).

Yang et al. (2011) concluded that increasing greenery cover in high-rise residential quarters of Shanghai improves summertime outdoor thermal comfort conditions. Using numerical simulation, a dense tree cover (Leaf Area Index (LAI) = 6.4) was added over a grass lawn and a hard pavement with an albedo of 0.2, resulting in a reduction of up to 15.0 °C and 20.0 °C in daytime PET, respectively.

Greenery in high-density settings could lead to higher reductions in PET than those in the open spaces (Kong et al. 2017). Therefore, tree canyons play a vital role in improving pedestrians' thermal comfort conditions. The effect of tree canopy on thermal comfort conditions depends on many factors such as tree canopy cover, Leaf Area Index (LAI), and aspect ratios (Sect. 2.2).

### 3.3 Green Roofs

Green roofs present a feasible design strategy for micro-climate amelioration and reducing energy consumption in cities (Liu et al. 2012). The impact of green roofs on the micro-climate affects both indoor and outdoor thermal comfort (Mutani and Todeschi 2020). Green roofs are an effective way of reducing buildings' heat gain in summer and heat loss in winter (Krebs and Johansson 2021). Therefore, adopting green roofs increases thermal comfort and reduces the energy consumption of buildings. Previous research in different climatic zones determined that an individual green roof reduces near-surface air temperature by 2–5 °C, roof surface temperature by 15–45 °C, and building energy consumption by up to 80% (Jim 2012).

The results of a study conducted in a school in Athens, Greece, indicated that the air temperature in a classroom with a green roof was 3.0 °C lower compared with the corresponding classroom under a concrete roof (Barmparesos et al. 2018) directly affecting students' thermal sensation and energy consumption. The effectiveness of green roofs to enhance thermal conditions is strongly dependent on the $U$-value of the roof and its characteristics (lightweight or massive, ventilated or unventilated, etc.) (Giuseppe and D'orazio 2015) in addition to the vegetation characteristics (Alexandri and Jones 2008).

Using a programmed two-dimensional prognostic micro-scale mode, the thermal effect of green roofs on the built environment was examined both inside the canyon and at roof level. The results determined that in the hot and arid climate of Riyadh, air temperature decreases on the roof and inside the canyon reaches up to 12.8 °C and 9.1 °C daytime average, respectively, confirming that green roofs can improve outdoors thermal conditions, both on the roof and street levels (Alexandri and Jones 2008).

By comparing different simulation models, it was concluded that green roofs can cause a decline by 0.3–3.0 °C in the average air temperature when applied on an urban scale (Santamouris 2014). Smith and Roebber (2011) found out that the air temperature during 19:00–23:00 is reduced up to 2.0–3.0 °C due to the higher albedo of the green roofs and its evapotranspiration effect. Microclimatic monitoring conducted on urban woodland intensive green roofs (IGR has > 150 mm substrate depth) revealed that the IGR improved thermal comfort by cooling mean PET by 10.9 °C during daytime in sunny scenarios (Lee and Jim 2019).

In contrast, Taleghani et al. (2016) showed that green roofs at the height of 6 m (two-story dwellings) led to minor changes in surface air temperature, MRT, and consequently thermal comfort, since this strategy modifies the energy balance at roof level, well above the pedestrians' height. Similarly, through analysing various scenarios in Melbourne, Australia, it was found that adding a green roof does not show improvement in PET at the pedestrian level and it should be implemented with other scenarios together (Krebs and Johansson 2021). Chen et al. (2004) conducted simulations to clarify the effect of high-rise buildings' green roofs on the pedestrian height thermal comfort conditions in Tokyo, Japan. They concluded that the effect of green roofs on pedestrians' thermal comfort conditions is negligible due to the height of the roofs.

It should be noticed that green roofs have a considerable effect on improving thermal comfort conditions of buildings, neighbourhoods, and micro-climate in severe thermal conditions of summer. However, a city-wide effect can be achieved by integrating a network of green roofs and other GIs within the urban fabric. As a result, in dense urban areas with a severe shortage of ground-level green spaces and intense UHI effects, large-scale installation of greenery on spatially concentrated roofs can compensate for the green-space deficit, mitigate urban climate, and improve humans' thermal comfort.

## 3.4 *Vertical Greenery*

Buildings' envelopes can easily be covered with vegetation and alter the micro-climate of the built environment contributing to an improvement in thermal comfort conditions. Façade greening induces a change in the building wall albedo, decreasing the heating of building walls and the reflection of incoming solar radiation causing an improvement in indoor and outdoor thermal comfort (Jänicke et al. 2015).

Several researchers have focused on the effects of vertical greenery on indoor thermal conditions confirming that greenery thermoregulates the superficial temperature of the wall and improves indoor thermal comfort (Jänicke et al. 2015). They have estimated that vertical greenery reduced the indoor air temperature by 0.5–5.0 °C during summer in different regions (Eumorfopoulou and Kontoleon 2009), resulting in the reduction of energy consumption by 5–90% (Zinzi and Agnoli 2012) confirming the positive effect of vertical greenery on indoor thermal comfort conditions during hot summer (Medl et al. 2017).

At the same time, vertical greenery can reduce heat gains and improve outdoor thermal comfort. Field studies conducted both before and after the implementation of green walls along a pathway in central Sao Paulo revealed that before the implementation of green walls, the air temperature was about 30.5 °C and the PET was 45.6 °C representing thermal discomfort (de Abreu-Harbich et al. 2016), whilst new studies carried out in the same site indicated that green walls can cool the surface temperature by up to 3.0 °C leading to an improvement of humans' comfort (Iuorio and Harbich 2019).

The results of a micro-scale CFD model of a neighbourhood in Hong Kong suggest that green facades are the second-best strategy for reductions in universal thermal comfort index (UTCI) after street trees (Jia and Wang 2021). Through a field measurement in the hot-humid climate of Guangzhou, China, Lin et al. (2019) determined that in a shaded area with a green façade and an unshaded area, the green façade reduced the average PET by 2.5 °C and 1.4 °C, respectively, compared to the outdoor environment. Bianco et al. (2017) found out that the maximum surface temperature of vertical greenery was measured 24.0 °C lower than that of a reference wall in Turin (northern Italy). The results of a study showed that greening facades improve pedestrians' thermal comfort with PET reductions varying from 0.2 to 1.4 °C depending on the arrangements of the greenery façade (Cui et al. 2022).

Various issues should be considered for analysing the magnitude of vertical greeneries' effects on thermal comfort, including the various climatic effects of plant species, the height and area of vertical greeneries, and their distribution in the urban environment as well as climatic features of the region (Hami et al. 2019). The combination of both green roofs and green walls leads to the highest mitigation of temperatures inside the canyon (Alexandri and Jones 2008) considering that the potential of green roofs and green facades to improve outdoor thermal comfort is restricted to the proximity of the building's facades and should be implemented as a valuable addition to other GIs.

For all climates examined, green walls have a stronger effect than green roofs inside the canyon. Nonetheless, green roofs have a greater effect at roof level and, consequently, at the urban scale.

## 4 GI for Mitigating Natural Hazards

Natural hazards are naturally occurring events that become disastrous when they create large fatalities and property damages, which hinder social and economic development (Debele et al. 2019; Kumar et al. 2020). Here, we provide more detailed insights about the impact of hydro-meteorological natural hazards such as floods, droughts, landslides, storm surges, and coastal erosions and their management through GI.

There is no doubt that the cities require actions and solutions for the alarming natural hazards also in addition to the development of urban quality of life (Kumar 2021). One possible solution could be the use of GI to mitigate the impact of natural hazards. For instance, GI as multi-purpose solutions (e.g. for urban flood mitigation) can help to reduce vulnerability through the inclusion of several elements that reduce dependency on only one measure.

GI is an emerging approach to making cities sustainable, healthy, and more viable to manage the impact of natural hazards. In recent decades, it has been seen by policymakers and urban planners as an effective planning tool to improve and support urban ecosystems, contribute to the conservation of biodiversity, decrease the effects of natural hazard, and improve the well-being of urban inhabitants. Here, we discuss how different types of GI can be used to tackle climate change and urbanisation-induced natural hazards such flooding, droughts, landslides, storm surge, and coastal erosion at different scales.

## 4.1 *Multi-hazards*

The multi-benefits provided by the GI have already been widely studied and recognised in the literature—impacts on human health and well-being (Kim and Miller 2019; Venkataramanan et al. 2019), stormwater management (Li et al. 2020), natural hazard management, reduction, and helping ecosystem services (Sahani et al. 2019; Debele et al. 2019; Kumar et al. 2020). Chávez et al. (2021) identified the environmental benefits of coastal GI as being mainly twofold—flood protection and defence against coastal erosion. The biophysical features of green spaces in urban areas help cities to adapt and cope with natural hazards by delaying the onset and propagation probability and thereby reduce the associated risks such as surface flood, drought, and landslide risks (Moos et al. 2018). Multi-functionality within the GI helps to integrate the environmental, social, cultural, and economic benefits of urban green areas and enables more efficient use of space. Best practices of GI in local governance, especially when combined with traditional grey or artificially built infrastructure, which is called the hybrid approach, could achieve greater urban sustainability and resilience and are recognised for their value in adapting and keeping pace with ongoing urbanisation and the impacts of future climate change (Debele et al. 2019). Debele et al. (2019) analysed 288 case studies that used GI as the main measures to mitigate five natural hazards (floods, droughts, heatwaves, landslides, and storm surges) and found that flooding is mostly managed by a hybrid approach and followed by the green approach whilst heatwaves are mostly managed by purely green approach. A hybrid approach is also the best measure to protect coastal areas from flooding and erosion risks, whereas a green approach is also a good solution to reduce the risks caused by drought and landslides. Moos et al. (2018) assessed the performance and efficacy of GI (forests) in decreasing damages to the mountainous area due to floods, landslides, snow avalanches, rockfalls, and debris flows, finding that forests are highly efficient and effective to reduce the associated risks by reducing the onset and propagation probability and intensity of hazards.

## 4.2 *Floods*

Newman et al. (2022) presented a case study through modelling to evaluate GI impacts on flood mitigation using the Sunnyside neighbourhood in Houston, Texas, USA, as a case study. The results showed that the economic benefit provided by GI could pay itself off in 33.5 years in indirect benefits. They also reported that the GI reduced the risk of flooding, reduced stormwater runoff contaminants, and provided a possible direction to protect vulnerable peoples. At the same time, the design can capture 2295.3 $m^3$ of surface runoff, producing USD 243,856 in annual green benefits (including reduced air pollutants—USD 133), $CO_2$ sequestration (USD 88), groundwater replenishment (USD 1231), reduced energy use (USD 39,596), reduced treatment benefits (USD 683), and compensatory value of trees (USD 202,125). It was also found that GI practices reduced runoff volume by 37% compared with the conventional/grey approach. Foster et al. (2011) analysed the advantages and cost-benefits of GI for flood protection. They found that: (i) the net present value to urban watersheds was ~ 10–14% higher than conventional roofs; (ii) trees provided US$1.3 billion in stormwater profits (based on US$0.66/cubic foot of storage) in Houston, and (iii) permeable pavement reduced storm-runoff/flood volume by 70–90% same as meadow or forest to manage the predicted future climate change impacts in a community in Canada. Nurmi et al. (2016) studied the cost-effectiveness of green roofs to provide ecosystem services in the city of Helsinki. The study focused on multiple benefits that green roofs provide such as energy savings, urban flood control, and health benefits. The total benefits for green roofs in floodwater management were 0.9–2.2 in 50% green-roof buildings' scenario and yielded benefits of ~ US$2.3 per 4.1 $m^2$ for the city-wide green-roof scenario.

## 4.3 Droughts

Keesstra et al. (2018) investigated the application of hybrid/mixed GI for drought risk management in various environmental conditions such as in Ethiopia, Mediterranean basin, Slovenia, Iceland, southern Portugal, and Sweden and concluded that GI is an efficient long-term approach in both managing natural hazard risk and benefit-to-cost ratio. Atampugre (2014) evaluated the benefits of GI (bench terraces, contour bunds, and Napier grass strips) for drought risk management and found that bench terraces provided the maximum benefits, followed by contour bunds and Napier grass strips to reduce agricultural risk and increase crop productivity regardless of the crop variety in the Saba sub-catchment of the Upper Tana catchment in Kenya. Mishra and Rai (2014) evaluated the benefits of GI such as agroforestry, vegetative barriers, terrace cultivations, etc. to reduce erosion and agricultural drought in rural watersheds in Sikkim Himalaya, India. The study found that vegetative barriers and agroforestry were the two most effective practices for reducing soil erosion and increasing crop productivity and income. Terraces were found to be more expensive to be developed than other practices.

## 4.4 Landslides

Petrone and Preti (2010) studied the species selection question, which was accomplished by conducting on-site tests on various native species, cuttings of which were applied to several types of GI (drainages with live fascine mattress, a live palisade, a vegetated live crib wall, a vegetative covering made of a metallic net and biotextile coupled with a live palisade made of bamboo) restoration works for landslide risk mitigation in Nicaragua, Italy. Among the species, *Gliricidia sepium* and *Tabebuia rosea* are adequate for landslide risk protection on slopes, whilst *Erythrina fusca* showed successful behaviour only in the live crib wall for riverbank protection. The most effective GI against shallow landslides is re-introducing vegetation cover on a slope together with grey engineering structures, such as live fascines or crib walls (Faucon and Lambers 2017). The other GI against landslides is cropland and forest management practices, road design, and the management of streams and rivers with respect to their hydrological effects (Dolidon et al. 2009).

## 4.5 Storm Surge and Coastal Erosion

Narayan et al. (2016) compared the importance of seagrass/kelp beds, mangroves, coral reefs, and salt marshes with submerged breakwaters in reducing flood wave height and coastal erosion using 69 case studies. The results of the comparison showed that the green approach (i.e. coastal habitats) reduced coastal flooding by between 35 and 71%; this effectiveness varied with the habitats and the locations. Coastline vegetation barriers act effectively to decrease erosion of coastline by limiting wave height in the range of 0.3–5% (Anderson et al. 2011). Similar effects can be derived from natural coral reefs (Reguero et al. 2018) or the construction of banks of oyster reefs. Fernández-Montblanc et al. (2020) studied the potential of GI (e.g. dune reconstruction and revegetation) in reducing coastal erosion and flooding under current and future extreme-weather and climate conditions in the in Bellocchio (Italy, Northern Adriatic Sea). The result showed an increase of 22% of the sediment budget of the dune-beach system under the current sea level, and o f57% and 82% under future sea level rise for low and high hazard scenarios, respectively. The maximum floodwater volume has decreased by 42% in the current sea level. The predicted increase of maximum floodwater volume under future high hazard and low hazard scenarios is reduced by 25% and 59%, respectively.

## 4.6 Design Principles of GI

GI performance and efficiency rely on a number of environmental and operational principles to provide the broad array of benefits and services. Regardless, the GI concept has gained attraction among researchers and decision-makers as a multi-functional approach to urban resilience and sustainability. But, to fulfil these multi-functional purposes, both the quantity and quality of urban GI must be addressed in planning processes, and the design and development of GI planning principles are fundamental (Monteiro et al. 2020). The designing and planning principles are the most important to identify the basic design procedures that make GI work and how design can inform site locations for GI implementation. GI principles are, actually, underlying grounds that support, guide, and facilitate the designing procedures of GI, in order to ensure that it contributes to a network of quality and functional green spaces, capable of meeting the needs of a determined urban area, contributing to the best way to the sustainability of a given area, depending on its scale (micro-, meso-, and macro-scale settings). For instance, to solve urban area challenges such as spatial conflicts and path dependencies, Kabisch et al. (2022) integrated and presented five principles for urban GI design, planning, and implementation. These five principles are developed to contribute to resilient urban futures, the design, planning, policy, and governance of GI. The principles could support overcoming governance gaps and advance the scientific discourse of urban GI towards a more effective and sustainable urban development.

## 5 Summary and Conclusions

Urbanisation and climate change are increasingly impacting human's health, well-being, and environment in various ways which includes air pollution, urban heat, flooding, droughts, etc. GI implemented as natural, sustainable solutions in cities can mitigate negative health and environmental impacts of these processes.

GI in near-road environments reduces air pollution exposure by dispersion and deposition mechanism. The type, dimensions, and characteristics of GI determine the positive or negative impact on air quality in near-road environments. Generally, GI in open-road conditions reduces air pollution exposure, whilst GI in street canyons requires complex site-specific considerations based on the built-geometry, meteorology, and characteristics of GI for improving air quality. The available GI recommendations and design tools provide useful information to the public for planning GI barriers in near-road environments.

All of the GI strategies improve indoor and outdoor thermal comfort conditions by shading and evapotranspiration. The magnitude of this effect varies depending on the urban geometry, GIs' characteristics, and also climatic conditions. Green roofs and vertical greenery are feasible strategies for urban dense areas with intense UHI effects where the severe shortage of ground-level green spaces is a limitation for applying other GIs. Some studies suggest that the effects of green roofs are more significant at roof level and negligible on street level, whilst other studies indicate a significant improvement of pedestrians' thermal comfort on street level. Therefore, a network of different GIs should be integrated in order to achieve the highest level of thermal comfort for urban dwellers considering urban geometry, the effectiveness of each GI strategy in improving thermal satisfaction, and other parameters.

GI implementation and maintenance for natural hazard mitigation almost certainly become increasingly important as urbanisation and climate change increase. GI as multi-purpose measures can be used to tackle the occurrence of multi-hazard and reduce vulnerability through the inclusion of several elements that reduce dependency on only one system and avoid the associated damages whilst simultaneously providing social, economic, and environmental benefits. For instance, GI implemented to manage landslide risks such as contour ditches, drainage trenches can simultaneously reduce flooding and agricultural drought through decreasing the surface runoff and increasing infiltration to the soil. Of the types of GI covered here, green roofs (~ 10–14%) and permeable pavement (70–90%) appear to be the most effective solutions against flooding. GI such as bench terraces, contour bunds, Napier grass strips, vegetative barriers, and agroforestry is the most effective practices to manage drought hazards. Landslides' risks were effectively managed by GI such as re-introducing vegetation cover on a slope together with grey engineering structures (hybrid), cropland and forest management practices, road design and the management of streams and rivers with respect to their hydrological effects, whereas the storm surge and coastal erosion were effectively managed by GI such as coastal habitats (35–71%), coastline vegetation barriers (0.3–5%), and dune reconstruction and revegetation (25–59%).

The effectiveness of GI depends on the environmental and operational principles. In this chapter, we highlighted that having a standard principle for delivering urban GI is the most important in order to help designers, owners, operators, and occupiers to design, deliver, and operate their assets in such a way that improves natural hazard resilience and environmental impact. In the era of global warming and urbanisation, the spatial prioritisation of implementing GI is of great importance for ecosystem restoration, climate adaptation, natural hazard mitigation, and urban resilience. GI provides numerous health, socioeconomic, and environmental benefits, with air pollution reduction being one of the most frequently studied. However, placing the wrong GI type or species at certain places can degrade the surrounding air quality by trapping the air pollutant or increasing the BVOC levels. But the holistic approach that can consider all aspects of pros and cons of GI urban planning is not yet developed. Most of these pros (air pollution mitigation, water and soil conservations, and UHI mitigation) and cons (BVOC/pollen emission, maintenance, and structure damage) are either studied alone or in few combinations. Therefore, planners and practitioners require clear and practical guidance in an easily digestible format to inform their decisions on GI selection and design of new GI urban planting projects, as well as managing existing green spaces to achieve positive outcomes. The development of decision support tools and services can help ensure a positive outcome based on universally recognised green infrastructure interventions and ensuring that a positive outcome in decision-making will help build confidence in driving top-down initiatives by national governments.

## 6 Declaration

All the figures used in this chapter are of the author's own work. These have been published as a part of their open-access articles that are duly cited and acknowledged in the captions of figures. No permissions for their use are therefore sought and required.

## References

Abhijith KV, Kumar P (2019) Field investigations for evaluating green infrastructure effects on air quality in open-road conditions. Atmos Environ 201:132–147. https://doi.org/10.1016/j.atmosenv.2018.12.036

Abhijith KV, Kumar P (2020) Quantifying particulate matter reduction and their deposition on the leaves of green infrastructure. Environ Pollut 265:114884. https://doi.org/10.1016/j.envpol.2020.114884

Abhijith KV, Kumar P, Gallagher J, McNabola A, Baldauf R, Pilla F, Broderick B, Di Sabatino S, Pulvirenti B (2017) Air pollution abatement performances of green infrastructure in the open road and built-up street canyon environments—a review. Atmos Environ 162:71–86. https://doi.org/10.1016/j.atmosenv.2017.05.014

Akbari H, Pomerantz M, Taha H (2001) Cool surfaces and shade trees to reduce energy use and improve air quality in urban areas. Sol Energ 70:295–310. https://doi.org/10.1016/S0038-092X(00)00089-X

Alexandri E, Jones P (2008) Temperature decreases in an urban canyon due to green walls and green roofs in diverse climates. Build Environ 43:480–493. https://doi.org/10.1016/j.buildenv.2006.10.055

Anderson ME, Smith JM, McKay SK (2011) Wave dissipation by vegetation. https://apps.dtic.mil/sti/pdfs/AD1003881.pdf. Accessed on 14th Apr 2022

Aram F, Solgi E, Baghaee S, García EH, Mosavi A, Band SS (2020a) How parks provide thermal comfort perception in the metropolitan cores; a case study in Madrid Mediterranean climatic zone. Clim Risk Manag 30:100245. https://doi.org/10.1016/j.crm.2020.100245

Aram F, Solgi E, Garcia EH, Mosavi A (2020b) Urban heat resilience at the time of global warming: evaluating the impact of the urban parks on outdoor thermal comfort. Environ Sci Eur 32:1–15. https://doi.org/10.1186/s12302-020-00393-8

Atampugre G (2014) Cost and benefit analysis of the adoption of soil and water conservation methods. Kenya Int J Sci Res Publ 4:1–8. https://www.ijsrp.org/research-paper-0814/ijsrp-p3247.pdf

Baldauf R (2017) Roadside vegetation design characteristics that can improve local, near-road air quality. Transp Res Part D Transp Environ 52:354–361. https://doi.org/10.1016/j.trd.2017.03.013

Barmparesos N, Assimakopoulos MN, Assimakopoulos VD, Loumos N, Sotiriou MA, Koukoumtzis A (2018) Correction: Barmparesos et al. Indoor Air Quality and Thermal Conditions in a Primary School with a Green Roof System. Atmosphere 9(11):417

Barnes TK, Brade AR, MacKenzie JD, Whyatt DJ, Carruthers J, Stocker X, Cai CN (2014) Hewitt Spatially-varying surface roughness and ground-level air quality in an operational dispersion model. Environ Pollut 185:44–51

Baró F (2015) A multi-scale assessment of regulating ecosystem services in Barcelona. In: Ecosystem services: concepts, methodologies and instruments for research and applied use, pp 87–100. https://doi.org/10.13140/RG.2.1.1108.8243

Barwise Y, Kumar P (2020) Designing vegetation barriers for urban air pollution abatement: a practical review for appropriate species selection. Clim Atmos Sci 3:12. https://doi.org/10.1038/s41612-020-0115-3

Barwise Y, Kumar P, Tiwari A, Rafi-Butt F, McNabola A, Cole S, Field BCT, Fuller J, Mendis J, Wyles K (2021) The co-development of HedgeDATE, a public engagement and decision support tool for air pollution exposure mitigation by green infrastructure. Sustain Cities Soc 75:103299. https://doi.org/10.1016/j.scs.2021.103299

Bianco L, Serra V, Larcher F, Perino M (2017) Thermal behaviour assessment of a novel vertical greenery module system: first results of a long-term monitoring campaign in an outdoor test cell. Energ Effi 10:625–638. https://doi.org/10.1007/s12053-016-9473-4

Cabaraban MTI, Kroll CN, Hirabayashi S, Nowak DJ (2013) Modelling of air pollutant removal by dry deposition to urban trees using a WRF/CMAQ/i-Tree Eco coupled system. Environ Pollut 176:123–133. https://doi.org/10.1016/j.envpol.2013.01.006

Calfapietra C, Fares S, Manes F, Morani A, Sgrigna G, Loreto F (2013) Role of biogenic volatile organic compounds (BVOC) emitted by urban trees on ozone concentration in cities: a review. Environ Pollut 183:71–80. https://doi.org/10.1016/j.envpol.2013.03.012

Chávez V, Lithgow D, Losada M, Silva-Casarin R (2021) Coastal green infrastructure to mitigate coastal squeeze. J Infrastruct Preserv Resilience 2:1–12

Chen H, Ooka R, Harayama K, Kato S, Li X (2004) Study on outdoor thermal environment of apartment block in Shenzhen, China with coupled simulation of convection, radiation and conduction. Energ Build 36:1247–1258. https://doi.org/10.1016/j.enbuild.2003.07.003

Chen T, Pan H, Lu M, Hang J, Lam CKC, Yuan C, Pearlmutter D (2021) Effects of tree plantings and aspect ratios on pedestrian visual and thermal comfort using scaled outdoor experiments. Sci Total Environ 801:149527. https://doi.org/10.1016/j.scitotenv.2021.149527

Cui D, Zhang Y, Li X, Yuan L, Mak CM, Kwok K (2022) Effects of different vertical façade greenery systems on pedestrian thermal comfort in deep street canyons. Urban For Urban Greening 72:1–13. https://doi.org/10.1016/j.ufug.2022.127582

De Abreu-Harbich LV, de Oliveira Morelli DD, Brocaneli PF, Labaki LC (2016) How do green façades mitigate the thermal stress under heat waves? Case of Minhocão Elevate Highway, São Paulo (Brazil), in Full Papers-PLEA 2016

Di Giuseppe E, D'orazio M (2015) Assessment of the effectiveness of cool and green roofs for the mitigation of the Heat Island effect and for the improvement of thermal comfort in Nearly Zero Energy Building. Architectural Sci Rev 58(2):134–143

Debele SE, Kumar P, Sahani J, Marti-Cardona B, Mickovski SB, Leo LS, Porcù F, Bertini F, Montesi D, Vojinovic Z, Di Sabatino S (2019) Nature-based solutions for hydro-meteorological hazards: revised concepts, classification schemes and databases. Environ Res 179:108799. https://doi.org/10.1016/j.envres.2019.108799

Di Marino M, Tiitu M, Lapintie K, Viinikka A, Kopperoinen L (2019) Integrating green infrastructure and ecosystem services in land use planning. Results from two Finnish case studies. Land Use Policy 82:643–656. https://doi.org/10.1016/j.landusepol.2019.01.007

Dolidon N, Hofer T, Jansky L, Sidle R (2009) Watershed and forest management for landslide risk reduction. In: Landslides–disaster risk reduction, vol 33. Springer, Berlin, Heidelberg, pp 633–649. https://doi.org/10.1007/978-3-540-69970-5_33

Eumorfopoulou EA, Kontoleon KJ (2009) Experimental approach to the contribution of plant-covered walls to the thermal behaviour of building envelopes. Build Environ 44:1024–1038. https://doi.org/10.1016/j.buildenv.2008.07.004

Fabbri K, Ugolini A, Iacovella A, Bianchi AP (2020) The effect of vegetation in outdoor thermal comfort in archaeological area in urban context. Build Environ 175:106816. https://doi.org/10.1016/j.buildenv.2020.106816

Faucon MP, Houben D, Lambers H (2017) Plant functional traits: soil and ecosystem services. Trends Plant Sci 22(5):385–394

Fernández-Montblanc T, Duo E, Ciavola P (2020) Dune reconstruction and revegetation as a potential measure to decrease coastal erosion and flooding under extreme storm conditions. Ocean Coast Manag 188:105075. https://doi.org/10.1016/j.ocecoaman.2019.105075

Foster J, Lowe A, Winkelman S (2011) The value of green infrastructure for urban climate adaptation, vol 750. Center for Clean Air Policy, pp 1–52. https://www.adaptationleader.org/wp-content/uploads/2022/07/THE_VALUE_OF_GREEN_INFRASTRUCTURE_FOR_UR.pdf

Gallagher J, Baldauf R, Fuller CH, Kumar P, Gill LW, McNabola A (2015) Passive methods for improving air quality in the built environment: a review of porous and solid barriers. Atmos Environ 120:61–70. https://doi.org/10.1016/j.atmosenv.2015.08.075

Gourdji S (2018) Review of plants to mitigate particulate matter, ozone as well as nitrogen dioxide air pollutants and applicable recommendations for green roofs in Montreal, Quebec. Environ Pollut 241:378–387. https://doi.org/10.1016/j.envpol.2018.05.053

Gromke C (2011) A vegetation modeling concept for building and environmental aerodynamics wind tunnel tests and its application in pollutant dispersion studies. Environ pollut 159(8–9):2094–2099

Gulati M, Scholtz L (2020) The case for investment in green infrastructure in African Cities. Available online at https://bit.ly/3jEqYwq. Accessed on 14th Apr 2022

Hami A, Abdi B, Zarehaghi D, Bin MS (2019) Assessing the thermal comfort effects of green spaces: a systematic review of methods, parameters, and plants' attributes. Sustain Cities Soc 49:101634. https://doi.org/10.1016/j.scs.2019.101634

Hefny MS, Heinke Schlünzen K, Grawe D (2015) Including trees in the numerical simulations of the wind flow in urban areas: should we care? J Wind Eng Ind. Aerodyn 144:84–95

Hermoso V, Morán-Ordóñez A, Lanzas M, Brotons L (2020) Designing a network of green infrastructure for the EU. Landsc Urban Plan 196:103732. https://doi.org/10.1016/j.landurbplan.2019.103732

Holst J, Mayer H (2011) Impacts of street design parameters on human-biometeorological variables. Meteorol Z 20:541. https://doi.org/10.1127/0941-2948/2011/0254

Imran HM, Jatin K, Anne N, Shobha M (2019) Impacts of future urban expansion on urban heat island effects during heatwave events in the city of Melbourne in southeast Australia. Q J R Meteorol Soc 145:1–17. https://doi.org/10.1002/qj.3580

Iuorio O, Harbich LA (2019) Green networks as a key of urban planning with thermal comfort and well-being BT. In: Lemes de Oliveira F, Mell I (eds) Planning cities with nature: theories, strategies and methods. Springer International Publishing, Cham, pp 97–109. https://doi.org/10.1007/978-3-030-01866-5_7

Jänicke B, Meier F, Hoelscher MT, Scherer D (2015) Evaluating the effects of façade greening on human bioclimate in a complex urban environment. Adv Meteorol 2015:1–15. https://doi.org/10.1155/2015/747259,1-15

Jeanjean APR, Monks PS, Leigh RJ (2016) Modelling the effectiveness of urban trees and grass on PM 2.5 reduction via dispersion and deposition at a city scale. Atmos Environ 147:1–10. https://doi.org/10.1016/j.atmosenv.2016.09.033

Jia S, Wang Y (2021) Effect of heat mitigation strategies on thermal environment, thermal comfort, and walkability: a case study in Hong Kong. Build Environ 201:107988. https://doi.org/10.1016/j.buildenv.2021.107988

Jim CY (2012) Effect of vegetation biomass structure on thermal performance of tropical green roof. Landscape Ecol Eng 8:173–187. https://link.springer.com/article/10.1007/s11355-011-0161-4

Kabisch N, Frantzeskaki N, Hansen R (2022) Principles for urban nature-based solutions. Ambio 51:1388–1401. https://doi.org/10.1007/s13280-021-01685-w

Karimi A, Sanaician II, Farhadi H, Norouzian-Maleki S (2020) Evaluation of the thermal indices and thermal comfort improvement by different vegetation species and materials in a medium-sized urban park. Energ Rep 6:1670–1684. https://doi.org/10.1016/j.egyr.2020.06.015

Keesstra S, Nunes J, Novara A, Finger D, Avelar D, Kalantari Z, Cerdà A (2018) The superior effect of nature-based solutions in land management for enhancing ecosystem services. Sci Total Environ 610:997–1009. https://doi.org/10.1016/j.scitotenv.2017.08.077

Khalili S, Fayaz R, Zolfaghari SA (2022) Analyzing outdoor thermal comfort conditions in a university campus in hot-arid climate: a case study in Birjand, Iran. Urban Clim 43:101128. https://doi.org/10.1016/j.uclim.2022.101128

Kim G, Miller PA (2019) The impact of green infrastructure on human health and well-being: the example of the huckleberry trail and the heritage community park and natural area in Blacksburg, Virginia. Sustain Cities Soc 48:101562. https://doi.org/10.1016/j.scs.2019.101562

Kong L, Lau KKL, Yuan C, Chen Y, Xu Y, Ren C, Ng E (2017) Regulation of outdoor thermal comfort by trees in Hong Kong. Sustain Cities Soc 31:12–25. https://doi.org/10.1016/j.scs.2017.01.018

Krebs LF, Johansson E (2021) Influence of microclimate on the effect of green roofs in Southern Brazil–A study coupling outdoor and indoor thermal simulations. Energ Build 241:110963

Kumar P (2021) Climate change and cities: challenges ahead. Front Sustain Cities 3:1–8. https://doi.org/10.3389/frsc.2021.645613

Kumar P, Druckman A, Gallagher J, Gatersleben B, Allison S, Eisenman TS, Hoang U, Hama S, Tiwari A, Sharma A, Abhijith KV, Adlakha D, McNabola A, Astell-Burt T, Feng X, Skeldon AC, de Lusignan S, Morawska L (2019) The nexus between air pollution, green infrastructure and human health. Environ Int 133:105181. https://doi.org/10.1016/j.envint.2019.105181

Kumar P, Debele SE, Sahani J, Aragão L, Barisani F, Basu B, Bucchignani E, Charizopoulos N, Di Sabatino S, Domeneghetti A, Edo AS (2020) Towards an operationalisation of nature-based solutions for natural hazards. Sci Total Environ 731:138855. https://doi.org/10.1016/j.scitotenv.2020.138855

Kumar P, Debele S, Khalili S, Halios CH, Sahani J, Aghamohammadi N, Andrade MFA, Athanassiadou M, Bhui K, Calvillo N, Cao SJ, Coulon F, Edmondson JE, Fletcher D, Freitas ED, Guo H, Hort MC, Katti M, Kjeldsen TR, Lehmann S, Locosselli GM, Malham SK, Morawska L, Parajuli R, Rogers CDF, Yao R, Wang F, Wenk J, & Jones L (2024). Urban heat mitigation by green and blue infrastructure: a review of drivers, effectiveness, and future needs. Innov 5:100588. https://doi.org/10.1016/j.xinn.2024.100588

Lee LSH, Jim CY (2019) Urban woodland on intensive green roof improved outdoor thermal comfort in subtropical summer. Int J Biometeorol 63:895–909. https://doi.org/10.1007/s00484-019-01702-4

Lee H, Holst J, Mayer H (2013) Modification of human-biometeorologically significant radiant flux densities by shading as local method to mitigate heat stress in summer within urban street canyons. Adv Meteorol 2013:312572. https://doi.org/10.1155/2013/312572

Levine JG, Brown AY, MacKenzie R (2021) Green infrastructure for roadside air quality (GI4RAQ) guidance & decision tree: an evidence-based approach to reducing roadside exposure to road transport pollution. Developed by the Birmingham Institute of Forest Research, University of Birmingham, and Transport for London. https://doi.org/10.25500/epapers.bham.00003398

Li L, Uyttenhove P, Van Eetvelde V (2020) Planning green infrastructure to mitigate urban surface water flooding risk—a methodology to identify priority areas applied in the city of Ghent. Landsc Urban Plan 194:103703. https://doi.org/10.1016/j.landurbplan.2019.103703

Lin H, Xiao Y, Musso F, Lu Y (2019) Green façade effects on thermal environment in transitional space: field measurement studies and computational fluid dynamics simulations. Sustainability 11:5691. https://doi.org/10.3390/su11205691

Liu KS, Hsueh SL, Wu WC, Chen YL (2012) A DFuzzy-DAHP decision-making model for evaluating energy-saving design strategies for residential buildings. Energies 5:4462–4480. https://doi.org/10.3390/en5114462

Medl A, Stangl R, Florineth F (2017) Vertical greening systems—a review on recent technologies and research advancement. Build Environ 125:227–239. https://doi.org/10.1016/j.buildenv.2017.08.054

Mishra PK, Rai SC (2014) A cost–benefit analysis of indigenous soil and water conservation measures in Sikkim Himalaya, India. Mt Res Dev 34:27–35. https://doi.org/10.1659/MRD-JOURNAL-D-12-00013.1

Monteiro R, Ferreira JC, Antunes P (2020) Green infrastructure planning principles: An integrated literature review. Land 9(12):525

Moos C, Bebi P, Schwarz M, Stoffel M, Sudmeier-Rieux K, Dorren L (2018) Ecosystem-based disaster risk reduction in mountains. Earth Sci Rev 177:497–513. https://doi.org/10.1016/j.earscirev.2017.12.011

Morakinyo TE, Kong L, Lau KKL, Yuan C, Ng E (2017) A study on the impact of shadow-cast and tree species on in-canyon and neighborhood's thermal comfort. Build Environ 115:1–17

Mutani G, Todeschi V (2020) The effects of green roofs on outdoor thermal comfort, urban heat island mitigation and energy savings. Atmosphere 11:123. https://doi.org/10.3390/atmos11020123

Narayan S, Beck MW, Reguero BG, Losada IJ, Van Wesenbeeck B, Pontee N, Sanchirico JN, Ingram JC, Lange GM, Burks-Copes KA (2016) The effectiveness, costs and coastal protection benefits of natural and 3 2 2 nature-based defences. PloS one 11(5):e0154735

Newman G, Sansom GT, Yu S, Kirsch KR, Li D, Kim Y, Horney JA, Kim G, Musharrat S (2022) A framework for evaluating the effects of green infrastructure in mitigating pollutant transferal and flood events in Sunnyside, Houston, TX. Sustainability 14:1–16. https://doi.org/10.3390/su14074247

Nowak DJ, Hirabayashi S, Doyle M, McGovern M, Pasher J (2018) Air pollution removal by urban forests in Canada and its effect on air quality and human health. Urban For Urban Green 29:40–48

Nurmi V, Votsis A, Perrels A, Lehvävirta S (2016) Green roof cost-benefit analysis: special emphasis on scenic benefits. J Benefit-Cost Anal 7:488–522. https://doi.org/10.1017/bca.2016.18

Ottosen TB, Kumar P (2020) The influence of the vegetation cycle on the mitigation of air pollution by a deciduous roadside hedge. Sustain Cities Soc 53:101919

Pace R, Biber P, Pretzsch H, Grote R (2018) Modeling ecosystem services for park trees: sensitivity of i-tree eco simulations to light exposure and tree species classification. Forests 9:1–18. https://doi.org/10.3390/f9020089

Park M, Hagishima A, Tanimoto J, Narita KI (2012) Effect of urban vegetation on outdoor thermal environment: field measurement at a scale model site. Build Environ 56:38–46. https://doi.org/10.1016/j.buildenv.2012.02.015

Pataki DE, Carreiro MM, Cherrier J, Grulke NE, Jennings V, Pincetl S, Pouyat RV, Whitlow TH, Zipperer WC (2011) Coupling biogeochemical cycles in urban environments: ecosystem services, green solutions, and misconceptions. Front Ecol Environ 9:27–36. https://doi.org/10.1890/090220

Petrone A, Preti F (2010) Soil bioengineering for risk mitigation and environmental restoration in a humid tropical area. Hydrol Earth Syst Sci 14:239–250. https://doi.org/10.5194/hess-14-239-2010

Pope CA, Ezzati M, Dockery DW (2009) Fine-particulate air pollution and life expectancy in the United States. N Engl J Med 360:376–386. https://doi.org/10.1056/NEJMsa0805646

Reguero BG, Beck MW, Agostini VN, Kramer P, Hancock B (2018) Coral reefs for coastal protection: a new methodological approach and engineering case study in Grenada. J Environ Manage 210:146–161. https://doi.org/10.1016/j.jenvman.2018.01.024

Sadrizadeh S, Yao R, Yuan F, Awbi H, Bahnfleth W, Bi Y, Cao G, Croitoru C, de Dear R, Haghighat F, Kumar P (2022) Indoor air quality and health in schools: a critical review for developing the roadmap for the future school environment. J Build Eng 57:104908. https://doi.org/10.1016/j.jobe.2022.104908

Sahani J, Kumar P, Debele S, Spyrou C, Loupis M, Aragao L, Porcu F, Shah MAR, Di Sabatino S (2019) Hydro-meteorological risk assessment methods and management by nature-based solutions. Sci Total Environ 696:133936. https://doi.org/10.1016/j.scitotenv.2019.133936

Salata F, Golasi I, Proietti R, de Lieto VA (2017) Implications of climate and outdoor thermal comfort on tourism: the case of Italy. Int J Biometeorol 61:2229–2244. https://doi.org/10.2167/jost549.0

Santamouris M (2014) Cooling the cities—a review of reflective and green roof mitigation technologies to fight heat island and improve comfort in urban environments. Sol Energ 103:682–703. https://doi.org/10.1016/j.solener.2012.07.003

Schwarz N, Schlink U, Franck U, Großmann K (2012) Relationship of land surface and air temperatures and its implications for quantifying urban heat island indicators—an application for the city of Leipzig (Germany). Ecol Indic 18:693–704

Smith KR, Roebber PJ (2011) Green roof mitigation potential for a proxy future climate scenario in Chicago, Illinois. J Appl Meteorol Climatol 50:507–522. https://doi.org/10.1175/2010JAMC2337.1

Spangenberg J, Shinzato P, Johansson E, Duarte D (2008) Simulation of the influence of vegetation on microclimate and thermal comfort in the city of São Paulo. Rev. da Soc. Bras. Arborização Urbana 3:1–19. https://doi.org/10.5380/revsbau.v3i2.66265

Susca T, Gaffin SR, Dell'Osso GR (2011) Positive effects of vegetation: Urban heat island and green roofs. Environ Pollut 159(8–9):2119–2126

Taleghani M, Sailor D, Ban-Weiss GA (2016) Micrometeorological simulations to predict the impacts of heat mitigation strategies on pedestrian thermal comfort in a Los Angeles neighborhood. Environ Res Lett 11:024003. https://doi.org/10.1088/1748-9326/11/2/024003

Tallis M, Taylor G, Sinnett D, Freer-Smith P (2011) Estimating the removal of atmospheric particulate pollution by the urban tree canopy of London, under current and future environments. Landsc Urban Plan 103:129–138. https://doi.org/10.1016/j.landurbplan.2011.07.003

Tiwari A, Kumar P (2020) Integrated dispersion-deposition modelling for air pollutant reduction via green infrastructure at an urban scale. Sci Total Environ 723:138078. https://doi.org/10.1016/j.scitotenv.2020.138078

Tiwari A, Kumar P, Baldauf R, Zhang KM, Pilla F, Di Sabatino S, Brattich E, Pulvirenti B (2019) Considerations for evaluating green infrastructure impacts in microscale and macroscale air pollution dispersion models. Sci Total Environ 672:410–426. https://doi.org/10.1016/j.scitotenv.2019.03.350

Tiwary A, Sinnett D, Peachey C, Chalabi Z, Vardoulakis S, Fletcher T, Leonardi G, Grundy C, Azapagic A, Hutchings TR (2009) An integrated tool to assess the role of new planting in PM10 capture and the human health benefits: a case study in London. Environ Pollut 157(10):2645–2653

Tomson M, Kumar P, Barwise Y, Perez P, Forehead H, French K, Morawska L, Watts J (2021) Green infrastructure for air quality improvement in street canyons. Environ Int 146:106288. https://doi.org/10.1016/j.envint.2020.106288

Tong Z, Baldauf RW, Isakov V, Deshmukh P, Max Zhang K (2016) Roadside vegetation barrier designs to mitigate near-road air pollution impacts. Sci Total Environ 541:920–927. https://doi.org/10.1016/j.scitotenv.2015.09.067

Venkataramanan V, Packman AI, Peters DR, Lopez D, McCuskey DJ, McDonald RI, Miller WM, Young SL (2019) A systematic review of the human health and social well-being outcomes of green infrastructure for stormwater and flood management. J Environ Manage 246:868–880. https://doi.org/10.1016/j.jenvman.2019.05.028

Vos PEJ, Maiheu B, Vankerkom J, Janssen S (2013) Improving local air quality in cities: to tree or not to tree? Environ Pollut 183:113–122. https://doi.org/10.1016/j.envpol.2012.10.021

Wang Y, Akbari H (2016) Analysis of urban heat island phenomenon and mitigation solutions evaluation for Montreal. Sustain Cities Soc 26:438–446

WHO (2016) Urban green spaces and health (No. WHO/EURO: 2016-3352-43111-60341). World Health Organization. Regional Office for Europe. https://apps.who.int/iris/handle/10665/345751. Accessed on June 2022

Yang J, Yu Q, Gong P (2008) Quantifying air pollution removal by green roofs in Chicago. Atmos Environ 42:7266–7273. https://doi.org/10.1016/j.atmosenv.2008.07.003

Yang F, Lau SSY, Qian F (2011) Thermal comfort effects of urban design strategies in high-rise urban environments in a sub-tropical climate. Archit Sci Rev 54:285–304. https://doi.org/10.1080/00038628.2011.613646

Yin S, Shen Z, Zhou ZX, Che S, Wang W (2011) Quantifying air pollution attenuation within urban parks: an experimental approach in Shanghai, China. Environ Pollut 159:2155–2163. https://doi.org/10.1016/j.envpol.2011.03.009

Zhou B, Rybski D, Kropp JP (2017) The role of city size and urban form in the surface urban heat island. Sci Rep 7(1):1–9

Zinzi M, Agnoli S (2012) Cool and green roofs. An energy and comfort comparison between passive cooling and mitigation urban heat island techniques for residential buildings in the Mediterranean region. Energ Build 55:66–76. https://doi.org/10.1016/j.enbuild.2011.09.024

**Prashant Kumar** is a Chair Professor in Air Quality and Health, the founding Director of the University of Surrey's Global Centre for Clean Air Research (GCARE) and Co-Director of the Institute for Sustanability. His cross-disciplinary research is focused on the interfaces of clean air engineering, human health and smart/sustainable living in cities/megacities. He contributes this chapter with Sisay Debele (research fellow at the University of Surrey's Global Centre for Clean Air Research), Arvind Tiwari (research fellow at the University of Surrey's Global Centre for Clean Air Research), Abhijith Kooloth Valappil (research fellow at the University of Surrey's Global Centre for Clean Air Research), Jeetendra Sahani (research fellow at the University of Surrey' Global Centre for Clean Air Research), and Soheila Khalili (PhD candidate at the University of Surrey's Global Centre for Clean Air Research).

# Urban Flood Resilience

**Dawei Han**

**Abstract** Floods threaten cities in various forms whose magnitude and frequency are likely exacerbated by climate change and increasing urbanisation. This chapter explains their causes, consequences, observation methods, hazard modelling techniques, and mitigation infrastructures relevant to urban flood resilience. Effective decision support tools are important in gathering information to facilitate transparent decision-making process in order to evaluate and plan urban resilience to flooding.

**Keywords** Urban flood resilience · Flood modelling · Hazard modelling · Flood · Policy

## 1 Introduction

Floods threaten cities in various forms (pluvial, fluvial, coastal, groundwater, etc.). Climate change and increasing urbanisation will likely exacerbate the magnitude and frequency of future urban flooding. Urban flood hazards may occur individually or jointly (e.g. compounding flooding). In the past, flood monitoring and modelling efforts have historically focused on fluvial and coastal flooding. Urban pluvial flooding is relatively less observed and modelled due to complex urban terrains and fast-changing storms. As the main driving force in urban pluvial flooding, rainfall over cities provides key information for flood hazard modelling. Monitoring rainfall can be carried out using rain gauges and weather radars either individually or jointly. It is important to understand their respective strengths and weaknesses and how to merge them to achieve the improved accuracy. Modelling urban floods should include flood disaster-causing factors, flood disaster-prone environments, and flood disaster-hit bodies. Agent-based models are increasingly developed in modelling the interaction between urban flooding and human society. Flood resilience depends on modern infrastructures (includes grey, green, and blue types). A major obstacle in wide adoptions of integrated grey/green/blue infrastructures in flooding resilience

D. Han (✉)
School of Civil, Aerospace and Design Engineering, University of Bristol, Bristol, UK
e-mail: d.han@bristol.ac.uk

 
R. Yao (ed.), *Resilient Urban Environments*, Cities and Nature,
https://doi.org/10.1007/978-3-031-55482-7_12

is a lack of agreed consistent standards of methodologies and adequate metrics for their performance evaluation. Resilience represents the capacity of a system to absorb disturbance without substantially changing its structure or functions. Understanding the overall system is essential to evaluating its resilience. Urban flood resilience management seeks both to reduce flood hazards to prevent the system shifting to an undesirable state and to keep the system in a preferred state. The planning of a flood-wise city requires understanding of the potential consequences from flooding hazard, designing for hybrid infrastructures to reduce these consequences, and preparing the local community to withstand potential flooding impact. It is a huge challenge to integrate various infrastructures' co-benefits into economic appraisal of mitigation measures and how to address resilience from a system-of-system perspective.

Due to the complexity and ever-changing nature of urban flood resilience, it is not possible to provide an exhaustive coverage of all aspects of the topic. Instead, key concepts of urban flood resilience are presented, and the relevant studies carried out from the researchers and collaborators in the author's research group are used to illustrate the typical issues in this ever-changing and challenging field. Due to the limited room in a book chapter, only concise descriptions are provided as signposting to the relevant links for the interested readers to explore further. Clearly, there are still more questions than answers despite of the huge progresses made by the research community.

## 2 Urban Flooding Hazards

Flooding is an overflow of water that submerges land that is usually dry. It is a common, widespread, and frequent natural hazard worldwide that has potential to cause severe socio-economic loss and environmental damages. Flood risk is particularly severe in urban areas due to the change of land use/land cover (e.g. increase of impervious areas) and the high concentration of population. Flooding may occur as an overflow of water from water bodies (river, lake, or ocean), or it may occur due to an accumulation of rainwater or groundwater locally. Urban flood hazards include the following types which may act individually or jointly (Han 2011):

Fluvial flood (i.e. river flood): it occurs along major rivers and in lower tributaries of large catchments. Since such a flood takes a great deal of time to build up to its high stage, adequate warning time is usually available for local communities to make necessary preparations. As a result, river floods are a major threat to properties instead of human lives. A recent well-known urban flooding caused by fluvial flooding occurred in August 2002 in Central Europe (Kundzewicz et al. 2013). The flood reached a magnitude of unprecedented heights on record. Prague received significant damage from this extreme flood (the worst ever to hit the city with the flow of River Vltava reaching 5300 $m^3/s$). In particular, there was significant damage to the Prague Metro subway system. However, due to the sufficient early warning of high water levels, the evacuations were effective in minimising the human casualties.

Urban pluvial flood: urbanisation reduces vegetation cover in the city and replaces the original land cover with concrete, asphalt, and roofing materials. The increased impervious areas decrease rainwater infiltration into the soil profile, causing increased runoff into urban drainage systems. When increased runoff is greater than the drainage capacity, urban flooding occurs. In July 2021, London was flooded twice by extreme rainfall (a month's rain falling on some parts of the city in just one hour). About 1000 buildings (include homes, schools, businesses, hospitals, and tube stations) were flooded. The severity and scale of this urban pluvial flood exceeded the design limits of the local drainage system (JBA 2021).

Groundwater flood: when the groundwater table rises (due to recent high rainfall), groundwater flooding may occur. Increased groundwater may rise through building floors or basement rooms. In contrast to river flooding or urban pluvial flooding (appearing suddenly), groundwater flooding is much slower (e.g. in days, weeks or even months after heavy or prolonged rainfall). Groundwater flooding tends to occur in low-lying areas because during very wet periods, there is additional groundwater flowing towards these areas causing the groundwater table to rise up to the surface. A challenge in dealing with groundwater flooding is that many traditional flood mitigation measures may not be effective (because groundwater can rise up through the building floors over a long time). Currently, groundwater flooding is estimated to affect a hundred thousand properties in the UK (Environment Agency 2022).

Coastal flood: many cities have been developed along costal zones. When seawater levels rise due to storm surge, high tides, etc., inundated coastal areas may suffer from coastal flooding. The seawater can flood coastal cities via direct flooding or overtopping/breaching of a barrier. Coastal flooding may be exacerbated by human influence through changing the coastal environment. Climate change (causing sea level to rise) and human population movements along costal lines increase the risk of coastal flooding in future (Climate Change Post 2022). A devastating coastal urban flooding occurred in New Orleans by Hurricane Katrina in 2005. As a Category 5 Atlantic hurricane, it had caused over 1,800 fatalities and $125 billion in damage. About 80% of the city was inundated for weeks. It had damaged most of the city's communication and transportation facilities, causing thousands of people stranded with little access to basic necessities (Ripley 2005).

Flood by dam failures: cities downstream of dams may be threatened by flooding if the dams fail due to high water level, poor maintenance, earthquake, landslide, etc. A well-known case is the Vajont Dam in Italy in 1963. With a height of 262 m, it was one of the tallest dams in the world. On 9 October 1963, a megatsunami caused by a landslide in the reservoir pushed 50,000,000 cubic metres of water over the dam crest in a wave of 250 m. This brought massive flooding and destruction to the valley below, leading to the destruction of several towns and villages with fatalities between 1900 and 2500 (Environment and Society 2022).

## 3 Urban Rainfall Monitoring and Flood Modelling

### *3.1 Urban Rainfall Monitoring*

To effectively manage urban flooding risk and improve flooding resilience, it is important to monitor urban flooding. In the past, flood monitoring and modelling efforts have historically focused on fluvial and coastal flooding. Urban pluvial flooding is relatively less observed and modelled due to complex urban terrains and fast-changing storms. In view of the context of climate change, urban pluvial flooding is most likely to increase in severity and frequency (IPCC 2021). Although usually associated with shallow water depths, the impact of urban pluvial floods can be far-reaching and widespread. Therefore, the focus of this section is on monitoring rainfall over cities because it is the main driving force in urban pluvial flooding.

There are three main approaches in observing rainfall in urban areas: rain gauge, weather radar, and satellite. Rain gauge is an instrument used to gather and measure the amount of precipitation over a period of time at a single location (a direct rainfall observation method). Weather radar is a type of remote sensing used to observe rainfall using microwave radiation. Modern weather radars are mostly pulse-Doppler radars, capable of detecting both the motion of rain droplets and the intensity of the rainfall. Weather radar data can be analysed to determine the structure of storms and their potential to cause urban pluvial flooding. Similar to ground-based weather radars, satellites in space use remote sensing to observe rainfall using a variety of sensors (e.g. active microwave sensor (radar), passive microwave sensor (Imager), and visible and infrared scanner). Since satellite rainfall data have coarser spatial and temporal resolutions in relation to ground-based weather radars, they are less effective in urban rainfall observations due to the demand of high spatial and temporal resolutions needed in urban flood modelling (Srivastava et al. 2022).

The use of rain gauges is the most conventional method employed in measuring rainfall. Since a single rain gauge only measures point rainfall, an efficiently and appropriately designed rain gauge network is needed to capture the maximum rainfall information with a minimum number of gauges (therefore, to reduce the cost of operating and maintaining rain gauge networks). Currently, there are still knowledge gaps in rain gauge network design due to the complexity of the problem. Design requirements include working out the number of gauges and their locations, given the frequency in time of sampling with the minimised uncertainty of rainfall estimation. The high variability of rainfall in time and space over cities is a significant and difficult issue to address. In comparison with the conventional approach to analyse the limited rainfall-related information (e.g. rainfall spatial correlation and patterns) over a study area, the information from remotely sensed rainfall measurements may be utilised to aid rain gauge network design which can be more direct and efficient. Weather radars are extensively installed around the world. Although weather radar rainfall products cannot replace rain gauge measurements at present, the radar measurements are an ideal foundation to provide valuable information into rainfall patterns and local topographic influences on rainfall. Dai et al. (2016) have explored the new approach using the weather radar data in rain gauge network design. The rain gauge network design is based on a combination of principal component analysis (PCA) and variable selection criteria using a weather radar rainfall dataset to provide both the optimum rain gauge density and rain gauge locations. The proposed methodology has been applied to over hundred different storm events measured by a weather radar network and assessed against an existing dense rain gauge network in England. The scheme offers a novel approach into rain gauge network design and has potential to be explored and implemented in other city regions.

Since radars and rain gauges measure rainfall differently, there exist potentially large discrepancies between them which hamper their effective use in urban flood modelling. Radar-gauge rainfall discrepancies are usually considered to be mainly from radar rainfall measurements. This ignores the fact that radar observes rainfall aloft, but a rain gauge measures rainfall on the ground. Raindrops observed aloft by weather radars are conventionally assumed to fall vertically to the ground. This assumption obviously does not stand because raindrop falling location changes due to wind drift and raindrop size also changes (e.g. evaporation, condensation, etc.). However, both effects are usually ignored in radar and rain gauge data analysis. To overcome such a problem, Dai et al. (2019) propose a formulated scheme to numerically simulate both raindrop drift and evaporation in the air to reduce the uncertainties of radar rainfall estimation. The Weather Research and Forecasting (WRF) model is used to produce high-resolution three-dimensional atmospheric fields. The adjustment method is explored and implemented using the Hameldon Hill radar in England. The first experiment includes only raindrop drift, the second one considers only evaporation, and the last includes both aspects (Fig. 12.1). The results demonstrate that a large drift for small raindrops impacted by the wind advection, and noticeable loss of rainfall is observed from raindrop evaporation (Fig. 12.2). Overall, all the three considerations improved the radar-gauge correlation and reduced their discrepancy significantly with the best results achieved by considering both raindrop drift and evaporation. The results of this work demonstrate that the radar rainfall estimated error induced by the two microphysical processes (evaporation and raindrop drift) cannot be ignored. The study provides useful contributions to the improvement of quantitative precipitation estimation from weather radar and enhances our understanding on precipitation processes jointly observed by rain gauges and weather radars.

Since both weather radar and rain gauges have their strengths and weaknesses in measuring rainfall over cities, merging rain gauge and radar data has significant potential to improve the accuracy of precipitation estimation, which provides better rainfall data for urban flood modelling. This is particularly important for small urban areas where no rain gauges are located within. Since the rain gauge network around the ungauged urban catchment is fixed, a relevant question is how to decide the optimal merging area that could produce the best rainfall estimation inside the urban catchment. To solve such a problem, the radar-gauge merging has been incrementally performed by gradually increasing the distance from the centre of the study area (Fadhel et al. 2016). The adaptive merging scheme is applied to a small urban catchment in west Yorkshire, England. The performance of the scheme is evaluated using

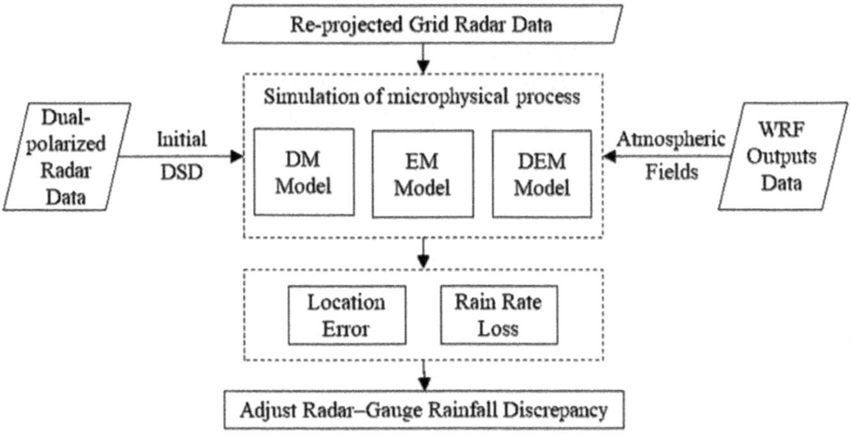


**Fig. 12.1** Adjustment procedure to resolve radar-gauge rainfall discrepancy using the WRF model and dual-polarisation radar (Dai et al. 2019) (*Note* The drift model (DM) omits E, which indicates that raindrop evaporation is ignored. Similarly, the evaporation model (EM) omits the raindrop motion. The drift-evaporation integrated model (DEM) retains all items and thus represents both effects)

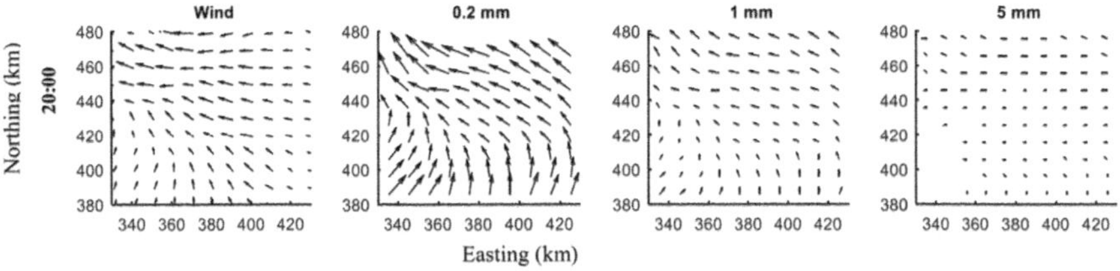


**Fig. 12.2** Maps of ground wind and raindrop drifts for sizes 0.2, 1, and 5 mm for different time steps for the 8 May 2015 event (Dai et al. 2019)

the experimental rain gauges temporarily installed inside the study area. The result demonstrates that an optimum radar-gauge merging area does exit, and consequently there is an optimum number of rain gauges for the best merged rainfall data. This study is important for urban flooding analysis because it provides much improved rainfall estimation for ungauged urban catchments.

## 3.2 *Urban Flood Modelling*

Heavy rain and other factors on the city may cause urban flood disasters in certain disaster-prone environments. Modelling urban floods should include the following parts: flood disaster-causing factors, flood disaster-prone environments, and flood disaster-hit bodies (Zhu et al. 2019). In term of flood disaster-causing factor analysis, modelling of fluvial and coastal floods has been well studied in past decades. These models greatly benefit from the increasing quality of topographic data and hydrological data. However, urban areas generally have complex and irregular topography with buildings, drainage network, and many other critical infrastructures. They pose substantial modelling challenges on the dynamics and nonlinear interactions of hydrological, hydrodynamic, and hydromorphological processes. The accuracy and efficiency of numerical flood models are often in conflict with each other. For the fast-simplified models, some key information is lost in its sacrifice for increasing speed (leading to less accuracy in modelling). On the other hand, complex models provide more accurate flood information but with the demand of more detailed data input and higher computational cost (an important limitation in real-time flood prediction). With the recent advances in computational power and the increase in the availability of data, the quality and complexity of these mathematical models have steadily increased. There are a wide variety of available hydrological and hydrodynamic models capable of simulating fluvial or pluvial flooding in urban environment, including MOUSE (Lindberg et al. 1989), HEC-HMS (Charley et al. 1995), MIKE-11 (Havnø et al. 1995), MIKE-SHE (DHI 2000), LISFLOOD (Bates and De Roo 2000), the Storm Water Management Model (SWMM) (Rossman 2015), HEC-RAS (Patel et al. 2017), etc. A major bottleneck for improving urban flood modelling is the limited quality of rainfall data as mentioned in the last section.

An important part of modelling urban floods is the exposure linked with the distribution of population and socio-economic resources. They are the key factors for evaluating urban flood impact due to the dynamic evolution of disaster-hit bodies and urban floods. Exposure to urban flooding at a certain time or period varies with the changes in the inundation areas and disaster-hit bodies. Social simulation techniques have been successfully used to represent the behaviours of organisations and humans. They include Bayesian networks, system dynamics, microsimulationm, event and fault trees, cellular automata, and agent-based models (ABMs). ABMs have been gradually introduced to the field of flood disaster risk modelling as they can simulate population movements as well as interactions among the population, hazard factors, and disaster-prone environments (Zhuo and Han 2020). A recent study (Zhu et al. 2019) has demonstrated a novel method based on the LISFLOOD-FP model and an ABM to simulate the urban flood inundation and exposure of urban populations, buildings, and infrastructures. The modelling was carried out in simulating flooding under varying conditions. Several scenarios (including flooding progression and various responses of local residents) were considered. Additionally, improvement to the microexposure analysis was carried out by incorporating dynamic features of the real-world situations. This technique was subsequently applied to an urban area (Lishui City, China) (Fig. 12.3). Such exposure simulation is useful in estimating disaster vulnerability and assessing losses. The results with different scenarios are useful to relevant government agencies in assessing risk, issuing early warnings, and planning mitigation responses to future urban flooding disasters (Fig. 12.4). As a result, effective mitigation measures can be taken to minimise flooding casualties by considering the dynamic distribution of the population.

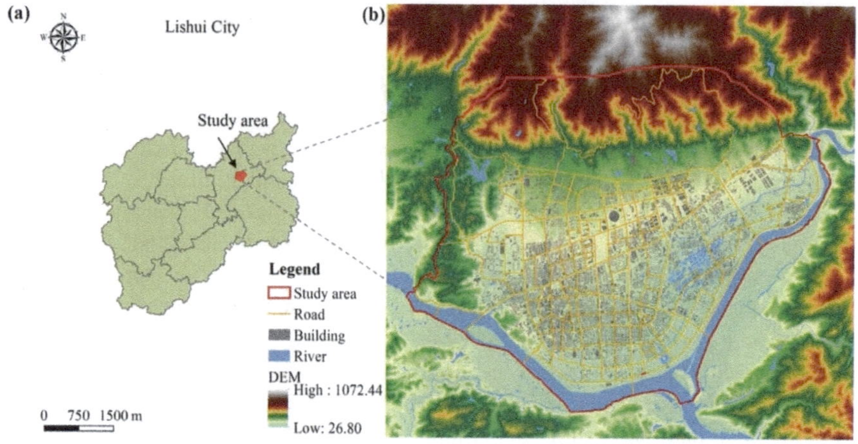


**Fig. 12.3** Location of the study area (**a**) and a digital elevation model indicating the specific details of the study area (**b**) (Zhuo et al. 2019)

(a) Road exposure (pluvial flood, T = 18:30) (b) Road exposure (fluvial flood, T = 18:30)

(c) Building exposure (pluvial flood, T = 18:30) (d) Building exposure (fluvial flood, T = 18:30)

**Fig. 12.4** Map of road and building exposures. T means time here

It is very challenging to verify the proposed modelling method in cities because human behaviour and psychological processes are difficult to observe. In the study by Zhu et al. (2019), the proposed method was verified indirectly. The traffic information (e.g. records from private and public transport) is collected and used to link with the simulated population movement results. Actual water inundation points were compared with the simulated water inundation results. Major uncertainties are related with the use and design of the ABM. The main issue is connected with the differences in the responses of local communities to flood disasters in the simulated scenario. Simplification of the behaviour patterns and disaster responses of local residents can result in discrepancies between the simulation and reality. Another limitation is the exclusion of the population movement beyond the city boundary (i.e. the exchange between the city and the outside). Moreover, the simulated water inundation points cannot be completely verified with the observed ones due to inadequacy in observations. There is a clear need for more abundant and accurate flooding hazard data to fine-tune the urban flooding simulations.

## 4 Flood-Wise Green, Blue, and Grey Infrastructure

An infrastructure consists of a network of man-made/natural systems that function cooperatively and synergistically to produce and distribute essential goods or services for human society and environment. Modern infrastructure related to flood resilience includes grey, green, and blue types. Grey infrastructure refers to the engineered infrastructure (e.g. drainage network, treatment plants, embankment, channelisation, storage tanks and reservoirs, etc.). Nowadays, nature-based solutions are increasingly introduced into urban flood infrastructures. Nature can be used to provide important complementary protection for local communities against flooding threats. When nature is harnessed as an infrastructural system, it is called "green infrastructure". Green infrastructure is at all scales. It facilitates stormwater management, as well as serving to provide an ecological service for social, economic, and environmental health of the surroundings, helping to improve air, soil, and water quality (CMCC 2021). Green infrastructure strategically uses networks of natural lands, landscapes, and other open spaces to conserve ecosystem functions and provide associated benefits to urban flooding resilience. Closely related to green infrastructure, blue infrastructure refers to urban infrastructure dealing with water (e.g. rivers, streams, ponds, lakes, etc.), which can be used to connect the benefits of green and grey infrastructure. Blue infrastructure aims to mimic green infrastructure's principles (e.g. following natural drainage paths), which may serve as a grey infrastructure's pre-treatment system. Rainwater is captured and stored where it falls, rather than being transported to the draining facility. Blue infrastructure can support unique aquatic biodiversity in urban areas. Due to their closeness, blue infrastructure and green infrastructure are commonly associated in urban environments and are referred together as "blue-green infrastructure" (such as rain gardens or reed beds). Coastal cities may also utilise pre-existing features of the coastline in flooding resilience design. Grey infrastructure (e.g. harbours, quays, piers, and other extensions of the urban environment) is also often integrated with the green infrastructure (the local marine environment) (Alberta Society 2016). In addition to flooding resilience, multifunctionality is at the core of blue-green infrastructures which are designed to reduce a variety of social and environmental vulnerabilities, as well as bringing co-benefits at the same time (e.g. improving human health and well-being, mitigating climate change, and providing business opportunities and jobs). The main components serve stormwater management, increasing biodiversity, reduction of heat stress, better air quality, food production, clean water, sustainable energy, and healthy soils. Anthropocentric functions include increased quality of life through recreation and the provision of shade and shelter in urban areas.

Alberta Society (2016) has promoted urban green–blue infrastructures with useful effects: (1) Urban forests: they are an important component of urban green infrastructure systems. Urban forests use appropriate tree and vegetation species to reduce/slow down the surface runoff and increase local soil water storage. Therefore, there is less need of maintenance and irrigation. Diversity of tree species has been considered in designing urban forests to avoid monocultures, which makes the developed urban forests more durable and resilient to pests and other harms; (2) Constructed wetlands: they work as a bio-filtration system (either as a subsurface flow system or a free water surface system) which contains wetland vegetation. They are mostly built on uplands and floodplains to avoid interruption or damage to natural wetlands and other aquatic resources. Appropriate planning and operations can help minimise possible harms to the wetlands (e.g. alteration of natural hydrology and introduction of invasive species) and increase flooding mitigation measures (flood retention); (3) Green and blue roofs: rainwater is detained locally to reduce its impact on the runoff drainage system. They also improve water and air qualities. The plants and soil provide more green space and insulation on roofs (hence reducing energy cost); and (4) Rain gardens: they are a form of stormwater management by capturing rainwater using shallow depressed areas in the landscape (covered with shrubs and plants) to collect rainwater from surrounding impervious surfaces and allow for the water to slowly infiltrate into the ground.

A major obstacle in wide adoptions of green/blue infrastructures in flooding resilience is a lack of agreed consistent standards of methodologies and adequate metrics for their performance evaluation. It is particularly challenging to evaluate the upscaled and interconnected infrastructure subsystems. There is a need to standardise cost-effective assessment and valuation in general economic and other assessments systems to make blue-green infrastructures appeal to funders. In comparison with green–blue infrastructures, grey infrastructure currently has clearer quantified benefit, asset life, depreciation, and return on investment which are needed for large engineering-focused flooding resilience infrastructure projects. Green–blue infrastructures present challenges in terms of measuring return on investment, risk management, and effectiveness in urban areas. They usually do not fit conventional flood risk management models. Therefore, there are inadequate standards or building/urban codes to govern how such projects should be implemented. As a largely untested concept, green–blue infrastructures also suffer from scientific knowledge gaps, socio-political uncertainties, and decision-making difficulties. It should be noted that green–blue infrastructures are not a substitute for grey infrastructure and vice versa because there will always be a requirement for grey infrastructure based on hard engineering. However, depending on the specific localised conditions and objectives, green–blue infrastructures can complement grey infrastructure in a hybrid approach to help increase flooding resilience in cities for the future. Nevertheless, various attempts have already been made to fill the gaps in this aspect using the following tools: (1) Green Infrastructure Valuation Toolkit: www.merseyforest.org.uk/services/gi-val. The Natural Economy Northwest programme (UK) and partners have developed a framework for assessing the potential economic and wider returns from investment in environmental improvements and green infrastructure; and (2) Green Value Calculator: greenvalues.cnt.org/index.php. This calculator is developed by the Centre for Neighbourhood Technology (CNT) in the US for evaluating performance, costs, and benefits of green infrastructure and low impact development solutions for stormwater management.

## 5 Assessing and Planning Urban Resilience to Flooding

A system is defined as a set of interacting components forming a complex whole. A system is delineated by its boundaries, surrounded by its environment, and characterised by its structure and functions. Systems may change from one state to another, which may result in large, abrupt, long-lasting changes to their structure or functions. Resilience represents the capacity of a system to absorb disturbance without substantially changing its structure or functions. Understanding the overall system is essential to evaluating its resilience. Urban flood resilience management seeks both to reduce flood hazards to prevent the system shifting to an undesirable state and to keep the system in a preferred state. Mao et al. (2016) have identified three types of human–water couplings with their related resilience framings: (1) the water subsystem, dealing with hydrological flood resilience to anthropogenic hazards; (2) the human subsystem, foregrounding social resilience to hydrological flood hazards; and (3) the coupled human–water system, exhibiting socio-hydrological resilience. These three subsystems may help the clarification and evaluation of different urban flooding resilience management challenges. The first two subsystems address hydrological and social states, while the third one emphasises the feedbacks and interactions between human and water components within complex flood resilience systems subject to internal or external disturbances. While the former two focus on intrinsic hazard–subsystem relations, the latter covers these subsystem relations and the broader and more iterative interplay between them. Coevolution and feedbacks of coupled human–water systems play an important role in defining a set of systematic properties (absorptive, adaptive, and transformative capacities). Assessing and planning urban flooding resilience in a socio-hydrological context is challenging because of different framings of water-related resilience, including hydrological flood resilience to anthropogenic disturbances, social resilience to hydrological flood disasters, and socio-hydrological resilience. Currently, there is limited agreement in characterising and assessing urban flooding resilience. Hence there is a need to review, compare, and classify existing resilience indicators in order to propose new quantification and assessment methods for different resilience framings, or even develop mathematical tools to quantitatively describe the resilient processes of the capacities (Mao et al. 2016). Urban resilience to flooding covers multiple dimensions, including physical and environmental factors such as extreme precipitation and land use/land cover, socio-economic factors such as governance regimes, and technical factors such as information tools and early warning systems for the local communities. A holistic evaluation of resilience requires addressing the entire flooding disaster process from preparedness and response to mitigation and recovery. There is a need to consider multiple aspects of systemic resilience, including absorptive, adaptive, and transformative capacities. It is highly challenging to quantify the complex and multifaceted resilience notions to design/plan strategies for future flooding resilience management and governance. The recent advances in systems analysis, flood simulation modelling, risk assessment, remote sensing, and data science (include big data) have offered new opportunities for assessing urban resilience to flooding under change (Chen and Han 2016). There is increasing research effort on identifying interdependencies between different infrastructure systems, which may lead to scenarios that have the potential to be critical from a system-of-systems perspective. This can then be used for designing/planning/improving appropriate measures for urban flooding resilience. Infrastructure improvements should form key part of a city-wide resilience strategy, to ensure that existing and newly constructed lifelines are able to cope in the event of future floods (Pregnolato et al. 2020).

Since a sound understanding of the overall system is an essential step to evaluating its resilience, effective and powerful decision support tools are needed to address the aforementioned three subsystems for assessing and planning urban resilience to flooding. Two attempts are introduced here. Dai et al (2020) developed a hazard-human coupled model (HazardCM) to assess city dynamic exposure to rainfall-triggered natural hazards (covering the aforementioned Subsystem 1 and Subsystem 2). It includes four components: urban environment module, agent-based model, city–hazard coupler, and dynamic exposure assessment (Fig. 12.5). Rainfall-triggered urban flood hazards under extreme hydrometeorological events are modelled in Lishui City of China. Scenarios covering different magnitudes, timings and locations, and return periods of flood hazards are investigated to derive the spatial distribution and evolution of human exposure. It offers a new way to analyse flooding exposure's space–time characteristics while considering the dynamic nature of both humans and floods. In term of Subsystem 3, the city itself is an integrated and complex system consisting of heterogeneous and interconnected subsystems pertaining to both physical and social structures (among them humans, infrastructures, organisations, and economy), which are connected by nonlinear, multiple interactions (network examples are shown in Fig. 12.6). To deal with such a complex system, an integrated decision support platform for city-regional resilience development planning has been explored by Resilience Brokers using a platform called "resilience.io" (resilience-brokers.org/resilience-io-2/) which acts as a core to aggregate information under the guidance of a holistic thinking framework. Such a system encourages participation of stakeholders and enables transparent decision-making to increase the urban flood resilience. Once the exposure and vulnerabilities are assessed (Subsystems 1 and 2), planning should embrace a holistic thinking framework to understand the interconnections between the key actors and sociotechnical processes and their interactions with nature which have shaped the city's flooding disaster risks to derive future scenarios. Such a system simulator as a decision support tool is key in linking all the drivers so that multi-sector resilience development planning will be assessed holistically in relation to urban flooding. It has potential to facilitate the city region to make choices between possible alternatives of adaptation and mitigation measures related to flood hazards under different scenarios, and consequently sound decisions can be made. A community cockpit encourages the transparent decision-making by active participation of the concerned community to view alternative planning options and provide feedback to the decision makers.

**Fig. 12.5** Model structure (Dai et al. 2020)

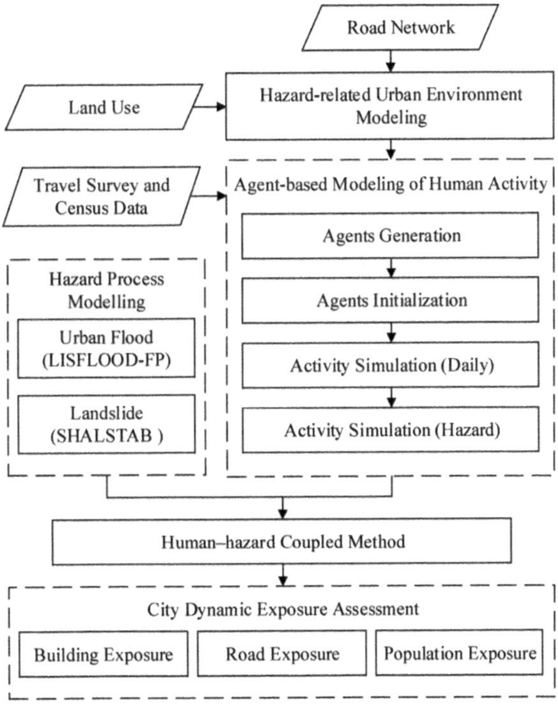

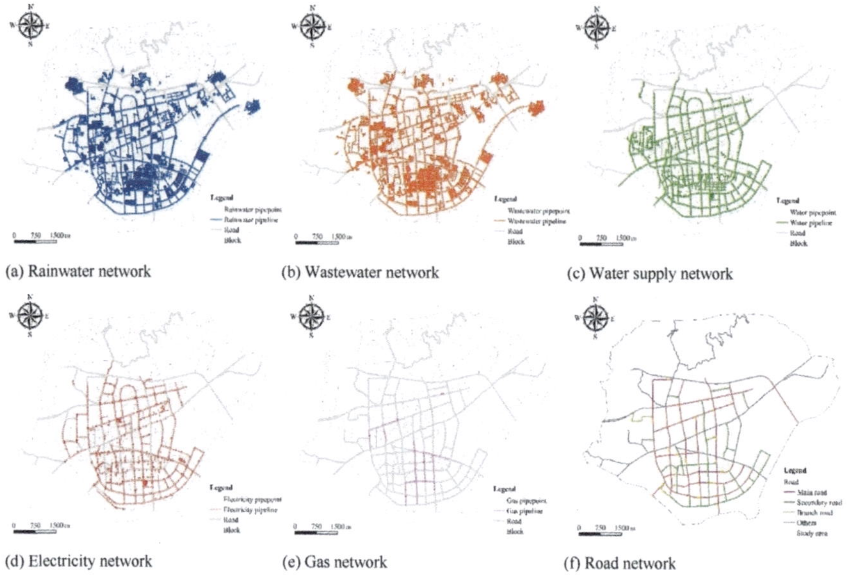


**Fig. 12.6** Map of the networks (Dai et al. 2020)

# 6 Summary

Flooding risk to cities has to be managed through resilient and sustainable planning, especially in fast-developing areas. There is a need to develop an integrated and widely applicable framework for assessing urban resilience to flooding. The planning of a flood-wise city requires understanding of the potential consequences from flooding hazard, designing for hybrid infrastructures (grey, green and blue) to reduce these consequences, and preparing the local community to withstand potential flooding impact. It is a huge challenge to integrate various infrastructures' co-benefits into economic appraisal of mitigation measures and how to address resilience from a system-of-system perspective. Resilience concept should be integrated with technical methods (e.g. high-resolution flood simulation and flood risk assessment) in order to understand the impact of flooding on various urban sectors. Integrated decision support systems are essential for efficiently and effectively assessing and planning city flooding resilience.

# References

Alberta Society (2016) Introduction to green infrastructure and grey infrastructure. https://albertawater.com/green-vs-grey-infrastructure/

Bates PD, De Roo APJ (2000) A simple raster-based model for flood inundation simulation. J Hydrol 236:54–77. https://doi.org/10.1016/S0022-1694(00)00278-X,2000

Chen Y, Han D (2016) Big data and hydroinformatics. J Hydroinf. https://doi.org/10.2166/hydro.2016.180
Climate Change Post (2022) Coastal flood risk United Kingdom. https://www.climatechangepost.com/united-kingdom/coastal-floods/
CMCC (2021) Foresight, Nature-Based Solutions. https://www.climateforesight.eu/seeds/nature-based-solutions/
Charley W, Pabst A, Peters J (1995) The hydrologic modeling system (HEC-HMS): design and development issues, Technical Paper No. 149, Hydrological Engineering Center, US Army Corps of Engineers, USA
Dai Q, Bray M, Zhuo L, Islam T, Han D (2016) A scheme for raingauge network design based on remotely-sensed rainfall measurements. J Hydrometeorol. https://doi.org/10.1175/JHM-D-16-0136.1
Dai Q, Yang Q, Han D, Rico-Ramirez MA, Zhang S (2019) Adjustment of radar-gauge rainfall discrepancy due to raindrop drift and evaporation using the weather research and forecasting model and dual-polarization radar. Water Resour Res. https://doi.org/10.1029/2019WR025517
Dai Q, Zhu X, Zhuo L, Han D, Liu Z, Zhang S (2020) A hazard-human coupled model (HazardCM) to assess city dynamic exposure to rainfall-triggered natural hazards. Environ Model Softw. https://doi.org/10.1016/j.envsoft.2020.104684
DHI—Danish Hydraulic Institute (2000) MIKE SHE Water movement user manual. DHI Water & Environment, Denmark
Environment Agency (2022) Flooding from groundwater. https://www.gov.uk/government/publications/flooding-from-groundwater
Environment and Society (2022) Expecting disaster: the 1963 landslide of the Vajont Dam. https://www.environmentandsociety.org/arcadia/expecting-disaster-1963-landslide-vajont-dam
Fadhel S, Rico-Ramirez M, Han D (2016) Exploration of an adaptive merging scheme for optimal precipitation estimation over ungauged urban catchment. J Hydroinformatics. https://doi.org/10.2166/hydro.2016.022
Havnø K, Madsen MN, Dørge J (1995) MIKE 11 – a generalized river modelling package. Computer models of watershed hydrology. Water Resources Publications, Colorado, pp 733–782
Han D (2011) Flood risk assessment and management. eISBN: 978-1-60805-047-5. Bentham Science Publishers
IPCC (2021) IPCC report six. https://www.ipcc.ch/report/sixth-assessment-report-working-group-i/
JBA (2021) A retrospective look at London surface water flash floods. https://www.jbarisk.com/flood-services/event-response/a-retrospective-look-at-summer-2021-london-flash-floods/
Kundzewicz ZW, Pińskwar I, Brakenridge GR (2013) Large floods in Europe, 1985–2009. Hydrol Sci J 58(1):1–7
Lindberg S, Nielsen JB, Carr R (1989) An integrated PC modelling system for hydraulic analysis of drainage systems, in: Watercomp'89: the first Australasian conference on technical computing in the water industry. Institution of Engineers, Australia, p 127
Patel DP, Ramirez JA, Srivastava PK, Bray M, Han D (2017) Assessment of flood inundation mapping of Surat city by coupled 1D/2D hydrodynamic modeling: a case application of the new HEC-RAS 5. Natural Hazards. https://doi.org/10.1007/s11069-017-2956-6
Pregnolato M, Han D, Lo Jacomo A, De Risi R, Agarwal J, Huang J (2020) Resilient infrastructures for reducing urban flooding risks. In: Water-wise cities and sustainable water systems: concepts, technologies, and applications. https://doi.org/10.2166/9781789060768_0181
Ripley A (2005) Hurricane Katrina: how the coast guard got it right. Time Magazine
Rossman LA (2015) Storm water management model user's manual Version 5.1 EPA-600/R-14/413b[z], National Risk Management Laboratory Laboratory Office of Research and Development US Environmental Protection Agency
Srivastava P, Gupta DK, Islam T, Han D, Prasad R (2022) Radar remote sensing: applications and challenges—earth observation. Elsevier Science Publishing Co Inc

Zhu X, Dai Q, Han D, Zhuo L, Zhu S, Zhang S (2019) Modeling the high-resolution dynamic exposure to flooding in a city region. Hydrol Earth Syst Sci (HESS). https://doi.org/10.5194/hess-23-3353-2019

Zhuo L, Han D (2020) Agent-based modelling and flood risk management: a compendious literature review, 2020. J Hydrol. https://doi.org/10.1016/j.jhydrol.2020.125600

# Energy Demand and Cities: Understanding the Complexity of Reduction Potential

**Stefán Thor Smith**

**Abstract** This chapter introduces the approach of complexity science to the study of cities in relation to energy demand. Energy demand reduction has been demonstrated to play a significant role in national emission reduction targets but achieving necessary reductions in energy demand has many (sometimes conflicting) implications for other UN Sustainable Development Goals. Whilst cities, and more broadly urbanisation, have been highlighted as offering potential opportunities for enabling energy demand reduction, they can also represent opportunities for demand growth. As complex adaptive systems, cities are made of many different elements and relations in which many different pathways exist for the emergence of energy demand reduction or growth. The interconnectivity (relations) and heterogeneity (many) in cities are discussed in relation to energy demand in cities along with ideas of causation. These fundamental properties of complex city systems are then briefly discussed in relation to proposed national energy demand reduction scenarios in the UK.

**Keywords** Energy demand · Complex adaptive systems · Interconnectivity · Heterogeneity · Causality

## 1 Introduction

The global success of mitigating climate change is dependent on transitions to net zero carbon energy systems. Whilst this is often understood as the ubiquitous dominance of renewable energy in national and regional energy generation mixes, the past few decades of energy policy have also given consideration to energy demand as part of this 'net zero' transition. For example, Directives of the European Commission have guided national policy on energy efficiency and energy technology adoption targets for its member states for over 20 years: The Energy Performance in Buildings Directive was first published in 2002 and has been instrumental in setting national standards related to energy use in buildings, both for new builds and retrofits across

S. T. Smith (✉)
School of the Built Environment, University of Reading, Reading RG6 6DF, UK
e-mail: s.t.smith@reading.ac.uk

 
R. Yao (ed.), *Resilient Urban Environments*, Cities and Nature,
https://doi.org/10.1007/978-3-031-55482-7_13

Europe. Although there is clear recognition that demand (more specifically a particular form of demand reduction) should be part of all pathways to a net zero carbon society, there is concern amongst the energy demand research community that current thinking on energy demand is too heavily focused on adoption of new technologies for providing efficiencies and electrification of demands leading to demand reduction (Shove 2020).

By definition, efficiency improvements in technology will, in isolation, lead to a reduction in energy consumed by the technology for a given task (or specified service). Put the technology into a social context, however, and the efficiency improvements take on new meaning. The 'optimal' service provision of technologies may now be more easily attained as a result of increased efficiency (e.g. a warmer/cooler or more comfortable home), or efficiency savings in relation to one activity (or service) leading to cost savings that enable new energy demanding activity. This broad understanding of what is known as the 'rebound effect' (Bordón-Lesme et al. 2022; Beppler et al. 2023) highlights that in order to understand potential for energy demand reduction, energy demand itself needs reframing to a collection of connected demands in society for services and goods: Improving 'accessibility' to one service through efficiency measures or new technology can increase capacity for demanding other services/goods or quality of service/goods at the expense of any true reduction in overall energy consumption. This raises questions of how reduction in energy use can fit with socially informed life-style goals, broader issues of basic need and well-being, and infrastructure development for provision of individual and societal needs.

Many of the 17 United Nations (UN) Sustainable Development Goals (SDGs) can be linked to this socio-technical framing of energy demand that help to expose some of the key challenges of energy demand reduction. Unless resource (and specifically energy in this chapter) redistribution strategies are integrated with measures for tackling poverty and health (SDG1, SDG3, SDG6), inequality (SDG5, SDG10), infrastructure growth (SDG9), and infrastructure resilience (SDG9), then addressing these fundamentally important issues of global sustainable development (as part of green growth agendas) suggests an increase in energy consumption will likely surpass any potential efficiency savings associated with current energy demanding activity (Lange and Berner 2022). Yet, when these SDGs are compared alongside sustainable consumption (SDG12), there is a clear, and difficult, challenge laid out in finding congruence between meeting sustainable development targets and energy demand reduction. For example, addressing inequality in relation to energy demanding services is less about demand reduction than it is about increased access to those services and the associated energy consumption. How the SDGs, and tensions therein, are addressed is often put as context specific—global issues to be addressed with locally informed action (Balk et al. 2022; Leavesley et al. 2022). The extent and manifestation of the challenges in meeting the SDGs will vary across the world; the implications to energy demand, therefore, vary across the world. Despite this recognition of global variation in the challenges of sustainable development and energy consumption, there is at least one common (and increasingly so) contextual factor in the manifestation of challenges and solutions: that of urbanisation.

The increasing urbanisation of global populations has put cities and towns under the spotlight for addressing the SDGs and demand reduction as part of SDG12. Research on cities, such as by Bettencourt et al. (2007), suggests that cities act as hubs of intensification ('reactors') in society, where innovation, culture, science, etc., can all be seen to evolve and thrive but also where issues raised in the SDGs are amplified (e.g. inequality, health and well-being, and resource consumption). Efficiencies have also been observed, with a reduction in infrastructure requirements associated with the same level of service provision in less urban regions. These intensifications and efficiencies scale with the size (population density) of the city. Efficiencies in size are seen in a sub-linear scaling of physical infrastructure requirement, and potential social and technological solutions for SDGs and problems associated with SDGs scale super-linearly (Bettencourt et al. 2007).

Applying the 'city as reactor' to the case of energy consumption, urban regions can offer efficiencies in service and resource use through shared demands. However, as part of this intensification and associated scaling, urban regions also act as incubators for new demands for services and associated resources. Energy use can have pathways for reduction (city efficiency) and pathways for growth (city producing new demands through incubation of social and technological innovation). This is important to the whole premise that cities are fundamentally important to meeting sustainability targets; how these sub-linear and super-linear scaling patterns combine, as well as which innovations take hold, will ultimately inform the ability for demand reduction to result from city development.

It is, therefore, important that the configuration of cities—the relationships between the social and technical (as a socio-technical system) that cities are made up of—can be understood in relation to different SDGs (specifically energy demand reduction in this chapter) so as to help guide future incubation/innovation/change of (and within) cities towards sustainable urban development targets (Rutherford and Coutard 2014).

To highlight some of the challenges associated with transitions of urban energy systems, this chapter first poses the idea that it is useful to consider cities as systems. As systems, cities are made of interacting elements, elements that can themselves be seen as subsystems of the whole city; system hierarchy exists in the city. The many different elements and interactions within and between them provide system complexity that makes it difficult (or impossible) to predict how the system will evolve. This does not mean that there is no value in considering future pathways to energy demand reduction in cities but rather that it requires comprehensive study of the many systems to help inform understanding of why, or why not, certain energy reduction pathways can be successful in a city context. This is covered in Sect. 2.

The final section of the chapter (before conclusions) looks at recent work conducted on setting out scenarios of low energy demand futures for the United Kingdom. These scenarios are then reflected on in the light of understanding of cities as complex systems. Whilst the examples used in this chapter are focused on the United Kingdom, the relevance of issues raised are global—although admittedly not comprehensively representative. Whilst many fundamental system elements of cities are universally shared (e.g., cities as hubs of socioeconomic activity and cities comprised of buildings, road infrastructure, etc.) the realisation of those system elements can differ significantly. For example, inequality in energy access is predominantly an economic issue in developed cities whilst in cities with informal settlements this is also about lack of access to energy infrastructure.

Rather than present a (selection of) model(s) used in the study of urban energy demand, this chapter sets out some of the core issues raised through complexity science that should, and do, inform energy demand studies and energy demand modelling activities. It is hoped that the chapter also helps highlight why energy demand (reduction) research is truly a multidisciplinary endeavour.

## 2 Cities as Complex Systems of Energy Demand

### *2.1 City System Complexity*

Batty's work on cities and complexity presents a paradigm of urban spatial form and structures (social, economic, physical) being better recognised as transient manifestations of processes rather than formal, designed, constant entities (Batty 2005). This is a challenge to the use of past theories of urban static spatial structures as containing explanatory power for recent urban form, function and development as these past theories fall short in the face of current observations of urban environments. Failing to account for the influence of dynamics on current form and current functioning of social, economic and physical structures is tantamount to a failure to understand urban development. This idea of dynamics in a city implies change(s) to a process or system under the presence of force(s). This dynamical systems view leads to the structures mentioned above not being described as static objects of description but as processes, systems of interactions, that make up the city.

The idea that cities are best understood as systems of interactions, raises questions such as: What systems make up a city? What are the interactions within and between systems? How are system boundaries to be understood? What ontological framework of systems can be applied to a city? In response to most of these questions, the answer is many. Many systems can be imagined in the city with numerous parts (or subsystems), therefore many opportunities for drawing boundaries, and with several varying interactions and many undetermined consequences. What the systems are is a matter of ontology, where systems might be drawn with distinct boundaries along established academic disciplinary lines (engineered systems vs. social systems vs. economic systems) or, as in the case I put forward in the introduction, a less clearly bounded system of discipline (e.g. where social and technical systems are indistinct elements of the same—socio-technical—system). The idea of many interactions and the indeterminate nature of the city system is what has led many to study cities and energy through complexity science (Allen 1982) and more specifically as complex adaptive systems (CAS) (Batty 2008).

Gell-Mann in (Cowan et al. 1994) and Holland (1992) present the adaptive nature of complex systems as more than just a knowable response to a given condition. Adaptation can take place at different levels in a system, with different temporal characteristics associated with the realisation of the adaptation and the permanence of the adaptation. Holland identified adaptive character in terms of "evolution, aggregate behaviour, and anticipation" and Gell-Mann presented adaptation as hierarchical in terms of complexity: "prescribed" adaptations by the system itself due to held knowledge in the system, "changed" systems due to learning from previously realised prescribed adaptations, and "survival" of systems (or subsystems) that are best able to be self-sustaining under their environmental conditions. Although their terminology differs, Holland and Gell-Mann's conceptual framings of complex adaptation share common ground: Gell-Mann's survival of the fittest is akin to Holland's evolutionary character of adaptation; prescribed system responses are the result of shared, collective, knowledge in the system and is comparable to the simplest example of anticipation (where the anticipated outcome of a response to an environmental/system change results from knowledge conditioning system function); and Holland points out that actual vs. anticipated outcome leads to new knowledge that in turn leads to change in system conditions. This adaptive capacity leads to emergent behaviours and is itself an emergent feature in CAS, setting it apart from simple (or direct) adaptations of systems.

The direct adaptations are simple in nature, part of a system's current functioning with expected response to some expected environmental factors and with limited interaction and learning. An example of energy demand here would be the provision of heat (or cooling) to a building according to local climatic conditions. In isolation this sort of adaptation is straightforward, the heating (cooling) system switches between states of on or off to meet the required environmental conditions as set by the building occupiers or managers. However, in the context of electrification of heat,[1] this otherwise simple adaptation is becoming complex. Increasing electrical (in place of energy vectors such as gas and oil) heat provision places new concerns on grid management that has instigated the development of new (smart) electricity meters, new heating control systems, and new energy tariffs and business models—all designed to control aggregated heat provision. The system of heating control has changed (or rather continues to change) and now raises questions such as: Who can access heating, and when, and for how long? How can issues of fuel poverty and inequality be addressed? What flexibility exists, or can be integrated into, heating provision?

Admittedly, this example of complexity is not singularly bounded to cities, suggesting the complexity is not that of a city but of the features of the energy system. There are, in fact, examples of studies framing complexity of energy demand and energy systems, in terms of technology and social organisations other than the city (see Rylatt et al. 2013; Bale et al. 2015 for example). However, the intensification of demand, and the efficiencies of service provision associated with the city are why this

[1] Electrification of energy demand in the United Kingdom (as in many other countries) is seen as one of the required routes to energy demand reduction and decarbonisation.

is a useful framing for understanding energy demand reduction. The issue of electrification of heat in a city context is subject to greater intensification and variation in demand.

The city as a complex adaptive system recognises two important things for understanding any sustainable development target, and energy demand (reduction) is no exception. These are: (i) that many means diversity (great and small) in properties and interactions that lead to many possible responses (adaptations) to the forces[2] of change and (ii) that the possible qualitative and quantitative properties of any evolution of the city (i.e. of a sustainable development pathway) are many, as well as realisable in multiple ways, and so the emergent features of the system are unpredictable. This leads, therefore, to a recognition that an emergence of energy demand reduction is dependent on the trial of multiple (again, the many) and differing innovations associated with energy demand. These innovations can, for example, be in relation to the social (Wittmayer et al. 2020), to technology (Grubb et al. 2021), and of urban morphology (Khavarian-Garmsir et al. 2023).

To understand the properties of the energy demand reduction challenge in cities requires an understanding of some of the common ideas associated with the explanation of complex systems.

## 2.2 Complexity Properties of Cities and Associated (Energy) Demand

Systems can (in general terms) be described as a set of relations between a set of elements (Ropohl 1999). These relations can be seen as transformational functions to a set of attributes as well as a set of relations between a set of structural components of a system. These functional and structural sets that make up the system can also be structural components of some higher order system, therefore seen as subsystems and part of a hierarchy of systems.

The interactive nature of city systems, the variability in system interactions, and the variation in the system components are fundamental to the idea of system complexity. Capturing types of interactions and types of system component does not tell us how, when, or where they might manifest in a city. Further, it does not tell us how, when, or where adaptations occur that inform the emergent features of a system. These properties[3] of interconnectivity, heterogeneity (i.e., variability) and causality of a complex system need to be studied to understand the emergent features of current and possible (not predicted) future energy demands of a city.

---

[2] Forces here to be interpreted broadly as such things as policy implementation, new technology innovation, environmental change, etc.

[3] Note that this is not an exhaustive exploration of complexity theory and complex systems. rather, these issues are highlighted as of particular importance to understanding energy demand (reduction) in cities.

### 2.2.1 Interconnectivity

The interconnected nature of problems and solutions in cities reflects the interlinkages that exist between the different activities, different infrastructure, and different communities of cities. The driver(s) of connection are arguably socially driven (based on organisation of work, leisure, religion, health, etc.) but are supported by, or realised through, technology (e.g., computers, modes of transport, health/leisure facilities). The system connections are multiple (e.g., coordination both spatially and temporally between work, home, leisure, etc.) and the options for connections between different system elements are also multiple (e.g., different modes of transport, leisure and work connected through shared information and communication platforms)—making the city interconnected. It is the realisation of interconnectivity options (through technology) that informs the complexity of cities and associated energy demands.

Our understanding of the interconnectivity within cities, and the ability for cities to reconfigure due to this interconnectivity (therefore changing energy demand), has recently improved as a result of the many observed responses to COVID-19. Connections in cities have been reconfigured in different ways as various national and local policies to the pandemic took hold. Greater levels of home working and less commuting, population shifts 'out of the city', necessity to maintain the spatial and temporal nature of 'key worker' jobs, are just some examples of how responses to the pandemic differed within and between cities. The relation of these changes in connectivity to energy use have resulted in observed shifting of energy demand; a reduction in work- and transport-related demands, and an increase in home energy demand as work-related energy demands are relocated (as a result of information and communication technology) and increased heating/cooling and lighting coincide with greater amounts of time spent in the home (Rowe et al. 2022). The changes in demand under COVID-19 restrictions have been observed to vary for different cities. Under the easing of COVID-19 the advantages and disadvantages of the imposed new structures to everyday life are still being debated. This demonstrates that, within cities, connections are multiple and bidirectional, with multiple consequences to social organisation, use of technology, and energy demand. This raises an important question: Does reconfiguration of connections in cities (due to new or changing relationships with technology and multiple pathways of social reorganisation) help to achieve energy demand reduction?

The view of interconnectivity presented so far is one of a city as isolated, where use of technology and social organisation are confined by a virtual city wall working as a spatial constraint. This intra-city connectivity does not present the full picture. Cities are connected to other cities on both national and global scales, sharing resources, innovations, technology and practices that inform demand (Smith 2003). At this scale, the city can be seen as a node in a global network of cities such that potential for new energy demands, or demand reduction, in one city can be shared and repeated (or translated) in another city (Shove et al. 2014).

### 2.2.2 Heterogeneity

Measures of system function (e.g., socioeconomic or sociocultural dissimilarities in population) or spatial clustering of structural elements in cities (e.g., access to services such as public transport and shops, or landcover type) have been used in many different studies of heterogeneity. The differing units of analysis used reflect the many disciplinary interests associated with the city. Studying the heterogeneity of city land use has, for example, been carried out in relation to urban planning policy (Koster and Rouwendal 2012), urban ecology (Cadenasso et al. 2007; Alberti 1999) and urban meteorology (Yu et al. 2022) to name but a few. In these studies of structural heterogeneity, heterogeneity is considered as a spatial feature of the city where the scale of spatial unit influences the degree of heterogeneity observed. Time is, of course, relevant to structural heterogeneity with changing morphology as part of changes (adaptations) to planning and regulatory requirements, and new demands in society (i.e., the SDGs). Incorporating temporality is therefore important for understanding change in (the process of) structural heterogeneity.

Functional heterogeneity is perhaps more obviously a spatial and temporal feature of cities. The social and economic classifiers of population can be tied to spatial units in a similar fashion to that of structural heterogeneity (e.g., mix of economic function and cultural diversity in neighbourhoods) but there also exist temporal features of organisation in these functional system traits that are important to understanding the variation in city functions. Chatterton's (1999) study on university students in Bristol reveals both a spatial and temporal configuration of daily activity particular to this city population, as well as a type of seasonality associated to their function and influence in a city, due to being largely a migratory (rather than permanent) population. Spatiotemporal differences in system functions can also be ascribed to different urban populations; the movement of family units with (young) school-age children being strongly coupled to school hours, health workers tied to specific location and the temporal organisation of shift work, etc.

Both structural and functional system features can be linked to energy demand: Housing archetypes (Sokol et al. 2017) are often used in housing energy demand studies as their built form influences efficiencies for achieving indoor thermal comfort and can inform embodied energy (Stephan and Athanassiadis 2017); issues of energy inequality are related to socioeconomic factors that influence access to energy (Smith et al. 2022; Smith 2021); and proximity of work to home along with available infrastructure can influence travel mode choices with implications for amount of energy consumed per trip (Brand et al. 2014). The housing archetypes, socioeconomic types and proximity of place are just three examples (of many) of heterogeneous features of a city informing energy demand and profiles of consumption.

As with the concept of city interconnectivity, heterogeneity can take on meaning both within a city (as explored above) and in cross-comparison of cities. The study of Brelsford et al. (2017) also suggests city size can inform heterogeneity (such as in level of inequality) of a city and the capability cities have for achieving SDGs. With this in mind, Figs. 1 and 2 have been produced to provide an introductory quantitative look at how heterogeneity informs energy demand. Area-averaged annual metered

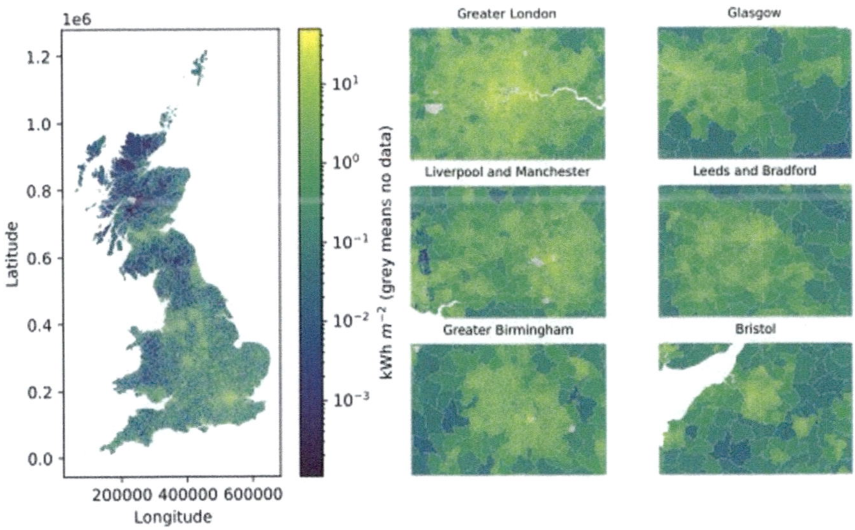


**Fig. 1** Postcode level annual metered domestic electricity consumption density May 2020–21 (BEIS 2020a)

domestic energy consumption for Great Britain (GB) has been spatially mapped at postcode sector level resolution for the two main energy vectors in GB (electricity and gas). These maps were created using Pope (2017) and national statistics on postcode level domestic consumption (BEIS 2020a, b). It is important to note that the definition of domestic in the data refers to annual consumption below a stated maximum and so includes some small business consumers and excludes some large domestic consumers. In the case of electricity consumption, meter and tariff types also inform the definition of domestic.

Figures 1 and 2 show a greater domestic energy use density in urban regions of Great Britain that highlights the importance of cities to energy demand reduction. Eight cities of differing size are captured in the six tiles. Liverpool and Manchester, Leeds and Bradford show how urban sprawl is blurring the boundaries of individual cities. The spatial distributions of annual domestic energy use density differ between all cities, although all display some central area of high demand. For the case of electricity demand, Greater London has multiple regions of high demand. For gas, all cities demonstrate multiple spatial clustering of higher demand. There is clear spatial heterogeneity in annual energy use density within each city, between cities and between the energy vectors (proxies for different energy demands) of electricity and gas.[4]

Whilst this brief spatial analysis of domestic annual energy consumption helps to link spatial heterogeneity with energy demand, it is by no means a comprehensive assessment of this form of heterogeneity. The use of total annual values by area

[4] Gas is the predominant energy vector for domestic heating in Great Britain.

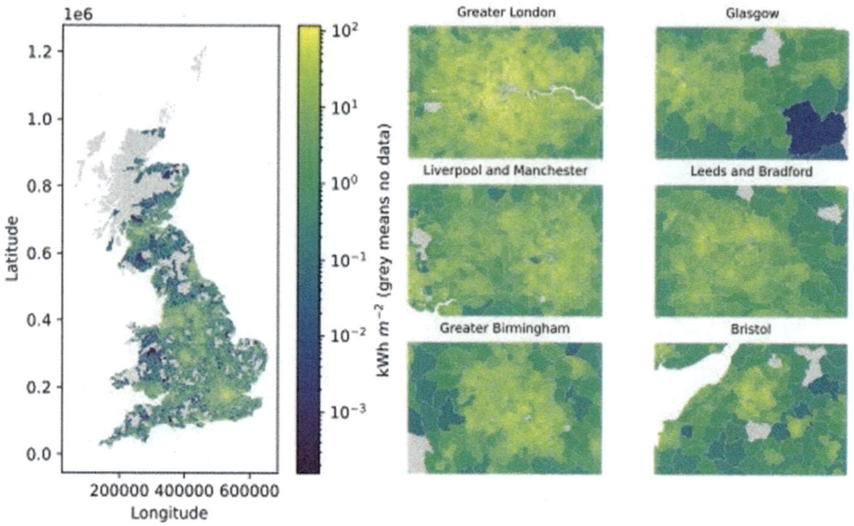


**Fig. 2** Postcode level annual metered domestic gas consumption density May 2020–21 (BEIS 2020b)

does not account for the distribution of consumption within each postcode sector—a limitation of the aggregate nature of the data used. More robust statistical measures of difference can be used (Brelsford et al. 2017), provided the data allows for such analysis. Averages of annual consumption, or regional totals of consumption provided in national statistics can show some degree of heterogeneity in demand within and between cities but ignore distributional effects. The heterogeneity in the interactions (the dynamics) of the city are difficult to assess through data alone. This motivates the development and use of models for studying urban energy systems (Keirstead et al. 2012; Capel-Timms et al. 2020).

### 2.2.3 Causality and Hierarchy

In the light of emergent features of a complex system, there is always a question of the causal nature of these features. For complexity theory, this comes down to questions of the relative scale of controlling system features: Are the observed city structures and functions caused by the bottom-up interactions between the elementary components of the system, or are the interactions a result of top-down influences of the social and physical environment? Key to the idea of causation is that complex systems are hierarchical, meaning the function at one level emerges from the influence of either a higher or lower system level.

For a city, we might think of individuals as interacting elements of a system, each with their own desires, goals, responsibilities, etc., that influence their spatial and temporal actions and interactions. The collective actions lead to emergent features at a higher level of social organisation, such as communities of religion, work, leisure, etc. This is bottom-up causation. However, the higher-level features can also be thought to influence lower-level elements and interactions by providing conditions that enable or constrain interaction. The higher-level system provides context that informs the lower level. This is top-down causation (Ellis 2012).

The direction of causation in relation to city energy demand is important to understanding how demand reduction scenarios might be achieved. To what extent are individual interactions between households (bottom-up) important to the diffusion of technology and low energy-use behaviours? How do national and local policy inform and constrain energy demand—are adaptations likely to be favourable to a sustained reduction in demand? The following section picks up on this last question and explores how cities as complex adaptive systems can inform proposed scenarios of national energy demand reduction.

Further reading on complexity and cities and, more broadly, complexity of the artificial, can be found in Portugali 2021 and Simon 1996.

## 3 Low Energy Demand (LED) Scenarios

Demand reduction is a challenge to a 170-year trend of sustained growth in energy demand, despite noted improvements in the economic intensity associated with energy use (Sorrell 2015). The LED scenarios (LEDS) for the United Kingdom (UK) (Barrett et al. 2021) are an assessment of the ability for different future demand profiles in the UK to meet climate mitigation targets. The assessment of four different demand reduction scenarios (Ignore, Steer, Shift, Transform) showed that demand reduction has a major role to play in achieving mitigation targets for 2050. The four scenarios are based on seven observed trends (see Table 1) in society that are impacting energy demand and are, therefore, key to the LEDS. The degree to which these trends evolve and inform future demand vary between scenarios. The scenarios of shift and transform look at significant changes to five identified energy demand sectors (see Table 1) in relation to the seven societal trends. A selection of demand reduction trends under each sector are summarised in Table 1 for these two scenarios.

The highlighted trends in Table 1 are the product of expert opinion from energy demand researchers in the UK and are thought to present trends most likely to have significant impact on national energy demand reduction. To understand the mitigation potential of these LEDS, the UK TIMES (UKTM) modelling framework was applied to assess mitigation against a baseline of 2010. The UKTM approach uses a system optimisation (cost minimisation) function where demands, energy flows and system balancing are considered at a national level. This optimisation approach assumes a system equilibrium can be associated with energy demand and the technology serving that demand.

The modelling approach used in the LEDS study provides comment on national scale demand reduction, but how these scenarios are realised and what the implication of the scenarios are for energy demand is not an issue that can be addressed by this

**Table 1** Matrix of the five demand reduction sectors and seven demand reduction trends as of the CREDS Low Energy Demand Scenarios

| Sectors trends | Nutrition | Mobility | Domestic buildings | Non-domestic buildings | Materials and products |
|---|---|---|---|---|---|
| Digitalisation | Optimise N3, N4 | Improved logistics M1 | D2 lighting, heating | Environmental control ND2 | MP1 |
| | Redistribution for N1 | Video conferencing M1, M2 | | | |
| | | Greater network integration M2, M3 | | | |
| Sharing and circular economy | N1 | Car share M2 leads to MP2 | D3 repurpose ND to D | Office sharing/ hot desk ND4 | MP2 by M2 and MP3 |
| | | | D3 shared occupancy | | |
| Energy efficiency | Reduce N4 | All non-HGV to M4 | Increase D1 | Increase ND1 | MP4 (e.g., zero-carbon steel) |
| | (i.e., reduce supermarkets) | | Heat-pump adoption (MP4) | ND2 | |
| Healthy society | N2 for personal health, reduces support services needed | M3 (personal health and improve air quality) | D1, D2 for fuel poverty and indoor environments | ND2 for improved indoor environment | MP4—fuel switching reduces air pollution |
| Environmental awareness | N2 | M1 (including aviation), M2 | Informing D1 and D4 | Organisational commitments to net zero | Motivates MP3 |
| Globalisation | N4 reconfigured less transportation and N1 | M1 (esp. flights) | – | – | Global standards for MP4 |

(continued)

**Table 1** (continued)

| Sectors trends | Nutrition | Mobility | Domestic buildings | Non-domestic buildings | Materials and products |
|---|---|---|---|---|---|
| Work and automation | Promotion of N2 in workplace | M1 by home working and new week structures (e.g. 4-day week) | Increased energy use from new working patterns | ND4 reduce energy use | – |

Captures the shift and transition pathways of the LEDS. The reduction pathways for each sector are: *Nutrition (N) N*1 reduce food waste, *N2* change diet to less energy/carbon intense, *N3* food production, *N4* supply chains; *Mobility (M) M1* reduce travel demand, *M2* reduce private car use, *M3* increased access to public and active transport, *M4* new technology (i.e. electric vehicles); *Domestic buildings (D) D1* building fabric retrofit, *D2* system controls, *D3* reduce housing demand, *D4* behavioural change to improve efficiency; *Non-domestic buildings (ND) ND1* building fabric retrofit, *ND2* new technology adoption, *ND3* behavioural change to improve efficiency, *ND4* reduce demand for ND property; *Materials and products (MP) MP1* reduce material waste by connecting industries (new resource flows), *MP2* reduce new goods/space production/construction, *MP3* increase reuse and increase length of life, *MP4* new technology giving efficiencies in material use and shift/transition away from $CO_2$ intensive materials/products

approach. This chapter has made the case that cities are fundamentally important to the landscape of energy demand (reduction) and, as complex adaptive systems, energy demand is an emergent feature of a system far from equilibrium, working on scales of neighbourhoods all the way up to networks of globally connected cities. It is therefore essential to our understanding of demand reduction potential that futures of low energy demand, as proposed in the shift and transform scenarios, are studied in relation to the complex city contexts where they are to emerge. The case of reduced demand in private transport (Table 1: M) is taken up to highlight some of the issues to be studied.

## *3.1 Example of Digitalisation and Demand Reduction in (City) Mobility—Exposing Complexity*

The digitalisation of working practices has opened up the possibility of greater flexibility in where and when people work. Greater home working does suggest less commuting and sustained shifts/transitions in the use of these digital technologies does open up the possibility for a sustained reduction in travel demand (Table 1: M1) and possibly reduced car ownership (Table 1: M2). Further, digitalisation is expected to increase access to public and active transport (Table 1: M3) by, for example, increasing ability to plan routes across different public transport modes/networks, purchase of tickets, etc.

According to the UK National Travel Survey of 2022, 52% of private vehicle trips were associated with shopping and leisure activity and 15% for commuting. The dominant mode of transport for trips above 1 mile in length being private vehicles and vans. However, in the case of London, prior to the COVID-19 pandemic, public transport and walking accounted for 60% of all surveyed trips (TfL 2022).

These two survey results suggest that the interconnectivity of activity will matter greatly for M1 and M2 to be sustained, even for trips less than 10 miles in length (length scales that are relevant for cities). Choice of mode of trip and ownership of cars is not likely to be a result of digitalising work activities alone. The coordination (interconnectivity) between these activities matter. Interconnectivity informs fixed relations like home–work location/distance that persist and lead to routine demands (Krygsman et al. 2004).

In the case of larger cities like London, trips on public and active transport networks might already be the dominant mode. Access to services (such as public transport) is not necessarily equally shared in cities and can be a question of spatial proximity (structural heterogeneity) rather than improved service coordination through digital means (M3). Whilst the efficiency and access of service has been shown to scale super-linearly with city size (suggesting growth of cities may also be a trend that could support energy demand reduction) strong interconnectivity in transport can lead to reliance on transport that requires energy (i.e., energy demands locked in and susceptible to growth).

Exploring the complexity of cities in relation to other trends and sectors in the LEDS can help further demonstrate the implications of city context to national energy demand reduction pathways. I invite you to consider other demand reduction scenarios presented in Table 1 as they might apply to cities as complex adaptive systems.

## 4 Conclusion

Cities have been presented in this chapter as complex adaptive systems that are characterised by hierarchies of interconnected, heterogeneous, structural and functional systems. The types of connection and heterogeneity that can be drawn out from the study of these city systems can help to inform how energy demanding services emerge and, therefore, how demand reduction might be achieved. How demand reduction pathways can be realised depends on the ability within city systems to support diversity and encourage adaptation. This has, however, an inherent danger of adaptations leading to emergent features of the city that do not, ultimately, lead to energy demand reductions. It is, therefore, important to gain understanding of the relationship of different causal pathways (top-down and bottom-up) to demand reduction in cities.

Causality along with interconnectivity and heterogeneity are properties of complex systems that are studied and modelled in relation to cities and energy demand. These properties can, and are, studied in isolation from each other and at times with no explicit reference to complexity science or the context of cities as complex adaptive systems.

The chapter started with positioning the challenge of energy demand reduction as a challenge to the integration of many of the UN Sustainable Development Goals, suggesting that cities (or more broadly, urbanisation) have a key role to play in addressing those challenges. Rather than positioning cities as fixed entities of given activity and demand, complexity science suggests cities are dynamic systems, always evolving and adapting. Development is not static, but the goals are not necessarily static either if cities are truly complex and adaptive. Setting targets for demand reduction are fundamentally important to planetary sustainability, but as energy-demanding services in cities are continually evolving, our approaches to sustained demand reduction in cities will also need to evolve. Complexity science applied to city structure and function provides a framework for thinking about the coevolving issue of energy demand reduction in our built environments.

## References

Alberti M (1999) Modeling the urban ecosystem: a conceptual framework. Environ Plan B Urban Anal City Sci 26(4):605–630. https://doi.org/10.1068/b260605

Allen PM (1982) Evolution, modelling, and design in a complex world. Environ Plan B Urban Anal City Sci 9(1):95–111. https://doi.org/10.1068/b090095

Bale CSE, Varga L, Foxon TJ (2015) Energy and complexity: new ways forward. Appl Energy 138:150–159. https://doi.org/10.1016/j.apenergy.2014.10.057

Balk D, Tagtachian D, Jiang L, Marcotullio P, Cook EM, Jones B, Mustafa A, McPhearson T (2022) Frameworks to envision equitable urban futures in a changing climate: a multi-level, multidisciplinary case study of New York city. Front Built Environ 8(October):949433. https://doi.org/10.3389/fbuil.2022.949433

Barrett J, Pye S, Betts-Davies S, Eyre N, Broad O, Price J, Norman J, Anable J, Bennett G, Brand C, Carr-Whitworth R, Marsden G, Oreszczyn T, Giesekam J, Garvey A, Ruyssevelt P, Scott K (2021) The role of energy demand reduction in achieving net-zero in the UK. Centre for Research into Energy Demand Solutions, Oxford

Batty M (2005) Cities and complexity: understanding cities with cellular automata, agent-based models, and fractals. MIT Press, Cambridge, Mass

Batty M (2008) The size, scale, and shape of cities. Science 319(5864):769–771. https://doi.org/10.1126/science.1151419

BEIS, Department for Business, Energy & Industrial Strategy (2020a) Official statistics. Postcode level electricity statistics 2020 (Experimental). https://www.gov.uk/government/statistics/postcode-level-electricity-statistics-2020-experimental

BEIS, Department for Business, Energy & Industrial Strategy (2020b) Official statistics. Postcode level gas statistics: 2020 (Experimental). https://www.gov.uk/government/statistics/postcode-level-gas-statistics-2020-experimental

Beppler RC, Matisoff DC, Oliver ME (2023) Electricity consumption changes following solar adoption: testing for a solar rebound. Econ Inq 61(1):58–81. https://doi.org/10.1111/ecin.13031

Bettencourt LMA, Lobo J, Helbing D, Ku C, West GB (2007) Growth, innovation, scaling, and the pace of life in cities. Proc Natl Acad Sci USA 104(17):7301–7306. https://doi.org/10.1073/pnas.0610172104

Bordón-Lesme M, Freire-González J, Rosa EP (2022) Do household energy services affect each other directly? The direct rebound effect of household electricity consumption in Spain. Energ Effi 15(7):47. https://doi.org/10.1007/s12053-022-10048-7

Brand C, Goodman A, Ogilvie D (2014) Evaluating the impacts of new walking and cycling infrastructure on carbon dioxide emissions from motorized travel: a controlled longitudinal study. Appl Energy 128:284–295. https://doi.org/10.1016/j.apenergy.2014.04.072

Brelsford C, Lobo J, Hand J, Bettencourt LMA (2017) Heterogeneity and scale of sustainable development in cities. Proc Natl Acad Sci USA 114(34):8963–8968. https://doi.org/10.1073/pnas.1606033114

Cadenasso ML, Pickett STA, Schwarz K (2007) Spatial heterogeneity in urban ecosystems: reconceptualizing land cover and a framework for classification. Front Ecol Environ 5(2):80–88. https://doi.org/10.1890/1540-9295(2007)5[80:SHIUER]2.0.CO;2

Capel-Timms I, Smith S T, Sun T, Grimmond S (2020) Dynamic anthropogenic activities impacting heat emissions (DASH V1.0): Development and evaluation. Geosci Model Develop 13(10):4891–4924. https://doi.org/10.5194/gmd-13-4891-2020

Chatterton P (1999) University students and city centres the formation of exclusive geographies. Geoforum 30(2):117–133. https://doi.org/10.1016/S0016-7185(98)00028-1

Cowan G A, Pines D, Meltzer D (1994) Complexity: metaphors, models, and reality. In: Santa Fe Institute studies in the sciences of complexity. Proceedings vol 19. Reading, Mass: Addison-Wesley.

Ellis GFR (2012) Top-down causation and emergence: some comments on mechanisms. Inter Focus 2(1):126–140. https://doi.org/10.1098/rsfs.2011.0062

Grubb M, Drummond P, Poncia A, McDowall W, Popp D, Samadi S, Penasco C, Gillingham KT, Smulders S, Glachant M, Hassall G, Mizuno E, Rubin ES, Dechezleprêtre A, Pavan G (2021) Induced innovation in energy technologies and systems: a review of evidence and potential implications for $CO_2$ mitigation. Environ Res Lett 16(4):043007. https://doi.org/10.1088/1748-9326/abde07

Holland J H (1992) Complex adaptive systems. Daeldalus 121(1):17–30

Keirstead J, Jennings M, Sivakumar A (2012) A Review of urban energy system models: approaches, challenges and opportunities. Renew Sustain Energy Rev 16(6):3847–3866. https://doi.org/10.1016/j.rser.2012.02.047

Khavarian-Garmsir AR, Sharifi A, Sadeghi A (2023) The 15-minute city: urban planning and design efforts toward creating sustainable neighborhoods. Cities 132(January):104101. https://doi.org/10.1016/j.cities.2022.104101

Koster HRA, Rouwendal J (2012) The impact of mixed land use on residential property values. J Reg Sci 52(5):733–761. https://doi.org/10.1111/j.1467-9787.2012.00776.x

Krygsman S, Dijst M, Arentze T (2004) Multimodal public transport: an analysis of travel time elements and the interconnectivity ratio. Transp Policy 11(3):265–275. https://doi.org/10.1016/j.tranpol.2003.12.001

Lange S, Berner A (2022) The growth rebound effect: a theoretical–empirical investigation into the relation between rebound effects and economic growth. J Clean Prod 371:133158. https://doi.org/10.1016/j.jclepro.2022.133158

Leavesley A, Trundle A, Oke C (2022) Cities and the SDGs: realities and possibilities of local engagement in global frameworks. Ambio 51(6):1416–1432. https://doi.org/10.1007/s13280-022-01714-2

Pope A (2017) GB Postcode Area, Sector, District, [PostalSector.shp]. University of Edinburgh. https://doi.org/10.7488/ds/1947

Portugali J (2021) Cities, complexity and beyond. In: *Handbook on cities and complexity*. Edward Elgar Publishing. https://doi.org/10.4337/9781789900125.00008.

Ropohl G (1999) Philosophy of socio-technical systems. Soc Philos Technol Quart Electron J 4(3):186–194. https://doi.org/10.5840/techne19994311

Rowe F, Robinson C, Patias N (2022) Sensing global changes in local patterns of energy consumption in cities during the early stages of the COVID-19 pandemic. Cities 129:103808. https://doi.org/10.1016/j.cities.2022.103808

Rutherford J, Coutard O (2014) Urban energy transitions: places, processes and politics of socio-technical change. Urban Studies 51(7):1353–1377. https://doi.org/10.1177/0042098013500090

Rylatt M, Gammon R, Boait P, Varga L, Allen P, Savill M, Snape R, Lemon M, Ardestani B, Pakka V, Fletcher G, Smith S, Fan D, Stathern M (2013) Cascade: an Agent Based Framework for Modeling the Dynamics of Smart Electricity Systems. Emerg Complex Organ 15(SPL. 2):1–13
Shove E (2020) Time to rethink energy research. Nat Energy 6(2):118–120. https://doi.org/10.1038/s41560-020-00739-9
Shove E, Walker G, Brown S (2014) Transnational transitions: the diffusion and integration of mechanical cooling. Urban Stud 51(7):1506–1519. https://doi.org/10.1177/0042098013500084
Simon HA (1996) The sciences of the artificial, 3rd edn. MIT Press, Cambridge
Smith RG (2003) World city actor-networks. Prog Hum Geogr 27(1):25–44. https://doi.org/10.1191/0309132503ph411oa
Smith S (2021) Mobilizing heterogeneity: understanding the dynamic qualities and diversity of energy access in Maputo. Urban Geogr. (August) 1–22. https://doi.org/10.1080/02723638.2021.1959769.
Smith S, Monstadt J, Otsuki K (2022) Enabling equitable energy access for Mozambique? Heterogeneous energy infrastructures in Maputo's growing urban periphery. Energy Res Soc Sci 90(August):102684. https://doi.org/10.1016/j.erss.2022.102684
Sokol J, Cerezo Davila C, Reinhart CF (2017) Validation of a Bayesian-based method for defining residential archetypes in urban building energy models. Energy Build. 134(January):11–24. https://doi.org/10.1016/j.enbuild.2016.10.050
Sorrell S (2015) Reducing energy demand: a review of issues, challenges and approaches. Renew Sustain Energy Rev 47(July):74–82. https://doi.org/10.1016/j.rser.2015.03.002
Stephan A, Athanassiadis A (2017) Quantifying and Mapping embodied environmental requirements of urban building stocks. Build Environ 114:187–202. https://doi.org/10.1016/j.buildenv.2016.11.043
TfL (2022) Transport for London. Travel in London Report 15
Wittmayer JM, de Geus T, Pel B, Avelino F, Hielscher S, Hoppe T, Mühlemeier S, Stasik A, Oxenaar S, Rogge KS, Visser V, Marín-González E, Ooms M, Buitelaar S, Foulds C, Petrick K, Klarwein S, Krupnik S, de Vries G, Wagner A, Härtwig A (2020) Beyond instrumentalism: broadening the understanding of social innovation in socio-technical energy systems. Energy Res Soc Sci 70(December):101689. https://doi.org/10.1016/j.erss.2020.101689
Yu T, Sützl BS, van Reeuwijk M (2022) Urban neighbourhood classification and multi-scale heterogeneity analysis of Greater London. Urban Anal City Sci 1–25. https://doi.org/10.1177/23998083221140890

# Clean and Renewable Energy

**Soteris A. Kalogirou**

**Abstract** In this chapter, various possible applications of renewable energy in cities and communities are presented. Being a clean form of energy, without any harmful emissions to the atmosphere, renewables effectively improve sustainability in the near and far future. Successful applications of renewables presented in this chapter include solar water heating, space heating and cooling systems, and photovoltaics and hybrid PV/T systems which provide both thermal and electrical energy. Another form of renewable energy systems used in buildings is the geothermal energy systems which employ ground heat exchangers in combination with heat pumps. These can supply both heating and cooling in buildings and communities. Besides the building application of the various renewable energy systems these are nowadays used more and more in district heating and cooling networks to satisfy either a district, a neighborhood, or a complete village or small city. Finally, the current status of these systems in the world is also presented to show the prospects for the future.

**Keywords** Solar water · Space heating · Hybrid photovoltaics · Geothermal · District networks

## 1 Introduction

Sustainability is a very broad term which has to do with a societal goal that broadly aims to allow human beings to co-exist on earth for a long time. Three different areas of sustainability are normally considered, environmental, social, and economic. Here we are interested particularly on the environmental area which leads to the avoidance of the depletion of natural resources in order to maintain an ecological balance. These are unavoidably related also to energy issues and in particular to the use of fossil

S. A. Kalogirou (✉)
Department of Mechanical Engineering and Materials Sciences and Engineering, Cyprus University of Technology, 50329, 3603 Limassol, Cyprus
e-mail: soteris.kalogirou@cut.ac.cy

Cyprus Academy of Sciences, Letters, and Arts, Nicosia, Cyprus



R. Yao (ed.), *Resilient Urban Environments*, Cities and Nature,
https://doi.org/10.1007/978-3-031-55482-7_14

fuels to satisfy the energy needs of humanity and their destructive effects that these have on the environment which led to climate change effects. Another issue is that fossil fuels resources are finite and when burned are very polluting. Therefore, the solution is to use sources that are friendly to the environment. This is mostly done today with renewable energy sources that are constantly supplied with energy from the sun directly or indirectly (wind, waves, biomass, etc.) at such rates so that they are considered practically inexhaustible and capable of substituting many of the conventional energy sources.

Nowadays a large number of renewable energy systems have been installed in various parts of the world. The total amount of renewable power capacity by the end of 2021 is 3146 GW. From these, solar PV constitutes 942 GW, a 175 GW increase from the 2020 figure, wind power is equal to 845 GW, compared to 743 in 2020, and solar thermal is equal to 522 $GW_{th}$, compared to 501 $GW_{th}$ for 2020 (REN 2022). The estimated renewable energy share for the global final energy consumption is 78.4% non-renewable and 12.6% renewable energy (9% other) whereas the figure for renewable electricity production is about 28.3% (REN 2022).

Cities, and in a more general term communities, are flooded with thousands to millions of people. Their needs among others require substantial quantities of energy with their consequent harmful emissions to the environment and its consequences on the climate which affect the quality of life in general. In view of this, renewable energy systems play an important role in achieving better living conditions. The various systems analyzed in this chapter are for direct use in buildings.

## 2 Solar Water Heating

Solar water heating is one of the most effective applications as after the system is installed it provides hot water to a building for many years without requiring any attention or significant maintenance. Basically, there are two main types of these systems, thermosiphon units or passive systems, shown schematically in Fig. 14.1, which operate without a pump to circulate the fluid, and active systems, shown schematically in Fig. 14.2, which employ a pump to do so which require a small amount of electrical energy (Kalogirou 2014). These systems are installed on the roof of the building or house. The main type of collector used in both types of systems is the flat plate collector. Usually, two or more such collectors are used. In the thermosiphon units the hot-water storage tank needs to be located above the top side of the collector to allow water circulation which is performed with the help of the small density difference between the hot and cold water. In active systems, as a pump is employed, the hot-water tank can be located indoors, usually in a plant room, and a differential thermostat is used to compare the temperature at the exit of the collector (top side) and the tank to switch on the pump when a defined amount of temperature difference is achieved. It switches off the pump when the difference is reduced to 2–3 °C. In air systems, a fan is used to circulate the air through the collector and an air-to-water heat exchanger to transfer the heat to the water circulated through the storage tank.

Two variations of active system are the direct and indirect systems.

A schematic diagram of a direct circulation system is shown in Fig. 14.2. In this system, a pump is used to circulate potable water from storage to the collectors

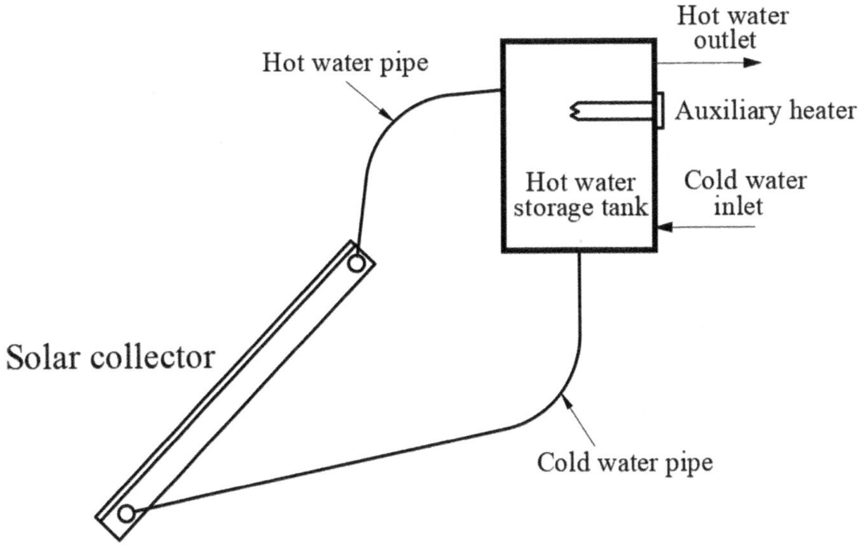


**Fig. 14.1** Schematic of a thermosiphon solar water heater

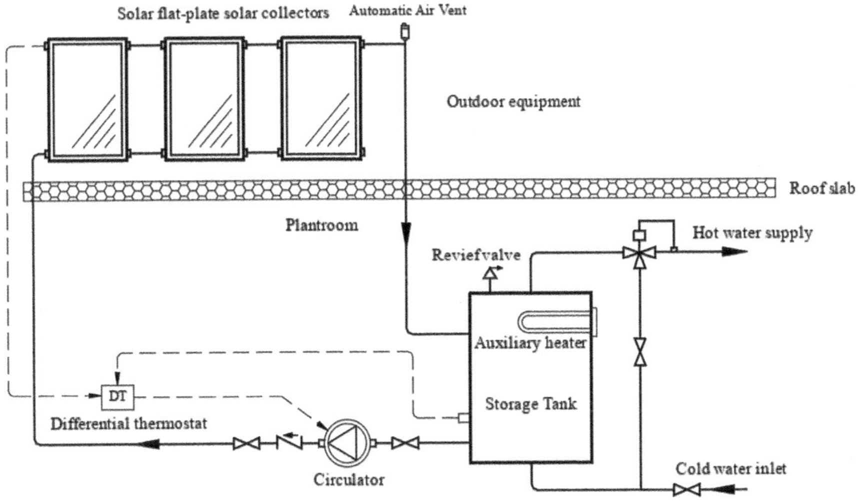


**Fig. 14.2** Direct circulation solar water-heating system schematic

when there is enough solar energy to increase its temperature and return the heated water to the storage tank where it stays until it is needed, and thus there is a need for the storage tank to be well insulated. The pump is controlled with a differential thermostat that measures and compares the temperature at the collector outlet and the storage tank and operates the pump whenever a certain temperature difference exists, as indicated above.

Because a pump is used to circulate the water, the collectors can be mounted either above or below the storage tank. Direct circulation systems can be used with water supplied from a cold-water storage tank or connected directly to city water mains. If the city water pressure is greater than the working pressure of the collectors, pressure-reducing valves and pressure relief valves are required to be installed. Direct water-heating systems should not be used in areas where the water is extremely hard or acidic, because calcium-scale deposits that will inevitably be formed may clog or corrode the collectors.

Direct circulation systems can be used in areas where freezing is infrequent. For extreme weather conditions, a freeze protection is usually provided by circulating warm water from the storage tank to the collectors. This loses some heat but protects the system. A special thermostat that operates the pump when temperature drops below a certain value is used in this case. Such recirculation freeze protection should be used only for locations where freezing occurs rarely, i.e., a few times a year, since stored heat is dumped in the process. A disadvantage of this system occurs in cases when there is power failure, in which case the pump will not operate, and the system could freeze. In such a case, a dump valve can be installed at the bottom of the collectors to provide additional protection (Kalogirou 2014).

For freeze protection, a variation of the direct circulation system, called the drain-down system, is used, shown schematically in Fig. 14.3. In this case, potable water is also pumped from storage to the collector array, where it is heated. When a freezing condition or a power failure occurs, the system drains automatically by isolating the collector array and exterior piping from the make-up water supply with the normally closed (NC) valve and draining it using the two normally open (NO) valves, shown in Fig. 14.3. It should be noted that the solar collectors and associated piping must be carefully installed with a small inclination to drain the collector's exterior piping when circulation stops. The one-way check valve shown on the top of the collectors in Fig. 14.3 is used to allow air to fill the collectors and piping during draining and to escape during a fill-up (Kalogirou 2014).

A schematic diagram of indirect water-heating systems is shown in Fig. 14.4. In this system, the pump also operates with a differential thermostat, and a heat-transfer fluid is circulated through the closed collector loop to a heat exchanger, where the collected heat is transferred to the potable water. The most commonly used heat-transfer fluids are water–ethylene glycol solutions and silicone oils. When non-potable or toxic heat-transfer fluids are used, double-wall heat exchangers or two heat exchangers in series should be employed for safety reasons. The heat exchanger can be a traditional external unit or can be located inside the storage tank or around the storage tank (tank mantle). As the collector loop is a closed circuit, an expansion tank and a pressure relief valve should be incorporated. According to the heat-transfer fluid

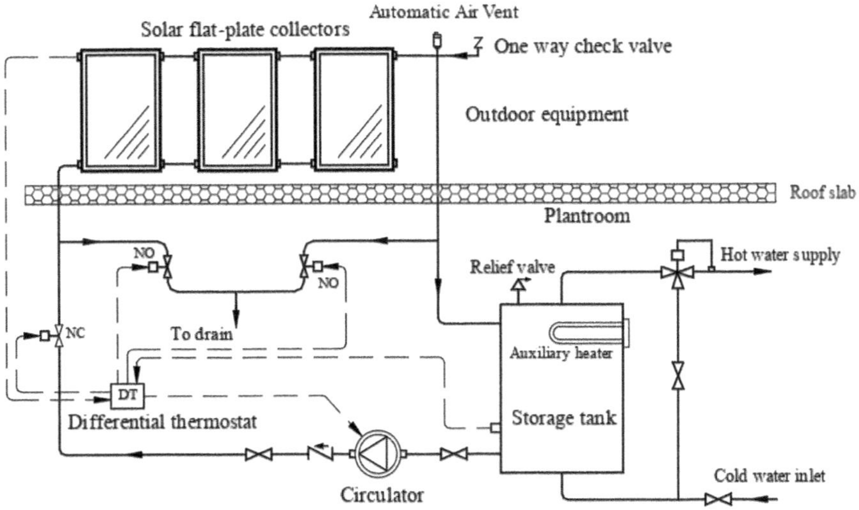


**Fig. 14.3** Drain-down solar water-heating system schematic

used, additional over-temperature protection may be needed to prevent the collector heat-transfer fluid from decomposing or becoming corrosive (Kalogirou 2014).

Systems of this type using water–ethylene glycol solutions are preferred in areas subject to extended freezing temperatures, because they offer good freeze protection. These systems are more expensive to construct and operate, since the solution should

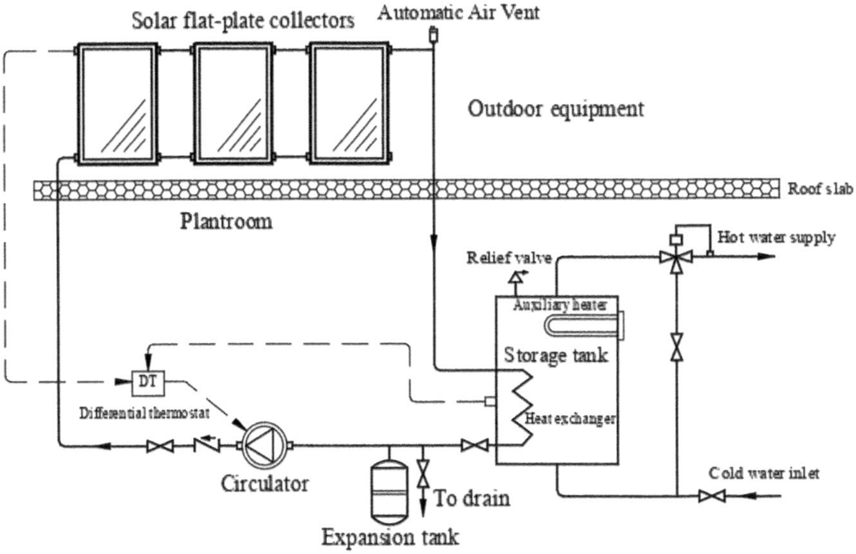


**Fig. 14.4** Indirect solar water-heating system

be checked every year and replaced every few years, depending on the quality of the fluid and the system temperatures achieved. A general rule of thumb to follow to size equipment is that the storage tank should be between 35 and 70 $l/m^2$ of collector area.

For freeze protection, a variation of the indirect water-heating system, called the drain-back system, can be used. These are indirect water-heating systems that normally circulate water through the closed collector loop to a heat exchanger, where its heat is transferred to potable water. Circulation continues when usable energy is available and the differential thermostat conditions are satisfied. Whenever the circulation pump stops, the collector fluid drains by gravity to a drain-back tank as shown in the schematic diagram of the system in Fig. 14.5. In this case, if the system is pressurized, the tank also serves as an expansion tank when the system is operating and thus must be protected with temperature and pressure relief valves. In the case of an unpressurized system the tank is open and vented to the atmosphere. The second pipe directed from the collectors to the top of the drain-back tank is to allow air to fill the collectors during a drain-back.

It should be noted that because the collector loop is isolated from the potable water, no valves are needed to actuate draining and scaling from a possible hard water is not a problem; however, the collector array and exterior piping must be adequately sloped as in drain-down systems to drain completely. As can be understood, freeze protection is inherent to the drain-back system because the collectors and the piping above the roof are kept empty whenever the pump is not running. A disadvantage of this system is that a pump with a high static head is required in order to fill the collector when the system starts up.

There is a good possibility that the collectors will be drained also during periods of insolation. Such a case can happen when there is no load requirement, and the

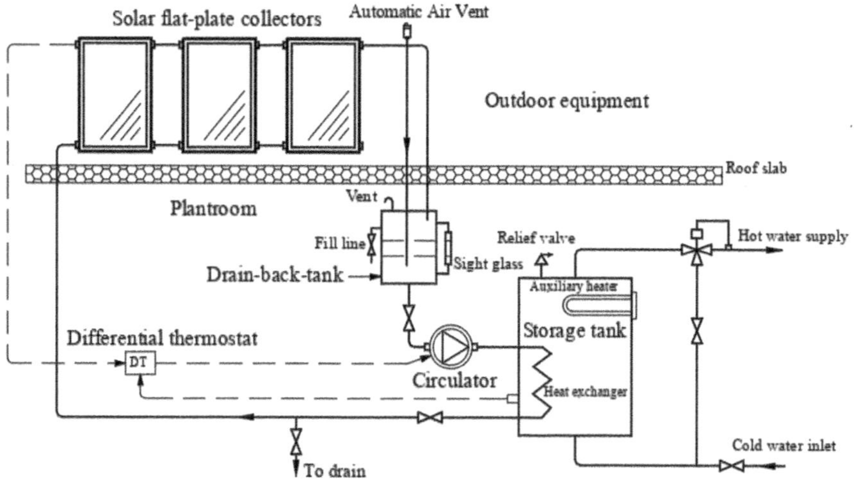


**Fig. 14.5** Drain-back solar water-heating system schematic

storage tank reaches a temperature that would not allow the differential thermostat to switch on the solar pump. It is therefore important to select collectors that can withstand prolonged periods of stagnation conditions.

An alternative design, which is suitable for small systems, is to drain the water directly in the storage tank. In this case, the system is open, and there is no need the have a separate drain-back tank and a heat exchanger; however, the system suffers from the disadvantages of the direct systems outlined before.

The most successful country in solar water-heating applications is Cyprus, where 92% of all houses are equipped with a solar water heater, mainly of the thermosiphon type. A typical neighborhood is shown in Fig. 14.6, where practically on the roof of every house there is a solar water heater installed and operating. In cities where usually high-rise buildings are built there is a restriction to the use of roof as there is not enough space to accommodate a system for each flat. In these applications building integrated systems are used, where the collectors are part of the building façade replacing also a traditional building component such as railings or balustrades (Kalogirou 2016) as shown in Fig. 14.7. Alternatively, spaces near a flat are employed; one such application in Hefei, China, is shown in Fig. 14.8. An additional advantage of using the collectors in this way is that large pipe runs are avoided.

**Fig. 14.6** Solar water-heating systems are affectively on the roof of every house in Cyprus (Photo by author)

**Fig. 14.7** BISTS railings and balustrades

**Fig. 14.8** BISTS in China high-rise buildings (photo by author)

## 3 Solar Space Heating and Cooling

An extension of the systems shown above, usually on a bigger scale, is solar space heating and cooling. In these systems, usually due to the larger number of collectors employed, active systems are used. These are effectively the same as the water-heating systems with their variations but applied on a bigger scale to satisfy the building thermal needs. Depending on the conditions that exist in a system at a particular time, the solar systems usually have five basic modes of operation:

- When solar energy is available and heat is not required in the building, solar energy is used to charge the storage unit.
- When solar energy is available and heat is required in the building, solar energy is used directly to supply the building load demand.
- When solar energy is not available, heat is required in the building, and the storage unit has stored energy, the stored energy is used to supply the building load demand.
- When solar energy is not available, heat is required in the building, and the storage unit has been depleted, auxiliary energy should be available and used to supply the building load demand.
- When the storage unit is fully heated, there are no loads to meet, and the collector is absorbing heat, solar energy must be discarded.

The last mode is achieved through the operation of pressure relief valves in fluid (water)-based systems. In the case of air collectors where the stagnant temperature is not harmful to the collector integrity, the flow of air is turned off, in which case the collector temperature rises to a level where the absorbed energy is dissipated by thermal losses.

In addition to the above operation modes, the solar energy system usually satisfies also the domestic hot water demand. The mode to select according to the prevailing conditions is usually controlled by thermostats. Therefore, depending on the load of heating, cooling, or hot water demand, the thermostat that is not satisfied gives a signal to operate a pump, provided that the collector temperature is higher than that of storage. Thus, by using the thermostats, it is possible to combine modes and operate in more than one mode simultaneously at a time. For simplicity some systems do not allow direct heating from a solar collector to the building heating units, but always transfer heat from collector to storage, which is used to satisfy the load, whenever this is needed. The auxiliary is also used to top up storage when there is no sunshine.

The solar cooling of buildings is an attractive idea because the cooling loads and availability of solar radiation are in phase. Additional advantage is that the combination of solar cooling and heating greatly improves the use factors of collectors compared to heating alone, when the collectors are used solely for hot water production.

Solar cooling can be achieved with two basic systems solar adsorption and solar absorption. Desiccants are sorbents, i.e., materials that have an ability to attract and hold other gases or liquids that have a particular affinity for water. The process of attracting and holding moisture is described as either absorption or adsorption, depending on whether the desiccant undergoes a chemical change as it collects moisture. Absorption changes the desiccant the way as it absorbs moisture; a typical example is table salt which changes from a solid to a liquid when moisture is absorbed. Adsorption, however, does not change the desiccant except the addition of water vapor weight, similar to the way a sponge is soaking up the water.

Therefore, absorption is the process of attracting and holding moisture by desiccants. During absorption, the desiccant undergoes a chemical change as it takes on moisture. Absorption machines are generally thermally activated, and they do not require high input shaft power. Therefore, where there is solar or other type of heat available, absorption machines could provide reliable and quiet cooling. Absorption systems are like vapor compression air-conditioning systems but differ in the pressurization stage. The pressurization is achieved by dissolving the evaporating refrigerant in the absorbent, in the absorber section on the low-pressure side (Fig. 14.9). For there the solution is pumped to a high pressure with an ordinary liquid pump. The addition of heat in the generator is used to separate the low-boiling refrigerant from the solution. In this way, the refrigerant vapor is compressed without the need for large amounts of electrical energy that the vapor compression air-conditioning systems require. The remainder of the system consists of a condenser, expansion valve, and evaporator, which function in a similar way as in a vapor compression air-conditioning system.

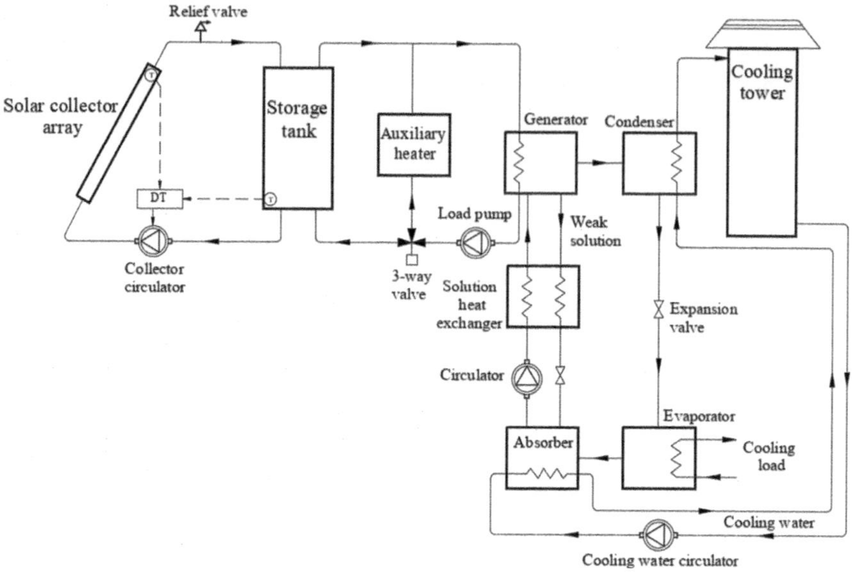


**Fig. 14.9** Li-Br system schematic

The most usual combinations of fluids include Li–Br–$H_2O$, where water vapor is the refrigerant, and ammonia–water ($NH_3$–$H_2O$) systems, where ammonia is the refrigerant.

## 4 Solar Photovoltaics

An important renewable energy application is photovoltaics. These are devises which produce electrical energy in the form of DC current directly from sunshine without involving any moving parts or noise. These are nowadays very popular systems as due to the worldwide scale of application which exceed 950 GW (REN 2022) their price per installed kW dropped substantially. This led to more and more applications each year, many of them installed on building roofs and facades. The usual application is the roof added one where the photovoltaics are installed on brackets on a flat concrete roof. In pitched roof the photovoltaics are installed using special supports and are esthetically more appealing as follow the roof inclination. In more advanced application photovoltaics can also be building integrated either on the roof or façade as shown in Fig. 14.10. In such a case a gap needs to exist at the back of the PV in which air is circulated to remove the thermal energy generated in order not to create unnecessary thermal loads to the building and increase the PV temperature and thus reduce its efficiency. During wintertime this thermal energy can be used to cover part of the thermal load of the building by directing hot air in the building.

**Fig. 14.10** BIPV systems on building roof and façade (photo by author)

The heat transfer phenomena in the gap are complicated, and natural convection is preferred to avoid the use of fans and the possibility of failure (Agathokleous and Kalogirou 2016; Agathokleous et al. 2018).

An alternative to the traditional PV is the hybrid PV/T systems in which some means of extracting the heat generated from sunshine not converted to electricity is used and employed to cover the thermal needs in several applications. They can be used effectively for converting the absorbed solar radiation into electricity and heat, therefore increasing their total energy output. In these systems, PV modules are coupled to heat extraction devices, in which water or air is heated and at the same time the PV module temperature is reduced to keep electrical efficiency at an appropriate level. Water-cooled PV/T systems are practical systems for water heating, and air-cooled PV/T system can be used in air-heating applications. These solar energy systems are of practical interest for many applications, as they can effectively contribute to cover both the electrical and thermal loads.

The additional thermal output provided from the PV/T systems makes them cost-effective compared to separate PV and thermal units of the same total aperture surface area. In PV/T system applications, the production of electricity is the main priority; therefore, it is more effective to operate the PV modules at low temperature so as to keep a PV cell electrical efficiency at the highest possible level.

Building-integrated PV, called BIPV, are already introduced above. When combined with heat extraction facilities they are called BIPV/T. Water or air can be used as a fluid circulation medium. When a BIPV/T heat removal uses water or other heat-transfer fluid, as the working fluid, the cost is generally higher due to the plumbing required and the need to build hydronic systems integration and greater weight. Although BIPV hydronic heat removal will improve PV efficiency, if the water circulation stops, as for example from a pump or power supply failure, high PV panel temperatures will be created which may affect PV durability and life. For a large façade area structural and cost constraints may also be a problem.

Air-heating BIPV/T collectors can also provide an optimal combination of both electrical and thermal conversion efficiencies. A BIPV/T facade could act as a thermosiphon photovoltaic-thermal air-heating collector which can provide preheated air in winter and also natural ventilation in summer and of course high electrical output throughout the year. A duct arranged behind the PV module allows air flow induced by a fan or naturally by buoyancy from the back of the PV panel, and this air movement is governed by a combination of natural convection and wind-induced flow (Batagiannis and Gibbons 2001).

For air flow in a BIPV duct section, wind blowing from one direction may assist the air movement and achieve greater cooling of the modules, whereas wind blowing from another direction may act against the required air flow direction and reduce the cooling potential (Batagiannis and Gibbons 2001). Furthermore, for a naturally ventilated BIPV/T, buoyancy forces are balanced by the pressure drop due to the friction at the entrance and exit (Brinkworth et al. 2000). For zero ambient wind velocity, the flow through the ventilated duct is driven only by buoyancy forces, whereas in all the other cases the flow inside the duct is due to a mixture of free and forced convection. Buoyancy induced air flow in a duct between the BIPV and the wall, even with a low mean air velocities can reduce BIPV operating temperature by about 15–20 °C giving also a 15% increase in electrical conversion efficiency (Brinkworth et al. 1997). For very high facades openings are required every 2–3 PV panel heights to avoid excessive air velocities in the duct.

Generally, the temperature attained by the BIPV/T systems depend on the incident solar energy, surface area, ambient air temperature, flow conditions, radiant surfaces, and the flow and temperature distribution (Tonui and Tripanagnostopoulos 2008).

## 5 Geothermal Systems

A completely different application of renewables in buildings is geothermal. This is used in combination with ground heat exchanger and a heat pump. Common heat pumps use an electrically driven compressor that compresses a refrigerant and raises its pressure and thus its temperature. The refrigerant has the ability to change state, from liquid to gas when heated and usually it boils at low temperatures. In the cooling mode, the common heat pump's refrigerant absorbs heat from the environment and becomes gas. Then the gas is compressed mechanically and has its temperature increased. The refrigerant at this stage is of high pressure and temperature and exchanges heat with a lower temperature medium (gas or liquid) as it passes through a condenser. Having its temperature dropped it returns to the liquid stage, and after passing through an expansion valve it becomes liquid at low temperature and pressure which is directed to the evaporator where it absorbs heat to evaporate, thus creating cooling. Then the process starts all over again with the refrigerant absorbing heat from the environment cooling it as it passes through an evaporator. In the heating mode the cycle is reversed.

Ground-coupled heat pumps or geothermal heat pumps are systems combining a heat pump with a ground heat exchanger for the heat exchange process, which improves the heat pump efficiency. Mainly, they are of two types; namely, the ground-coupled (closed-loop) system or the groundwater (open-loop) system. Both types of GHEs, open or closed loop, are pollutant-free since the only effect they have on the ground is the small increase or decrease of its temperature in a certain distance around the borehole.

Ground-coupled heat pumps exchange heat with the ground instead of the atmosphere. Effectively the coefficient of performance (COP) is increased by ejecting heat to ground instead of ambient air. To achieve this the condenser of the heat pump must be water cooled, and the thermal energy is dissipated in the ground at a higher temperature in winter, relative to ambient, and to a lower in summer again relative to the ambient air. This improves the thermodynamic efficiency of the heat pump cycle, increasing the COP, thus reducing the electrical energy requirements for the compressor of the heat pump.

The usual type of ground heat exchanger is a U-type plastic pipe, shown in Fig. 14.11, going all the way down the borehole and returns back. The inside of the heat exchanger pipe is corrugated, as shown in Fig. 14.11b, to assure turbulent flow and thus greater heat exchanger coefficients. The end of the U-pipe heat exchanger is shown in Fig. 14.11c. The usual depth of the borehole is around 100 m. This is also the major cost of any geothermal system because the borehole needs to be drilled with special machinery, install the U-type pipe, and then refill the borehole with the same material taken out when drilled or with bentonite which also improves the performance of the heat exchanger in case of horizontal flowing water stream. Alternative designs include two or three U-type heat exchanges in the same borehole. It should be noted that a lot of space is required around a borehole to work effectively without ground temperature saturation. This distance is around 6 m between nearby boreholes.

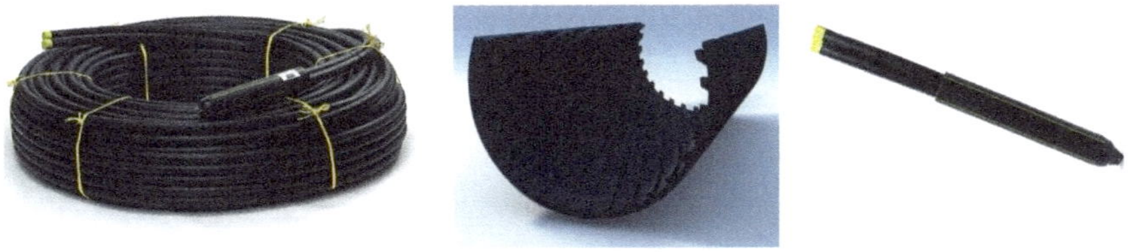

(a) The U-pipe heat exchanger role (b) The inside of the plastic pipe (c) The tip of the heat exchanger

**Fig. 14.11** Geothermal pipe heat exchanger

## 6 District Heating and Cooling

District heating and cooling distributes heating and cooling from a central energy system through underground installed pipes to the buildings connected to the system. These buildings can be a neighborhood, a block of buildings, campus buildings, etc. Therefore, individual buildings connected to the system do not need boilers, chillers, or cooling towers.

Customers use the hot water and chilled water to meet their space heating, water heating, and air-conditioning needs. Once the energy (hot water, steam, and chilled water) is used by the individual buildings, the water is returned to the central plant which is then either re-heated or re-chilled and then re-circulated throughout the underground, closed-loop pipes of the District Energy System.

A heating network generates and distributes heat in the form of hot water and superheated steam employing one or more generating units. They generally use a range of different primary energy sources for heat generation, including locally generated energy and renewables in the form of household waste incineration, biomass, biogas, solar, geothermal, combined heat and power (CHP) systems, heat recovered from wastewater, and any other form of energy that is available in the area either as a source or as a waste (e.g., from a local power plant or local industry). A cooling network is a centralized system that provides chilled water to supply an air-conditioning system. It generally includes chilled water production and distribution pipes to provide cooling services to all connected buildings. Operating as a closed circuit, the cooling network always includes two pipelines: one supplying chilled water to users and the other returning the used water to the production units (Rezaie and Rosen 2012).

A latest innovation in district heating and cooling systems is the use of low circulating temperature busted to the required temperature by a heat pump. This is also called fifth-generation district heating and cooling system.

The central energy plant is at the heart of every District Energy System. These plants operate at 90% system efficiency and are about 300% more efficient and environmentally friendly than typical single building plants.

## 7 Conclusions

A variety of systems are described in this chapter for the use of renewable energy in various aspects of the building to improve the prospects of sustainable energy future. These include solar water heating, space heating and cooling systems, and photovoltaics and hybrid PV/T systems which provide both thermal and electrical energy. Additionally, geothermal energy systems are presented which employ ground heat exchangers in combination with heat pumps. These renewable energy systems are used not only in buildings directly but nowadays these are also used in district heating and cooling networks to satisfy either a district, a neighborhood, a complete village, or small city. These systems are also more cost-effective nowadays, and this improves the prospects for a sustainable future.

## References

Agathokleous R, Kalogirou SA (2016) Double skin facades (DSF) and building integrated photovoltaics (BIPV): a review of configurations and heat transfer characteristics. Renew Energ 89:743–756

Agathokleous R, Kalogirou SA, Karellas S (2018) Exergy analysis of naturally ventilated building integrated photovoltaics/thermal (BIPV/T) system. Renew Energ 128(Part B):541–552

Batagiannis P, Gibbons C (2001) Thermal assessment of silicon-based composite materials used in photovoltaics. In: Conference proceedings of renewable energy in maritime island climates, Belfast, UK, pp 151–157

Brinkworth BJ, Cross BM, Marshall RH, Hongxing Y (1997) Thermal regulation of photovoltaic cladding. Sol Energ 61(3):169–178

Brinkworth BJ, Marshall RH, Ibarahim Z (2000) A validated model of naturally ventilated PV cladding. Sol Energ 69(1):67–81

Kalogirou SA (2014) Solar energy engineering: processes and systems, 2nd ed. Academic Press, Elsevier Science. ISBN 978-0-12-397270-5

Kalogirou SA (2016) Building integrated solar thermal systems—a new era of renewables in buildings. Bul Chem Commun 48(Special Issue E):102–108

REN (2022) Renewables 2022—global status report. REN 21

Rezaie B, Rosen MA (2012) District heating and cooling: review of technology and potential enhancements. Appl Energ 93:2–10

Tonui JK, Tripanagnostopoulos Y (2008) Performance improvement of PV/T solar collectors with natural air flow operation. Sol Energ 82(1):1–12

**Soteris Kalogirou** is at the Department of Mechanical Engineering and Materials Sciences and Engineering of the Cyprus University of Technology, Limassol, Cyprus. In addition to his Ph.D., he holds the title of D.Sc. He is a Fellow of the European Academy of Sciences and Founding Member of the Cyprus Academy of Sciences, Letters and Arts. He is Editor-in-Chief of Renewable Energy and Honorary Editor of Energy.

# Global Policies and Practices for Transforming Resilient City-Regions

**Peter Head**

**Abstract** In the recent 20 years, the world has coped with urbanisation, congestion, pollution, and climate change while trying to improve the quality of life. The evolution of policies to set frameworks and provide goals to help guide decision-making and investment globally is described, and examples of their application in cities are given. Tools and models to support transformation are similarly described. Case studies for leading examples of changes made in cities towards a more resilient, sustainable, and liveable future are provided from London, Shanghai, Accra, Singapore, Amsterdam, and Melbourne. These include urban forests, urban agriculture, water storage and re-use, flood management using nature-based solutions, walking, cycling, and electric public transport for mobility, and using waste as a resource.

**Keywords** Transformation · Well-managed urbanisation

## 1 Policies and Delivery Frameworks Sustainable Development Goal

In the 20 years between 2002 and 2022, the world was coping with urbanisation, congestion, pollution, and climate change while trying to improve quality of life. Policies were evolving to set frameworks and goals to help guide decision-making and investment outcomes, and some experience in their use was gained.

The landmark Brundtland Commission Report "Our Common Future" set out the "Sustainable Development" agenda, anticipating that almost half of humanity will live in cities by 2002. Continued population growth suggested that the developing world must massively increase its capacity to produce and manage its urban infrastructure, services, and shelter merely to maintain current inadequate conditions.

P. Head (✉)
CBE FREng FRSA 2022, London, UK
e-mail: peter.head@ecosequestrust.org



R. Yao (ed.), *Resilient Urban Environments*, Cities and Nature,
https://doi.org/10.1007/978-3-031-55482-7_15

Cities in developed economies like London faced problems, deteriorating infrastructure, environmental degradation, inner-city decay, poor schooling, and neighbourhood collapse with poor air quality leading to serious health problems. By the time of the London 2012 Olympics, the city had reinvented itself as a world-leading sustainable city, and this chapter will later explain how this was done.

The Brundtland report went on to say that "good city management for sustainable development requires decentralization of funds, political power, and personnel—to local authorities, which are best placed to appreciate and manage local needs".

In 2012 the United Nations initiated consultations on the goals and frameworks. The process to create the SDGs was initiated at the Rio +20 conference on sustainable development in Rio de Janeiro in 2012.

It was believed a city goal was critical. Cities and metropolitan areas were powerhouses of economic growth—contributing about 60% of global GDP. They also account for about 70% of global carbon emissions and over 60% of resource use. A focus on slum dwellers and the provision of adequate homes needed to be included, with associated critical issues like waste collection, water and sanitation systems, mobility, worsening air pollution, and unplanned urban sprawl. It was argued strongly that a systems approach to sustainability, resilience, and well-being was important, and that protection of natural and human heritage was included. Eventually, after a huge international effort, SDG Sustainable Cities and Communities was agreed with the set of objectives and targets to be used along with other goals. The goals were finally agreed upon at the UN in New York on 27 September 2015 (Fig. 15.1).

The focus is on disaster risk management and reduction as opposed to disaster management. New global objectives and principles were adopted:

**Fig. 15.1** Picture of all 17 sustainable development goals (UNDP 2023)

- The reduction of disaster risk as an expected outcome;
- A goal focused on preventing the new risk;
- Reducing existing risk and strengthening resilience which aligns with SDG11 in cities;
- The primary responsibility of states is to prevent and reduce disaster risk, with all-of-society and all-of-State institutions engagement;
- The scope of disaster risk reduction is broadened significantly to focus on both natural and man-made hazards and related environmental, technological, and biological hazards and risks;
- Health resilience is strongly promoted throughout.

Climate change and global warming continued to accelerate, and the Conference of the Parties COP meetings continued each year throughout this period. A legally binding international treaty on climate change was eventually agreed in 2015—The Paris Agreement. It was adopted by 196 Parties at COP 21 in Paris on 12 December 2015 and entered into force on 4 November 2016. Its goal is to limit global warming to well below 2, preferably to 1.5 °C, compared to pre-industrial levels.

As these global frameworks were coming together, the world's urban community were working to create guidance on how well-planned and well-managed urbanisation could enable sustainable development and resilience for developing and developed countries. The New Urban Agenda was adopted at the United Nations Conference on Housing and Sustainable Urban Development (Habitat III) in Quito, Ecuador, on 20 October 2016. It is a useful document, laying out qualitative standards and principles for the planning, construction, development, management, and improvement of urban areas.

All these frameworks and legal agreements needed to be brought together in cities, to focus on risk-informed sustainable development, achieving better resilience and long-term health and well-being as shown in the diagram (Fig. 15.2) from UNDRR GAR 19. This chapter explores how this complex process can be done.

A new report from UNDRR GAR 2022 called "Our World at Risk" has a sobering assessment since 2015.

> Risk creation is outstripping risk reduction looking at disasters, economic loss and the underlying vulnerabilities that drive risk, such as poverty and inequality. At the same time ecosystems and biospheres that support all life are at risk of collapse.

This is a grim worrying assessment. Global systems are becoming more connected and therefore more vulnerable in an uncertain risk landscape.

Local risks such as COVID-19 quickly become global risks, and these, like climate change, are having major impacts in every locality. Indirect, cascading impacts of these local and global problems like droughts and food insecurity are being observed as different tipping points being crossed.

The assessment in GAR2022 is clear. Unless we build resilience to systemic risk in city regions, the SDGs cannot be achieved. An ever-growing percentage of GDP is being spent on disaster recovery, leaving less and less for health, education, and other basic services to improve SDG performance. Therefore, SDG metrics are going

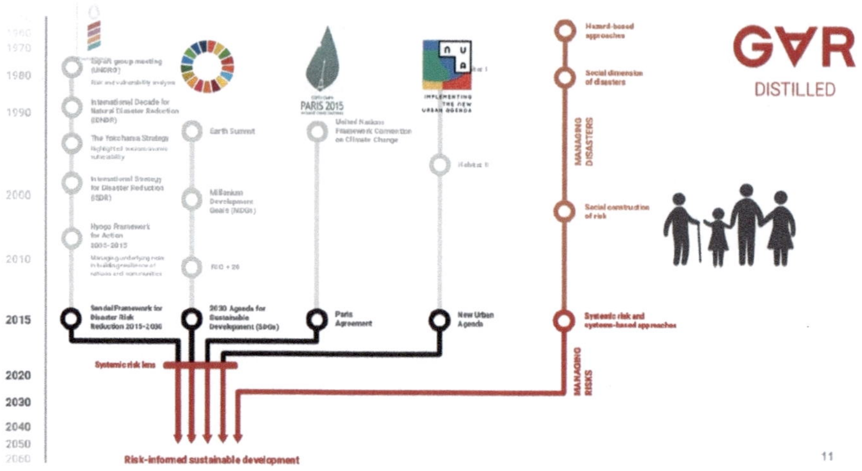


Fig. 15.2 UNDRR GAR 2019 spell out this acronym

backwards. This is the huge challenge the world now faces in implementation up to 2030.

## 2 The Systems Approach

This section discusses the **practical implementation of SDGs Framework in** London.

In 1999, in response to the Brundtland Commission Report, the UK government set out a strategy to help "deliver a better quality of life" through sustainable development. It was used to create a set of indicators which were rolled out quickly across all government departments including energy, transport, construction, environment, and water and became "Securing the Future in 2005". Devolution of power to city regions was a key policy too in line with Brundtland, and the re-establishment of the Greater London Authority GLA with a population of 4 million was a priority. A new Mayor was elected to run the authority in 2000 and the Business pressure group London First decided to explore how businesses could support sustainable development in practice. A new Sustainability Unit and worked quickly with businesses to explore what could be done to implement sustainable development in the City. In 2002 the new Mayor Ken Livingstone decided to set up an independent London Sustainable Development Commission, and a 20 strong Commission was appointed. The Commission set about commenting on the London Spatial Plan and other policies, pointing out where sustainable development was not being followed. It was decided to create a Sustainable Development Framework to set out the vision of a "World Class" city of the future with the greatest possible sense of physical,

**Fig. 15.3** London sustainable development commission framework

emotional, intellectual, and spiritual well-being and a guide for policy development and appraisal of projects. It was called the 4R's (Fig. 15.3).

A set of metrics and Quality of Life Indicators were used to monitor progress annually. The level of detail and data evidence underpinning each indicator were very impressive as shown in the report for 2017, and these metrics were used to inform the evolution of policies across the GLA to support positive change.

The London congestion charge scheme was introduced in 2003 and is one of the largest congestion charge zones in the world. The charge not only helped to reduce high traffic flow in the city streets but also reduced air and noise pollution in the central London area and raised investment funds for London's transport system (the proposed extension is highly controversial and opponents predict a significant decline in small business, huge opposition in Oxford and elsewhere, so need to acknowledge there are conflicting interests and therefore a political element to these measures, particularly if this book is presented as a manual of what to do). By introducing an increase in the number of buses there was an unprecedented mode shift from cars to buses compared to other cities—from 37% use of public transport in 2001 to 47% in 2012. Space in streets was freed up to build dedicated cycle lanes and super-bikeways (Fig. 15.4). This also helped reduce London's primary $CO_2$ emissions over the same period by 14%.

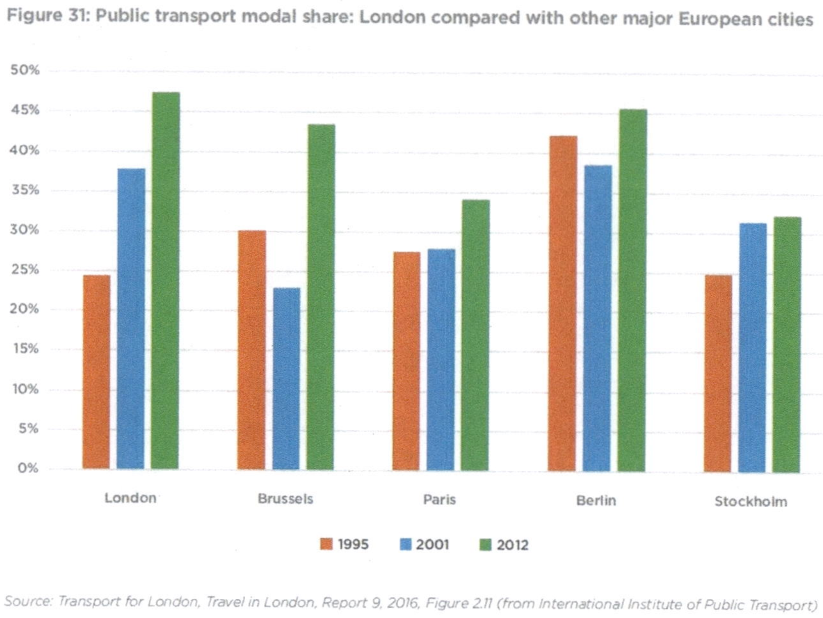


**Fig. 15.4** Mode share change in London from the congestion charge

The team at the Greater London Authority used the Sustainable Development Framework to improve and integrate sector policies, which became mutually supportive. Education was radically improved, with London moving from worst to best in UK regions, water quality in the Thames was improved and fish returned, and green spaces were enlarged and accessibility to them improved. Biodiversity indicators showed improvement while new hospitals and health facilities were rolled out with improved bus access. All this was achieved despite London reeling from the 2008 financial crisis with soaring unemployment and the economy shrinking.

In 2002 London decided to bid to host the 2012 Olympics, and the Sustainable Development Framework was used to train the bid team, who decided to aim to create the most sustainable Games ever staged. The bid was won on this basis, and implementation was monitored by the London Sustainable Development Commission, who made sure that sustainable procurement was at the heart of delivery and creating a sustainable city legacy.

Procurement objectives for the sustainability of Games delivery were set; the Commission was asked to verify their validity. They advised they were good "as long the leaders made sure they were actually delivered" because by 2012 they would be the norm as the industry progressed and to their credit, this was what happened.

London 2012, which cost £9.3 billion, led to the regeneration of the eastern part of the city, which turned a 500 acre (200 hectare) area of mostly toxic wasteland into the Olympic Park, where most events were staged and which would see the construction of 11,000 homes after the Games ended. Many more residents in the East of London now have direct access to large green spaces and new leisure facilities.

A very big downside of all this success came after 2008, in the lead-up to the Olympics, when London was seen to be a great attractive future sustainable city and property investors all over the world brought their money to the city. At the same time, the new mayor lowered expectations of building affordable homes as investment poured in. The new affordable homes promised in East London did not happen. House prices and rents rose very quickly, and homes became unaffordable to everyone, with house price-to-salary ratios rising to 25:1. Housing costs have driven up child poverty, and London now has the worst level of child poverty in the UK. This is not sustainable or resilient and is a lesson for resilience building. It is important to tackle all social and environmental aspects at the same time. Another growing problem is that a lot of new development around London has not included porous surfaces, and flooding has become more of a problem as rainfall intensities increase with climate change.

## 3 New Modelling Approaches to Deliver Systems Transformation

The New Urban Agenda sets out the paradigm shift that is needed to achieve sustainability and resilience based on the science of cities; it lays out standards and principles for the planning, construction, development, management, and improvement of urban areas along its five main pillars of implementation: national urban policies, urban legislation and regulations, urban planning and design, local economy and municipal finance, and local implementation.

Integrated urban and territorial planning is recognised as being critical, based on the principles of equitable, efficient, and sustainable use of land and natural resources. This needs to be backed by data collection, mapping, modelling analysis, and evidenced-based knowledge sharing.

The systems problems we are facing, climate change, ecological destruction, and pollution, need systems solutions. An opportunity that is available comes from the rapid evolution of open data, satellite data, artificial intelligence, and computing power to analyse complex city systems and find resilient pathways for the future.

Local data and systems modelling tools and technologies, embedded in a generic platform, can be used for real collaboration to enable the integration of human and ecological systems with economics, all centred around improving human well-being. Such innovative, open-source tools can enable the use of local open data and are now being developed for this purpose. They model resource flows in different scenarios for development and can therefore be used to steer urban decoupling and a circular economy. They can also enable different policy and tax regimes to be assessed and associated revenues to be calculated for these scenarios, and these calculations can be linked to municipal finance tools. Models (Fig. 15.5) can be set up at national,

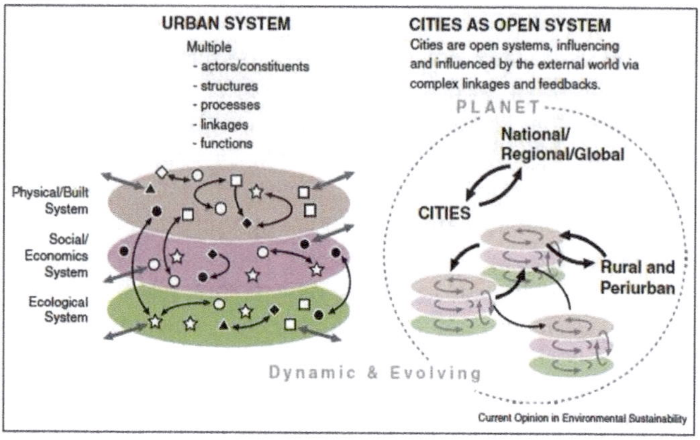


**Fig. 15.5** Open systems modelling of cities

regional, and local scales, used to assess financial relationships between them and can then inform project selection frameworks developed by the banking community for designing sustainable infrastructure. In this way, national infrastructure planning can be integrated with regional planning and forms of monitoring.

An example of new emerging models is the integrated systems resource modelling toolkit—resilience.io designed and prototyped by The Ecological Sequestration Trust.

The toolkit design (Fig. 15.6) allows temporal and spatial economics of any region of the world to be operationalised by simulating a functioning society, its relevant building blocks, dynamic entities, and transfers.

The entire model design is centred on depicting all relevant natural and human-operated processes in an economy (or a sector) as a computation of material and energy balances, driven by labour inputs where appropriate, such that the resource basis of societies becomes fully transparent.

By representing the economy in such a resource-based manner, the economy is captured as a system of systems, to assert interrelations between all economic sectors, including the population as a driver of processes, as well as exploring the benefits of alternative interventions (technology, policy, economic). This is inclusive of activities normally not captured in economics including the informal or "not registered" economy, which can be identified on a similar basis of resource flows with inputs and outputs driven by humans and their activities.

This enables the transition to a circular economy to be explored and transition policies and strategies to be put in place. Urban decoupling and exposure to climate and disaster risks can be assessed and implemented in every investment decision. The Platform can be used to explore different scenarios as part of the design process. This approach has been demonstrated with a limited prototype of the resilience.io platform in the Greater Accra Metropolitan Area, Ghana, and has been used in Australia and the UK.

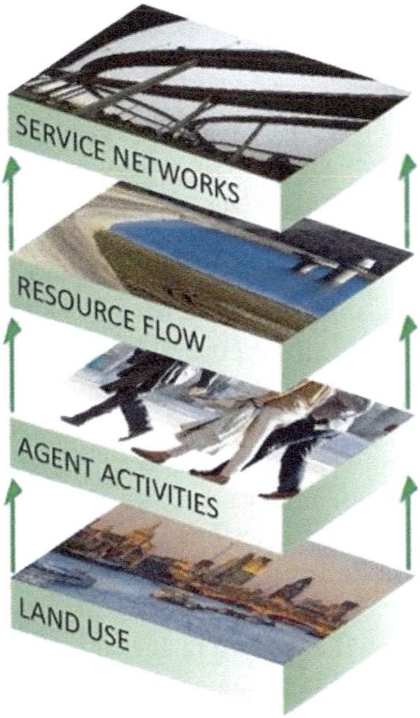


**Fig. 15.6** *Resilience.io* integrated systems modelling toolkit

Platforms like this enable multiple stakeholders, including communities and micro-, small- and medium-sized private enterprises, to collaborate for the best possible ecological-social-economic systems health outcomes. The aim is that systems modelling can be used by academia, for policy development, setting regulations and creating governance structures. There are real benefits and cost savings in such an approach.

A recent assessment from the World Bank shows that significant savings are possible using such an integrated systems approach to create a policy mix that will enable countries to achieve Sustainable Development Goals while limiting spending on new infrastructure systems from an estimated 8% of Gross Domestic Product per year to 4.5% of Gross Domestic Product per year, a full 40% reduction. This mix includes universal access to water, sanitation, and electricity; greater mobility; improved food security; better protection from floods using nature-based solutions; and eventually full decarbonisation.

## 4 New Approaches to Impact Financing

Cities are human creations, places in which we aspire to enable inhabitants to lead peaceful, healthy, prosperous, and free lives with full respect for human rights for all. They are places in which we aim to achieve gender equality, empower women and girls, reduce poverty, create jobs, and generate equitable prosperity. Cities present an opportunity for the inhabitants to commit to sharing resources and space in a way that ensures the lasting protection of the planet and its natural resources. Human settlements are the embodiment of the human spirit, where we determine our rights and responsibilities, both as individuals and collectively.

Enhanced performance requires a new relationship, or "smart partnership", between national and local governments, supported by the private sector, with a clearer recognition of the role of cities in the achievement of national development goals, but also a more effective division of labour at the local level which can strengthen the efficiency and integration of urban public services by municipalities and local governments. Strengthened urban management by the public sector can create new incentives, and an enabling environment, for private sector investment towards the provision of critical infrastructure which also supports citizens' well-being and resilience, the bedrock of the Sustainable Development Goals and Sendai Framework.

Public sector finances have been depleted by the COVID-19 emergency and are also being increasingly challenged by the need for funding climate change impact disaster recovery. Private sector investment is therefore increasingly being sought to support improved resilience and human and ecological health.

The establishment of a transparent and sound regulatory framework for investment is a prerequisite for private financing. Such a framework ensures that contracts are upheld, that local governments are protected from expropriation, and that commercial disputes can be arbitrated. In the absence of this framework, there is a lack of confidence in municipal capital investments among institutional and individual investors with negative implications for domestic debt markets. Unlocking the transformative power of financial resources requires a key which can be found in the untapped potential of human and social capital. This key unlocks constraints which hinder the flow and development of human talent, creativity and innovative capacity, which is seriously needed to shift away from financing business as usual.

A consortium has been working in 2021/22 to establish Trust2Impact which is bringing together a system financing process that allows existing financial innovation to be plugged and played into a portfolio of interventions in a city region and to capture the "hidden" cash flow value of social and environmental improvement interventions. Leading entrepreneurs are identified to accelerate transformational progress towards the desired local social and environmental outcomes. In this process, social entrepreneurs retain sweat equity in the upscaling of their mission and benefit from the greater value created by the collaborative systems approach. This value also accrues to local businesses, particularly through larger market opportunities linked to clean-up and better community health, and they would be invited to contribute to the overall funding, contingent on these benefits being realised.

This will enable emerging economy communities to move from being charitable beneficiaries to being both empowered consumers and auditors of social goods. Local communities will be encouraged to audit outcomes through access to open data.

As many growth markets have shown, development can be facilitated by legal and technology innovation including data and modelling systems such as those described in this chapter. This can be used to create an enabling environment that facilitates a measurable, higher, replicable socio-economic impact while reducing both risk and unit cost. In turn, this should lead to a scaled, virtuous impact circle producing broader revenue and predictability, while at the same time "hardwiring" social stakeholders' agendas. In essence, what is being proposed is a complete "re-think" which offers a system where the quicker you solve the social and connected environmental issues, the higher the return is produced. This could potentially unlock local pension funds and other sources of capital, well in excess of the 4.5% Gross Domestic Product needed.

United Nations Development Programme is following this thinking, proposing the introduction of "Systemic Funds", financing structures that have the mandate to develop systemic portfolios and flexibly deploy financial resources to them. "Systemic Fund designs and invests into entire portfolios at once, rather than single project interventions. It is multi-asset in nature, able to deploy different types of financing depending on the nature of the option to be invested in, from short-term grants to impact- and return-driven equity or credit investments".

## 5 Negative Impacts of the COVID-19 Pandemic and Climate Change

Data from 2019 shows that the impacts of climate change, and associated increasing inequality across and within countries, are undermining progress on the sustainable development agenda, threatening to reverse many of the gains made over the last decades that have improved people's lives. Extreme weather conditions, more frequent and severe natural disasters, and the collapse of ecosystems are causing increased food insecurity and are all worsening people's safety and health, forcing many communities to suffer from poverty, displacement, and widening inequalities.

Countries are having to spend an increasing sum of 2–4% of Gross Domestic Product per annum on disaster recovery as countries are hit by droughts, floods, and heatwaves. The investment levels required to deliver Sustainable Development Goals in cities are estimated to be at least 4.5% of Gross Domestic Product per annum, so investing nearly half of this in climate disaster recovery will have a very big and increasing impact on the ability to move the Sustainable Development Goals forward in many countries.

The Sustainable Development Goals Report 2020 brings together the latest data to confirm that, before the COVID-19 pandemic, progress remained uneven, and the world was not on track to meet the Goals by 2030. Some gains were visible: the share of children and youth out of school had fallen; the incidence of many communicable diseases was in decline; access to safely managed drinking water had improved; and women's representation in leadership roles was increasing.

At the same time, the number of people suffering from food insecurity was on the rise, the natural environment continued to deteriorate at an alarming rate, and dramatic levels of inequality persisted in all regions. Change to deal with these issues was not happening at the speed or scale required. The impacts of environmental deterioration are taking a toll on people's lives. Extreme weather conditions, more frequent and severe natural disasters, and the collapse of ecosystems are causing increased food insecurity and are all worsening people's safety and health, forcing many communities to suffer from poverty, displacement, and widening inequalities.

COVID-19 is another unprecedented health, economic, and social crisis and is now threatening lives and livelihoods, making the achievement of goals even more challenging. No country has been spared. Health systems in many countries have been driven to the brink of collapse. The livelihood of half the global workforce has been severely affected. More than 1.6 billion students have been out of school, and tens of millions of people are being pushed back into extreme poverty and hunger, erasing the modest progress made in recent years.

The Ukraine War is threatening food security further, so the situation is very serious indeed.

Climate change and COVID-19 have made the mounting debt crisis in developing countries much worse. Global debt levels have continued to set new records and grew to 247 trillion US dollars in July 2019, up from 168 trillion US dollars in 2008 at the start of the financial crisis. Levels of external indebtedness are at an all-time high in developing countries, which have seen both their public and private debt burdens grow by one-third over the last decade. External debt stocks in developing countries now amount to approximately 10 trillion US dollars. Amid the uncertain global conditions described above, many developing countries are experiencing default, distress, or rapidly growing vulnerabilities. Therefore, private finance systems innovation is very important.

After all the positive feelings of progress and direction up to 2015, cities around the world are slipping, in the lead-up to 2030, towards a dangerous less resilient and less sustainable future, that must be addressed urgently.

## 6 Case Studies from China, Australia, UK, and Netherlands

### China Sponge Cities

Chinese cities have managed flooding with traditional drainage systems that take water into rivers or the sea. Increasing urbanisation, construction activity, and increased rainfall intensity connected to climate change have all meant that flooding risk in Chinese cities has increased significantly. Also, water demand has increased as cities expand, and water supply has been challenging.

Between 2008 and 2010, 137 cities were flooded more than 3 times a year, and 57 cities were flooded for more than 12 h. At the same time, rainwater that gets discharged into rivers or the sea leaves the city and does not replenish the groundwater supply, leading to scarcity of water in the city—a problem that affects 30 out of 32 Chinese megacities.

Chinese sponge cities are required to maintain 30% of the city as green space. Another 30% is dedicated to community space. This means there is enough space to create more ponds and water-absorbing parks that can capture vast amounts of water. In this 60% of the land in cities, the aim has been to use nature to retain water so it does not drain away, in ponds and wetlands and some underground storage tanks.

Existing city neighbourhoods were retrofitted one by one in each city as part of an overall water management and ecological transformation model. Swales were put into streets; pavements were made porous; and storage tanks were put underneath. Concrete riverbanks were softened with natural planting. Green areas were converted to store water and to accommodate river overflow and fitted with leisure facilities like paths and seating and increased biodiversity using native plants and species.

Sponge city's design aims to manage regular yearly intense rainfall, but not accommodate extreme less frequent storms. The retrofit systems have been tested in recent intense storms, one nominated sponge city Zhenzhou was overwhelmed, and 300 people were killed in a recent unprecedented storm. However, the economic benefits of reducing regular floods and the health benefits are significant and worthwhile.

**Melbourne Australia—Urban Cooling**

Heatwaves are getting longer, hotter, and more frequent in Australia because of climate change, with the largest impacts occurring in cities due to the urban heat island effect. Heating of hard surfaces during the day lifts the ambient air temperature in the city centre by between 2 and 6 °C compared to rural areas. The Australian Emergency Management Agency and other government organisations recognised that overheating poses a serious threat to health and well-being. Long-term exposure to extreme temperatures and heat may cause cardiovascular, respiratory, cramps, rashes, and heat stroke-related problems and affect cognitive and emotional abilities. The most at-risk groups include the elderly, children, pregnant women, patients with chronic diseases, people with physical and mental disabilities, and low-income communities. Between 1995 and 2009, the city of Melbourne suffered extreme hot weather resulting in severe drought, water shortage, and heatwaves that killed several hundred people. Lowering night-time temperatures is particularly important so that homes can be cooled down overnight.

The city subsequently implemented a detailed strategy to reduce the heat island effects and ensure the population has access to cooler city spaces on hot days.

The city produced an open space strategy and an urban forest strategy.

Since 2010, forty million Australian dollars have been invested in related initiatives including urban forests and shrubbery; green space and rainwater harvesting; permeable paving and protection of waterways; and wetlands. The following strategy was created using modelling:

(1) Urban Forest Strategy—doubling the municipal canopy cover from 20 to 40% and expanding the stormwater harvesting network to capture 50% of the required water.
(2) Open Space Strategy—increasing the permeability of the city and expanding the green space network by 7.6%.

(3) Changing footpath and roof materials to use lighter pigments.
(4) Minimise extensively glazed facades or curtain walls.
(5) Implementation of active misting in communal areas for evaporative cooling.

The goal has been to cool the city by 4 °C and to lower energy use for cooling. 15,000 trees have been planted, 40 streets retrofitted to improve permeability, and an in-road stormwater harvesting system started. A four-year citizen's engagement programme educates and mobilises citizens. They are able to record temperatures as they move around the city and inform success or otherwise of measures taken (Fig. 15.7). Feedback from the community has been very positive so far as the programme evolves.

**UK London National Park City**

The National Park City movement began with a simple idea in 2013: how might the population think differently about the way they live in our cities if they applied the principles and purpose of National Parks to urban areas?

In July 2019 that became a reality when the Mayor of London signed the Charter making London the first such National Park City, receiving the support of a majority of local ward councillors across every London Borough. There are many walks across the city called "Green Chain" walks, linking up the spaces across Boroughs.

The publicity says:

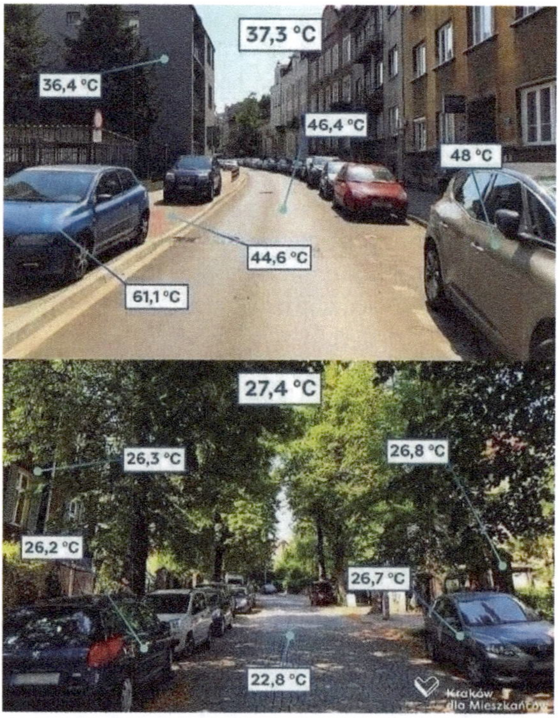


**Fig. 15.7** Measurements of temperatures in 2 adjacent streets at the same time in Warsaw

> We want more bird song, ultimate frisbee, hill-rolling, tree climbing, cycling, hedgehogs, volunteering, sharing, outdoor play, kayaking, clean air, otters, greener streets, outdoor learning, ball games, outdoor art and hilltop dancing in the city!

The overall London National Park City metrics are:

- 1572 $km^2$ in area
- 3.8 million gardens
- 8.6 million people
- 8.3 million trees
- 14,000 species
- 30,000 allotments
- 3000 parks
- 300 farms
- 50 + canoe clubs
- 1000 kms of signed footpaths
- 850 kms of rivers, canals and streams
- 14,000 wildlife species
- 2 Special Protection Areas
- 3 Special Areas of Conservation
- 4 UNESCO sites
- 37 SSSIs
- 142 Local Nature Reserves
- 1400 Sites of Importance for Nature Conservation
- 2 National Nature Reserves—Richmond Park and Ruislip Woods
- 47% green—parks, gardens, woodlands, grassland
- 2.5% blue—ponds, canals, rivers, lakes
- 9–10% of London is domestic buildings
- Roads make up 12% of London's space.

There are four aims for the London National Park City:

(1) Connect more people to nature and the outdoors
(2) Create more high-quality green and blue space
(3) Promote the rich cultural life enabled by the outdoors
(4) Inspire more cities to become National Park Cites.

The aim is to introduce more people to the nature around them and to celebrate the work carried out by so many conservation and grassroots organisations already working hard across London and beyond.

People will be able to live better lives in cities when they have access to good quality, wilderness, and outdoors. More green space has been shown to add value to walking and cycling routes, local economies, and children's development. Nature helps manage the climate, reduces pollution, and addresses many physical and mental health issues.

Research has shown significant mental health benefits from access to green spaces in London.

Using data from 25,518 people, research shows that Londoners who live within 300 m of green space have significantly better mental well-being. Proximity to green space was more important than lifestyle factors such as employment, income, and general health.

The Boroughs and communities are now working together to map and improve local spaces to link up and create this complete vision which is rooted in the original 2002 Sustainable Development Framework.

### Amsterdam Cycling Revolution

At the start of the twentieth century, bikes far outnumbered cars in Dutch cities and the bicycle was considered a respectable mode of transport for men and women. But when the Dutch economy began to boom in the post-war era, more and more people were able to afford cars, and urban policymakers came to view cars as the travel mode of the future. Entire Amsterdam neighbourhoods were destroyed to make way for motorised traffic. The use of bikes decreased by 6% every year, and the general idea was that bicycles would eventually disappear altogether.

All that growing traffic took its toll. The number of traffic casualties rose to a peak of 3300 deaths in 1971. More than 400 children were killed in traffic accidents that year.

After years of community activism, Dutch politicians became aware of the many advantages of cycling, and their transport policies shifted—maybe the car was not the mode of transport of the future after all. In the 1980s, Dutch towns and cities began introducing measures to make their streets more cycle-friendly. Initially, their aims were far from ambitious.

Nowadays the Netherlands boasts 22,000 miles of cycle paths. More than a quarter of all trips are made by bicycle, compared with 2% in the UK—and this rises to 38% in Amsterdam.

One of the great challenges is parking. Increased bicycle use means more bikes that need parking space. Amsterdam is now the only city in the world with more bicycles than inhabitants and has a lot of experience with parking and has created a helpful manual for designing parking spaces.

It has been found that bicycle parking in busy areas requires behavioural change. There is not enough space to park a bike just anywhere. Too often, routes for pedestrians, people with disabilities or emergency services are hindered by parked bicycles. In order to keep Amsterdam accessible, cyclists have to park with more consideration for their surroundings and other people, and this requires awareness training.

In order to address the influx of bicycles at train stations, a simple solution has been proposed. The idea entails that a commuter arriving at the station by bike in the morning, parks the bicycle used to access the station in the bicycle parking area, and then boards the train to work, school, university, or elsewhere. The bicycle is then available to someone else arriving later that morning by train. The commuter arriving by train can then utilise the bicycle in cycling to work, university, and other destinations in Amsterdam. In this way, the bicycles are used in a more efficient manner, and most importantly, less (expensive) bicycle parking spaces at the train station would be required.

There is now a pilot programme, and participants need to register and use an APP so that bikes can be made available for use. During the pilot, the use of the scheme was free for participants.

Research in 2015 showed that cycling prevents about 6500 deaths each year in the Netherlands as a whole, and Dutch people have half a year-longer life expectancy because of cycling. These health benefits correspond to more than 3% of the Dutch Gross Domestic Product. The study confirmed that investments in bicycle-promoting policies (e.g. improved bicycle infrastructure and facilities) are likely to yield a high cost–benefit ratio in the long term.

It took 40 years for Amsterdam to transform into a cycle-friendly and healthier city, mainly due to pressure from the population to change. Every city in the world could do this within one generation.

**Singapore Urban Farming**

The Greening Singapore strategy was developed in 2008–9, and the full 10-year plan was published in 2012. It included a plan to become water self-sufficient and resilient by storing and reusing water but almost no food was grown in the city and so food security was very poor. With the favourable climate, abundant roof spaces, gardens, and sunlight for vertical farms, special advisers argued strongly in 2009 that the government should encourage urban farming. This has now been encouraged, and Singapore now has a thriving urban farming community, including the latest technologies in vertical farms, rooftop farming, mushroom farming, and aquaponics for fish. The plan is to produce 30% of food by 2030, and already the market has grown to SIN$22m for arable farming and SIN$33m for fish.

A leading example is the "Edible Garden City" (EGC) which has designed, built, and maintained 200 food gardens. Farm-to-table is seen to be a great way to connect city dwellers and get them interested in urban farming in Singapore.

A typical 5000 sq. ft garden supports a 72-seat restaurant, with Okinawan spinach, purslane, red leaf hibiscus, and Ulam Raja, growing in the garden terrace with climate control storage refurbished from shipping containers.

"Onhand Agrarian" operates fish farm ponds using a closed-loop system Integrated Multi-Trophic Aquaculture System (IMTRAS). Waste from fish is recycled as food for crustaceans and in turn for other bottom feeders, which break down the waste into fertiliser for algae and other plants. In turn, they release oxygen into the water to nourish clams and other shellfish. The farm supplies fresh grouper, snapper, seabass, prawns, crabs, mussels, squid, and cuttlefish to a steady clientele of subscribers who pay a fixed monthly amount for a mixed bag of whatever they harvest fresh from the farms.

The activation of under-utilised spaces has been the core of Edible Garden City's operations (80 of its farms are on top of buildings) demonstrating how built-up spaces have the potential to become productive land.

With land being scarce, multi-storey vegetable farms with LED lighting and recirculating aquaculture systems can produce 10 to 15 times more vegetables and fish than conventional farms. "Citiponics" is one of Singapore's first rooftop farms. The farm is on top of a car park, a type of structure that services almost every neighbourhood in Singapore. And vertical farms lit by LEDs are emerging within buildings too.

## 7 Reflections and the Way Forward

In 2008 a research report was written on how the world could move to a more resilient model of The Ecological Age in which greenhouse gas emissions were reduced by 80%, and the ecological footprint was reduced by a factor of 4. Many of the case studies from all over the world, shared in this chapter, show examples now emerging of the practice set out in that paper.

However, all cities need to move much faster with the transformation using data-driven systems modelling and new systems finance to mobilise private capital and local pension funds.

Perhaps the common feature for success is transformation through a decentralised local community effort through social entrepreneurs, enabled by local government and supported by the private sector.

Just when the threats are greatest there is still time for communities and young people to come together and move the change process forward towards the Ecological Age. Urban forests, urban agriculture, water storage and re-use, flood management using nature-based solutions, walking, cycling and electric public transport for mobility, using waste as a resource and gathering and using energy efficiently.

## References

Adoption of SDGs at UN September (2015) https://sustainabledevelopment.un.org/sdinaction/newsletter/september2015

New Urban Agenda UN Habitat https://habitat3.org/the-new-urban-agenda/

Amsterdam Manual for Bicycle Parking Spaces. https://bikecity.amsterdam.nl/en/inspiration/a-manual/

Breaking the Silos E4H and Geneva https://global-geneva.com/building-bridges-for-geneva-and-the-world-breaking-from-the-silos/

Resilience Brokers resilience.io https://resiliencebrokers.org/resilience-io-2/

Brundtland Commission Report Our Common Future https://sustainabledevelopment.un.org/content/documents/5987our-common-future.pdf

China 12th 5 year Plan 2011–2015 https://policy.asiapacificenergy.org/node/37

Climate Resilient Melbourne Urban Sustainability Exchange https://use.metropolis.org/case-studies/4-c-cooler-melbourne#casestudydetail

LSDC Comments on the Draft London Plan https://www.london.gov.uk/about-us/organisations-we-work/london-sustainable-development-commission

Cooling South Melbourne Impact Assessment Modelling http://uhimitigationindex.be.unsw.edu.au/static/files/Cooling%20South%20Melbourne%2019-11-20.pdf

London Sustainable Development Framework https://www.london.gov.uk/sites/default/files/lsdc_-_sustainable_development_framework_2002.pdf

Bhopal Disaster (1984) https://en.wikipedia.org/wiki/Bhopal_disaster
Dongtan-Chenjia Town Ecological City http://www.srcb.com/html/zt/chenjiazhen/index.html
Dutch Cycling-Quantifying the Health and Economic Benefits https://www.ncbi.nlm.nih.gov/pmc/articles/PMC4504332/
Entering the Ecological Age Peter Head (2008) https://www.arup.com/perspectives/publications/speeches-and-lectures/section/entering-the-ecological-age-the-engineers-role
Hyogo Framework for Action 2005–2015 https://www.unisdr.org/2005/wcdr/intergover/official-doc/L-docs/Hyogo-framework-for-action-english.pdf
From funding projects to funding portfolios https://medium.com/@undp.innovation/from-funding-projects-to-funding-portfolios-b14c744f8adf
G20 Policy Brief on Integrated, Data Driven, Risk-Based Planning of Sustainable Infrastructure https://www.g20-insights.org/wp-content/uploads/2020/11/T20_TF3_PB4.pdf
Ghana SDG Baseline Report (2008) https://mofep.gov.gh/sites/default/files/news/Ghana%27s-SDG-Budget-Baseline-Report-Aug-09-18.pdf
Great Hanshin Earthquake (1995) https://en.wikipedia.org/wiki/Great_Hanshin_earthquake
IABSE Annual Meetings (1995) San Francisco https://www.tandfonline.com/doi/abs/10.2749/101686695791555862
Kyoto Protocol https://en.wikipedia.org/wiki/Kyoto_Protocol
London National Park City https://www.nationalparkcity.london/what-is-the-london-national-park-city
London Quality of Life Indicators (2017) LSDC https://www.london.gov.uk/sites/default/files/lsdc_-_qol_2017_evidence_report.pdf
London Sustainable Development Commission 2002–2022 https://www.london.gov.uk/about-us/organisations-we-work/london-sustainable-development-commission
London (2012) Olympics—Sustainable procurement https://assets.publishing.service.gov.uk/government/uploads/system/uploads/attachment_data/file/224038/pb13977-sustainable-procurement-construction.PDF
Overview of China Sponge City Programme https://growgreenproject.eu/wp-content/uploads/2021/01/Sponge-City-Programme-in-Wuhan-China.pdf
Report of World Summit on Sustainable Development Johannesburg 2002. https://documents-dds-ny.un.org/doc/UNDOC/GEN/N02/636/93/PDF/N0263693.pdf?OpenElement
Rozenberg J, Marianne F (2019) Beyond the gap: how countries can afford the infrastructure they need while protecting the planet. World Bank Group, Washington, DC. https://openknowledge.worldbank.org/handle/10986/31291
Sendai Framework for Disaster Risk Reduction 2015–2030 https://www.undrr.org/publication/sendai-framework-disaster-risk-reduction-2015-2030
Second Severn Crossing https://severnbridges.org/design-and-contract-preparations/
Shanghai Expo Legacy Better City Better Life https://www.bie-paris.org/site/en/focus/entry/not-just-better-city-better-life
Sichuan Earthquake A disaster on an immense scale BBC https://www.bbc.co.uk/news/science-environment-22398684
Singapore urban agriculture https://qz.com/1985399/3-ways-singapore-is-creating-food-security-with-urban-farms/
Green Space is Good for Your Health https://warwick.ac.uk/newsandevents/pressreleases/green_space_is/
The Paris Agreement UNFCCC (2016) https://unfccc.int/process-and-meetings/the-paris-agreement/the-paris-agreement
UK Government Securing Our Future https://www.gov.uk/government/publications/securing-the-future-delivering-uk-sustainable-development-strategy
UN Millennium Summit (2000) https://documents-dds-ny.un.org/doc/UNDOC/GEN/N02/636/93/PDF/N0263693.pdf?OpenElement
UN Sustainable Development Goals Progress Report (2019) https://www.un.org/sustainabledevelopment/blog/2019/07/sdg-progress-report-2019/

UN Sustainable Development Goals Progress Report (2020) https://unstats.un.org/sdgs/report/2020/
UNDP, United Nation Development Programme, What are the Sustainable Development Goals?, https://www.undp.org/sustainable-development-goals, Accessed 15th May, 2023
UNDRR GAR 2022 our world at risk https://www.undrr.org/gar2022-our-world-risk

# Zero-Carbon City and Community in Japan—Policies, Proposals, and Examples

**Hiroshi Yoshino**

**Abstract** This chapter discusses policy trends for zero-carbon cities (communities) in Japan, recommendations by academia, and examples of decarbonized city planning based on literature. Regarding policy trends, the overall movement since the 2050 carbon neutral declaration by the former Prime Minister in October 2020 is discussed. Further, the Plan for Global Warming Countermeasures in 2021, which intends to create more than 100 decarbonized leading areas by FY2030, is highlighted. Next, the proposals made by the Architectural Institute of Japan and the Science Council of Japan regarding zero-carbon cities and communities are explained and commented upon. In the latter half of this paper, an overview of the decarbonization efforts in 26 regions selected as decarbonization leading areas and specific examples of four regions is provided. In conclusion, the author summarizes the efforts regarding zero-carbon cities/communities via enhanced shelter performance by insulating and air-tightening buildings, improving facility efficiency, and introducing renewable energy sources.

**Keywords** Carbon neutral · Cities/communities · Policy trend · Decarbonization leading area · Thermal insulation and airtightness

## 1 Introduction

In October 2020, former Prime Minister Yoshihide Suga declared the goal of carbon neutrality by 2050 in Japan, which resulted in intensified efforts toward decarbonization. However, since the enactment of the Act on the Rationalization Energy Use (METI 1979) in 1979, the government has responded to global trends such as the holding of the Conference of the Parties to the United Nations Framework Convention on Climate Change (UNFCCC) and publishing reports from the Intergovernmental Panel on Climate Change (IPCC) by implementing various measures pertaining to energy conservation and decarbonization, including the Act on the Promotion of

---

H. Yoshino (✉)
Tohoku University, Sendai, Japan
e-mail: yoshino@sabine.pln.archi.tohoku.ac.jp

 
R. Yao (ed.), *Resilient Urban Environments*, Cities and Nature,
https://doi.org/10.1007/978-3-031-55482-7_16

Global Warming Countermeasures (MOE 1998), which has been revised every few years; the Plan for Global Warming Countermeasures (MOE 2021a); and the Strategic Energy Plan (METI 2021), which has been revised every few years.

At the city and community level, the Low Carbon City Promotion Act (MLIT 2012) was enacted in 2012. In 2017, Act on Improvement of Energy Consumption Performance of Buildings (the Building Energy Conservation Act) (MLIT 2015) was enforced, and in 2022, it was revised to impose an obligation to comply with the energy efficiency standards for all newly built houses and non-houses. Numerous studies have shown that enhancing the insulation of houses results in improved health and comfort, whereas improvement of equipment efficiency and the use of renewable energy in houses and non-houses result in enhanced resilience. Therefore, strengthening of building insulation, improvement of equipment efficiency and encouraging the use of renewable energy are the most basic aspects in promoting decarbonization and sustainability at the city level.

Meanwhile, the Architectural Institute of Japan (AIJ) and the Science Council of Japan have issued several proposals for decarbonized cities and buildings in response to the social movements. Although these proposals are targeted toward decarbonization, they also offer other benefits of decarbonization that will results in improved health and comfort and higher resilience.

In addition to reviewing these policies and proposals, as concrete examples of plans for decarbonizing cities and regions, the proposal plans for 26 regions selected as "decarbonization leading areas" (MOE 2022a) are introduced herein, and decarbonization trends in Japan are discussed.

## 2 Zero-Carbon City/Community Policy After the Carbon Neutral Declaration

### *2.1 2050 Carbon Neutrality Declaration*

On October 26, 2020, former Prime Minister Yoshihide Suga declared the goal for carbon-neutrality by 2050 in Japan. On April 22, 2021, he presented a speech at the Leaders' Summit on Climate hosted by the United States, where he urged the residents of Japan to reduce greenhouse gas emissions in fiscal year 2030 by 46% as compared with those from fiscal year 2013, and to strive toward achieving a further reduction by 50%.

Under the Paris Agreement adopted at the UNFCCC (COP21) in December 2015, Japan's target for fiscal year 2030 is 26%. The 46% reduction mentioned above is based on the carbon-neutrality declaration by the former Prime Minister, efforts expended by developed countries after the adoption of the Paris Agreement, and "Special Report on Global Warming of 1.5 °C" (IPCC 2018) by IPCC released in October 2018. Figure 1 shows the previous past $CO_2$ emission amounts (NIES 2022)

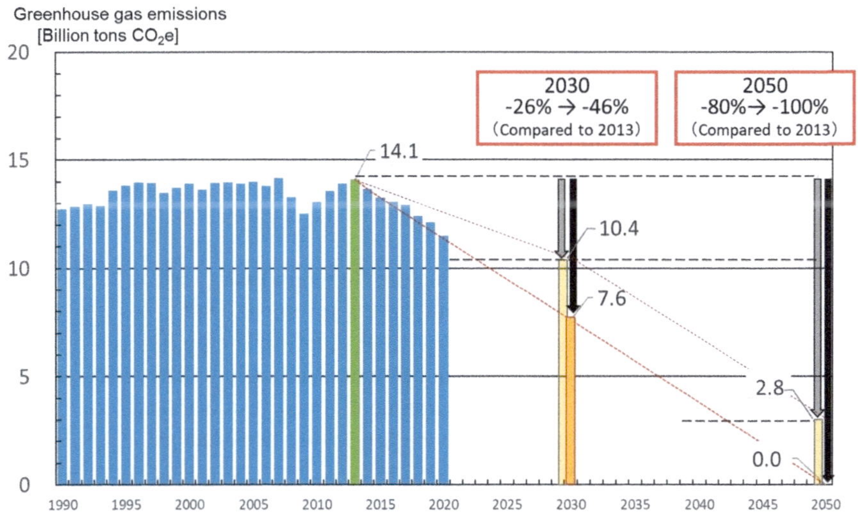


**Fig. 1** Trend of $CO_2$ emission and future reduction targets (NIES 2022). This figure was created based on Japan's GHG emissions data (FY1990—2020) by Greenhouse Gas Inventory Office of Japan, National Institute for Environmental Studies

and future reduction targets. Previous trends suggest that the targets for 2030 and 2050 are challenging.

## *2.2 Subsequent Movement After the Declaration*

In response to the carbon neutrality declaration by the former prime minister, efforts toward decarbonization intensified. In December 2020, the "2050 Green Growth Strategy Through Achieving Carbon Neutrality" (METI 2020) was announced, whose aim was to achieve zero emissions in infrastructure, urban spaces, etc. and a roadmap for promoting Zero Energy Building (ZEB) and Zero Energy Housing (ZEH)[1] was presented.

In addition, on October 22, 2021, the Plan for Global Warming Countermeasures (MOE 2021a) was revised for the first time in five years. The revised plan indicates that the $CO_2$ reduction rate by sector should be reduced by 51% and 66% for commercial and residential buildings respectively, as compared with those in fiscal year 2013. Additionally, the plan describes (1) the expansion of obligatory compliance with energy-saving standards for houses and buildings, (2) the use of renewable

[1] ZEB (net Zero Energy Building) and ZEH (net Zero Energy House) are defined as buildings that have a net annual primary energy consumptionConsumption of zero or approximately zero, which is achieved by improving the energy-saving performance of buildings and equipment, using energy across the board, and using on-site renewable energy.

energy such as solar power, which benefits local communities, (3) the local production and consumption of energy, (4) the wide-area utilization of energy, and (5) urban greening for absorbing $CO_2$. In response to the plan, the government intends to create more than 100 decarbonization leading areas by fiscal year 2030, and 26 areas have been selected by 2021. A decarbonized leading area achieves goals such as zero $CO_2$ emission associated with electricity consumption in the residential and commercial sectors based on regional characteristics. This will be discussed in detail in Chaps. 5 and 6.

Furthermore, on October 22, 2021, the 6th Strategic Energy Plan (METI 2021) was announced. In this plan, the following aspects are emphasized to promote energy policy (S+3E) while ensuring Safety: (1) a stable supply of energy (Energy security), (2) a low-cost energy supply by improving economic efficiency (Efficiency), and (3) compatibility with the environment (Environment). In the residential and commercial sectors, the government aims to ensure energy-saving performance at the level of ZEH/ZEB standards for newly built houses and buildings constructed after fiscal year 2030, and makes efforts to make it mandatory to comply with energy conservation standards under the Building Energy Conservation Act, and to enhance the level of Top Runner[2] equipment/building materials.

In addition, 823 local governments (45 prefectures, 476 cities, 20 special wards, 239 towns, and 43 villages), including Tokyo, Kyoto, and Yokohama, have announced that they will achieve "substantial zero $CO_2$ emissions by 2050" (MOE 2022b). The total population of the announced municipalities reached approximately 124.48 million, which corresponds to almost 100% of Japan's population.

## 3 Decarbonization Trend in Building and Communities

### *3.1 Energy Conservation Efforts for Buildings*

Energy conservation efforts in Japan began in 1979 after the Act on the Rational Use of Energy (METI 1979) was enacted in response to the oil crisis. All building owners are obliged to implement appropriate efforts, such as installing thermal insulation in buildings.

In 2015, the Building Energy Conservation Act (MLIT 2015) was enacted to further intensify energy conservation. Regulatory measures such as obligatory compliance with energy conservation standards for large-scale commercial buildings were implemented. Subsequently, in response to the carbon neutrality declaration, this Act was revised on June 2022. Under this law, the obligation to comply with the energy conservation standards was imposed on all newly constructed residential and commercial buildings constructed after 2025.

---

[2] Top-runner products are specific equipment and products that have the highest energy consumption efficiency among the products available on the market; they are determined based on their performance and future prospects in terms of technological development.

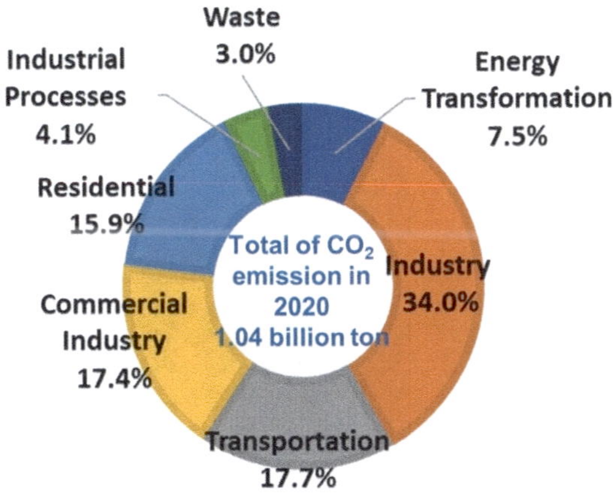


**Fig. 2** $CO_2$ emission in Japan (NIES 2022). This figure was created based on Japan's GHG emissions data (FY2020) by Greenhouse Gas Inventory Office of Japan, National Institute for Environmental Studies

## *3.2 Low Carbonization of Cities and Communities*

With regard to low-carbon cities and communities, in 2012 the Law Carbon City Act was enacted to address changes in energy supply and demand caused by the Great East Japan Earthquake, as well as increasing public awareness toward energy conservation and global warming. To provide context, Fig. 2 shows the amount of $CO_2$ emitted from urban activities, that are residential, commercial and transportation sectors, which constitutes approximately one-half of Japan's total $CO_2$ emission (NIES 2022). The Low-Carbon City Development Guidance (MLIT 2011) describes the basic concepts of a low-carbon city, particularly: (1) The basis of the development of a low-carbon city, and (2) The methodology to assess the effects of measures to develop a low-carbon city with required numerical data. Figure 3 shows the concept of the Law Carbon City Act (MLIT 2022) involving (1) compact city, (2) promotion of public transportation, (3) low-carbon buildings and (4) management of green-spaces and energy-use. Figure 4 shows the direct and indirect approaches for achieving low-carbon cities (MLTI 2011). By the end of June 2021, 26 cities had created low-carbon urban development plans under this law (MLTI 2021). These plans show that various initiatives have been introduced based on the characteristics of the cities.

## *3.3 Eco-Model City—"FutureCity"—SDGs Future City*

From 2008 to 2014, the national government selected 23 cities and regions as "Eco-Model Cities" (CAO 2023a), which were set with high goals and designated to pioneering efforts to realize a low-carbon society. In addition, six cities among the Eco-Model Cities and five cities among the cities affected by the Great East Japan Earthquake were designated as "FutureCities" (CAO 2023a), which are highly

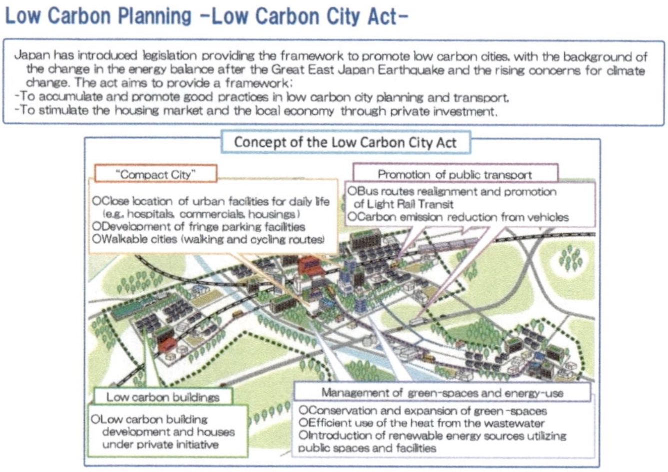


**Fig. 3** Concept of the low carbon city act (MLTI 2022). City planning division, city bureau, ministry of land, infrastructure, transport and tourism

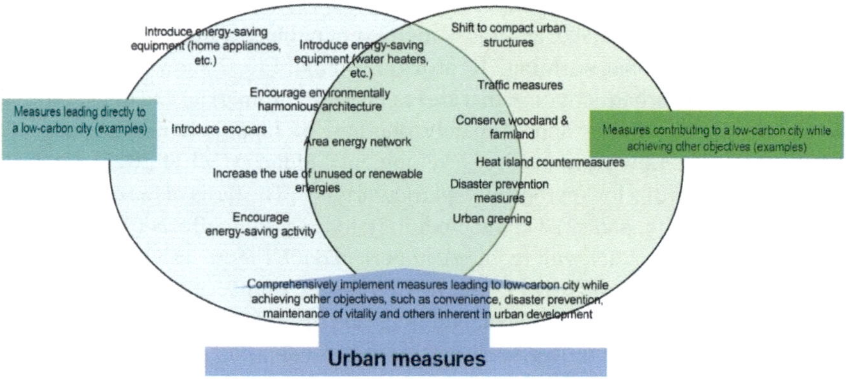


**Fig. 4** Correlation between urban and low-carbon measures (MLTI 2011)

sustainable cities that excel in the three aspects, i.e., the environment, society, and economy.

Furthermore, as its successor, the "SDGs Future City" concept (CAO 2023b) was introduced. These cities were selected from among the cities and regions selected to promote basic and comprehensive efforts that are consistent with the principles of the SDGs as well as those that can potentially realize sustainable development through the creation of new values for the economy, society, and environment. For "SDGs Future City" and "Municipal SDGs model project", 29 cities and 10 projects in 2018, 31 cities and 10 projects in 2019, 33 cities and 10 projects in 2020, 31 cities and 10 projects in 2021, and 30 cities and 10 projects in 2022 have been selected.

Municipal SDGs model projects were selected from among the SDGs Future Cities because the content of the planned implementation was an important initiative.

## 4 Proposals for Zero-Carbon Cities and Communities in the Academic Field

### 4.1 *Activities Related to Global Environmental Issues in the AIJ*

The AIJ has been established for 136 years and is one of the largest academic societies in Japan with 36,000 members. In recent years, the prevention of global warming has become a significant issue, and the Global Environment Committee has endeavored to address the issue. The Committee was established in 1995, and at the 3rd Conference of the Parties to the UNFCC held in Kyoto in 1997, the AIJ released the "Response to Climate Warming in the Field of Architectural Research (Chairman's Statement)" (AIJ 1997). It stated that $LCCO_2$ should be reduced by 30% for new buildings, and the useful life of buildings should be tripled.

In 2000, the "Architecture Charter for a Global Environment" (AIJ 2000) was published, which became a guide for addressing global warming.

### 4.2 *"Proposal of 2050 Vision: Building-Related Measures to Counteract Global Warming Towards Carbon-Neutralization" by 17 Building-Related Organizations*

The fourth IPCC report was published in 2007, and the first commitment period (2008 to 2012) of the Kyoto Protocol began in 2008. Subsequently, in December 2009, seventeen building-related organizations including AIJ published a leaflet titled "Proposal 2050 Vision: Building-related Measures to Counteract Global Warming Towards Carbon-Neutralization" (AIJ 2009).

This proposal presents the following basic concepts as methods of planning, designing, constructing, and maintaining buildings, cities, and regions toward carbon neutrality by 2050. (1) Design and operate buildings to minimize energy consumption; (2) design buildings to allow the use of renewable energy sources for providing the amount of energy consumed; (3) design and operate a building to extend its lifespan; (4) promote the utilization of eco-materials that reduce $CO_2$; and (5) promote efforts at urban and regional levels.

### 4.3 Proposal of AIJ Action Plan Toward a Low-Carbon Society in the Year 2050-Road Map for the Carbon Neutralization of the Built Environment of Japan (AIJ 2015)

In March 2015, the AIJ announced its own proposal as shown in the title, based on a proposal from 17 organizations. The proposal was coordinated by the author and other members of AIJ. The policy recommendations of this proposal are as follows: (1) Promote energy conservation and low environmental impact; (2) realize a long life and stock society; (3) promote the use of renewable energy; (4) promote the transition to a green economy; (5) conserve and promote the utilization of forest resources. In terms of research and development themes, 21 subjects related to new buildings, existing buildings and cities were proposed.

### 4.4 Establishment of the Council for Promoting Low-Carbon Society in Collaboration with 18 Building-Related Organizations and Publication of Proposals

Based on the recommendations of the proposal mentioned in Sect. 4.2, when the author was the president of the AIJ, he and the other members of AIJ urged 18 architecture-related organizations to establish the Council for Promoting Low-Carbon Society (CPLCS 2023) in July 2014. The purpose of this is to disseminate information and provide recommendations for the national government, local governments, and citizens to realize a low-carbon society.

In July 2020, the Council published "Three Proposals for Carbon Neutrality by 2050 -How to Implement Decarbonized and Resilient Human Settlements by Forming a Stock-based Society-" (CPLCS 2020). The following three recommendations were provided:

(1) Integrating energy and environmental plans into urban planning
(2) Renovating and making optimally using existing building stock
(3) Decarbonization and resilient recovery

### 4.5 Efforts Related to Low Carbonization by the Science Council of Japan

The Science Council of Japan (SCJ 2023) was established in January 1949 under the jurisdiction of the Prime Minister as a "special agency" that functions independently of the government, to reflect and expand scientific efforts within the government, industry, and citizens.

The discussion related to the decarbonization of cities and buildings has been continued since the 20th term (October 2005) by a subcommittee of the Committee on Civil Engineering and Architecture of Section III (Science and Engineering). In the 20th term (three years from October in 2004), the Construction and Society Subcommittee (Chairman: Shuzo Murakami, Professor at the Faculty of Science and Technology, Keio University (at that time)) compiled an external report entitled "Policy Proposals for Reducing Energy Consumption in Residential and Commercial Buildings."(SCJ 2004).

In the 3rd term (three years from October 2014), the "Subcommittee on Pathway to Realization of a Low-Carbon, Healthy Society, and Lifestyles/Consumer Behavior (Chairman: Hiroshi Yoshino)" compiled a proposal entitled, "Pathway to Low Carbon and Healthy Lifestyle and Cities/Buildings" (SCJ 2017). The basic ideas of the proposal are outlined as follows: (1) Economic growth must be separated from environmental degradation and avoid the paradigm of the 20th-century mass-production/mass-consumption civilization, which is evaluated by the GDP scale. (2) In the macro design of environmental problems, both the improvement in Q (i.e., the spatial performance of cities and regions) of (Q/L) and the reduction in L (i.e., the environmental load) must be considered. In terms of Q, not only economic benefits but also the sufficiency or satisfaction of life based on the quality of living, the natural environment, safety and security, amenities, etc. should be evaluated. (3) To realize a low-carbon and healthy living society, public awareness must be reformed, and a keyword is slimming down. (4) The ancient culture of "knowing sufficiency" should be regarded as a new social paradigm globally, and wisdom from the industry, government, academia, and citizens should be gathered and exploited to develop a low-carbon society.

## 5 Concept of "Decarbonization Leading Areas" and Characteristics of Selected Regions

According to the application guidelines (MOE 2021b), the target set for decarbonization leading areas is net-zero $CO_2$ emission from electricity consumption in the residential and commercial sectors by fiscal year 2030. In addition, these areas are to reduce other greenhouse gas emissions, such as those from transportation and heat use, by utilizing local resources, which is consistent with Japan's fiscal year 2030 target. The scope of the areas (MOE 2021b) includes administrative districts, villages, and facility groups that optimize power interchange and energy supply/demand using a single control technology. In particular, they include residential areas, business and commercial areas, natural areas, and facility groups.

The characteristics of the 26 regional projects selected for 2021 (MOE 2022a) are as follows:

(1) Frequently adopted decarbonization technologies include photovoltaic (PV) power and storage batteries. PV power generation systems are typically installed on the roofs of buildings or on premises within the area; however, in certain

cases, they are installed outside the area using the power purchase agreement (PPA) method.[3]

(2) Biomass power is also used in many areas, and it is generated using wood, garbage, and livestock manure.
(3) Other small scale energy sources are used, including wind power, water power, and digester gas.
(4) The promotion of ZEBs has been initiated for residential and commercial buildings. The author believes that this policy is highly demanded.
(5) New local electric power companies/public corporations, local energy management systems, local microgrids, large-scale demand responses, etc. are established in some areas.
(6) Decarbonization for transportation, conversion to electric vehicles (EVs), and adoption of fuel cell electric vehicles for public vehicles and buses have been realized.
(7) Projects featuring regional brands such as a famous castles, famous mountains, remote islands, and nickname of famous cities are conducted.
(8) One-third of the regions advocate the use of independently distributed energy and the strengthening of resilience through the construction of microgrids.
(9) Many projects have been conducted to improve health and comfort, which is a co-benefit of decarbonization.

Among these areas selected, many cities have been selected as Eco Model City and/or "FutureCity" and/or SDGs Future City. Decarbonization plans have been implemented in these cities.

## 6 Examples of Decarbonized Cities and Communities

In this chapter, four regions from among the "Decarbonization Leading Areas" are overviewed (MOE 2022b).

### *6.1 Yokohama City: Large City Decarbonization Model Challenged Through Public–Private Partnership in Minato Mirai 21 District*

Yokohama is a large city, with a population of 3.77 million. It was selected as an Eco-Model City in 2008, a "FutureCity" in 2011, and an SDGs Future City in 2018. All the

[3] PPA is an abbreviation for "Power Purchase Agreement"; it is a mechanism to provide electric power generated by a PV power generation system to electric power users of a facility for a fee. The system is installed by a company (PPA operator) that owns and manages solar power generation equipment in the space provided by the facility owner, such as the site and roof.

**Fig. 5** Yokohama Minato Mirai 21 District viewed from the bay (Photographed by the author)

plans devised target the entire city of Yokohama; however, the selected decarbonization-advanced areas include private and public offices and commercial facilities in the Minato Mirai 21 district. The Minato Mirai 21 district is a redevelopment area facing the Port of Yokohama, which is lined with skyscrapers, including the Yokohama Landmark Tower, and has a population of approximately 10,000. Figure 5 shows a view of the area from the bay. In 32 among 64 facilities in this district, decarbonization will be achieved via the following methods: (1) Using solar power generation equipment through off-site PPA in unused spaces in the suburbs of the city (rooftops of public housing and elementary/junior high schools, and regulating ponds); (2) installing solar power generation equipment on the walls; (3) utilizing existing waste power and wind power generation equipment; (4) expanding the use of renewable energy from other local governments through wide-area cooperation; (5) alleviating system pressure due to large-scale demand response. In addition, existing district heating and cooling facilities in the area will be renewed and strengthened, whereas food waste and plastic bottles in the area will be reduced or utilized appropriately (via biomass power generation, composting, reuse of plastic bottles, etc.).

## *6.2 Himeji City: Himeji Castle Zero Carbon Concept ~ Decarbonized Domino Starting from World Heritage and National Treasure "Himeji"*

Himeji City is a core city located in Hyogo Prefecture, west of Osaka, and has a population of approximately 770,000, including the population of the surrounding

**Fig. 6** Himeji castle (Photographed by the author)

municipalities. Himeji Castle, shown in Fig. 6 is located in a World Heritage Site, and is designated as an international convention and tourism city. Thirteen public facilities (Area A) in a designated historic site and 13 idle lands in the city (Area B) as renewable energy introduction areas in the suburb of Himeji City were targeted.

From the perspective of "reducing," "creating," and "replacing" energy, the content of the initiative was to achieve a "zero-carbon castle" by reducing $CO_2$ emissions associated with electricity consumption in Area A to virtually zero. Hence, the following are proposed or considered: (1) At public facilities in Area A, the power consumption is to be reduced by introducing LEDs and improving air conditioning. (2) Since the use of conventional photovoltaic power generation facilities is regulated from the viewpoint of landscape protection, next-generation PV cells such as film type cells are to be introduced. (3) A new PV power generation facility will be installed on idle lands owned by the city in Area B, and power will be supplied to public facilities in Area A through a corporate PPA. The surplus electricity generated at that time will be used to produce "green hydrogen" and to promote the decarbonization of business operators in the city. (4) The use of storage batteries to stabilize solar power generation is affected by the weather and other factors. (5) The introduction of environment-friendly next-generation vehicles for citizens and businesses in the city is supported. The relationship between Area A and B is shown in Fig. 7.

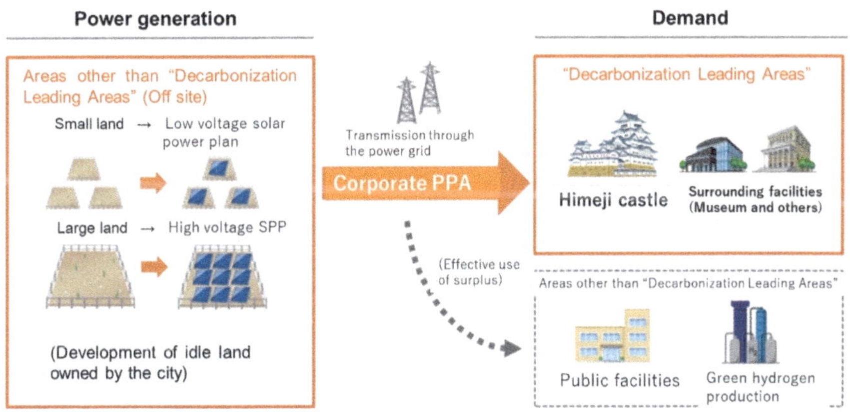


**Fig.7** The relationship between the A area(Demand: Right of Figure) and the B area(Power generation: Left of Figure (MOE 2022a). From the application materials for "Decarbonization Leading Areas" by Himeji City (jointly proposed by Kansai Electric Power Co.). (Author translated the original chart written in Japanese)

## *6.3 Kitakyushu City: Enhancement in Competitiveness of Local Industries through Introduction of Renewable Energy, Optimal Operation Model, and Horizontal Expansion in Public Facilities, etc*

Kitakyushu is the northernmost city in the Kyushu region and the second largest city in the region after Fukuoka. The population size is 920,000. It was selected as an Eco-Model City in 2008, as "FutureCity" in 2011 and SDGs Future City in 2018. The city has progressed significantly in terms of introducing the use of renewable energy. For example, in 2009, a 1.2 m pipeline was installed beginning from a hydrogen station to an urban area, and the world's first demonstration experiment at the community level was conducted to enable its use as an energy source for households, commercial facilities, and public facilities.

The decarbonization leading area targets public facilities in 18 cities and towns (1.36 million people) in the Kitakyushu metropolitan area and recycling companies in Kitakyushu Eco-Town. Through the introduction of self-consumption PV by the third party ownership (PPA), storage battery of EVs and energy efficient equipment, decarbonization will be completed for public facilities (3600 facilities) in the Kitakyushu metropolitan area (18 cities and towns) and for recycling companies in Kitakyushu Eco-Town. To reduce costs, a system for reusing and recycling used PV panels and EV batteries can be constructed via cooperation with recycling companies and automobile manufactures.

In addition, considering the characteristics of the Hibikinada area, where renewable energy is abundant, including wind power, regional management should be implemented through the use of hydrogen and the introduction of large-scale storage

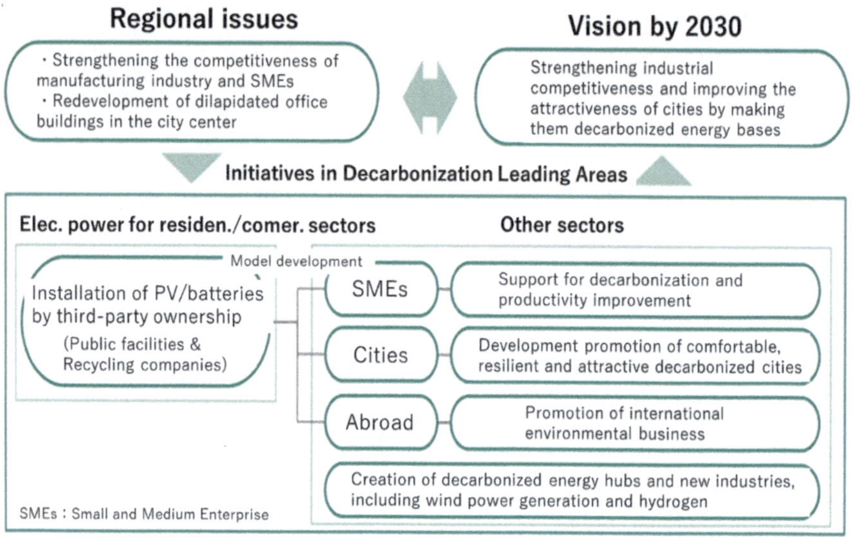


**Fig.8** Relationship between regional issues, goals to be achieved by 2030, and initiatives in the private sector and other sectors. From the internet of "Selected by the national government as a "Decarbonization Leading Area". Green Growth Promotion Division, Environment Bureau. " (Kitakyushu-city 2022) (Author translated the original chart written in Japanese)

batteries; furthermore, the use of renewable energy while reducing output control should be maximized. Figure 8 (Kitakyushu-city 2022) shows the relationship among regional issues, goals to be achieved by 2030, and initiatives in the private sectors and others.

Figure 9 shows Kitakyushu City's strategic vision (Kitakyushu-city 2021) for the future of energy beyond the decarbonization advanced area, which includes developing a wind power base, sophisticated energy storage system base, and hydrogen production and supply base.

## 6.4 Higashimatsushima City: A Future City Connected from Earthquake Reconstruction—A Future Circle Created by People, Energy, and the Region

Higashimatsushima city is located near Sendai City along the Pacific Ocean, which was devastated by the Great East Japan Earthquake in March 2011. It was selected as a "FutureCity" in 2011 and as an SDGs Future City in 2018. The decarbonization leading area includes the entire Nobiru District, which comprises 1071 houses and 71 offices in the commercial and industrial sectors. By utilizing the new local electric power operated by the Higashimatsushima Organization for Progress and Economy,

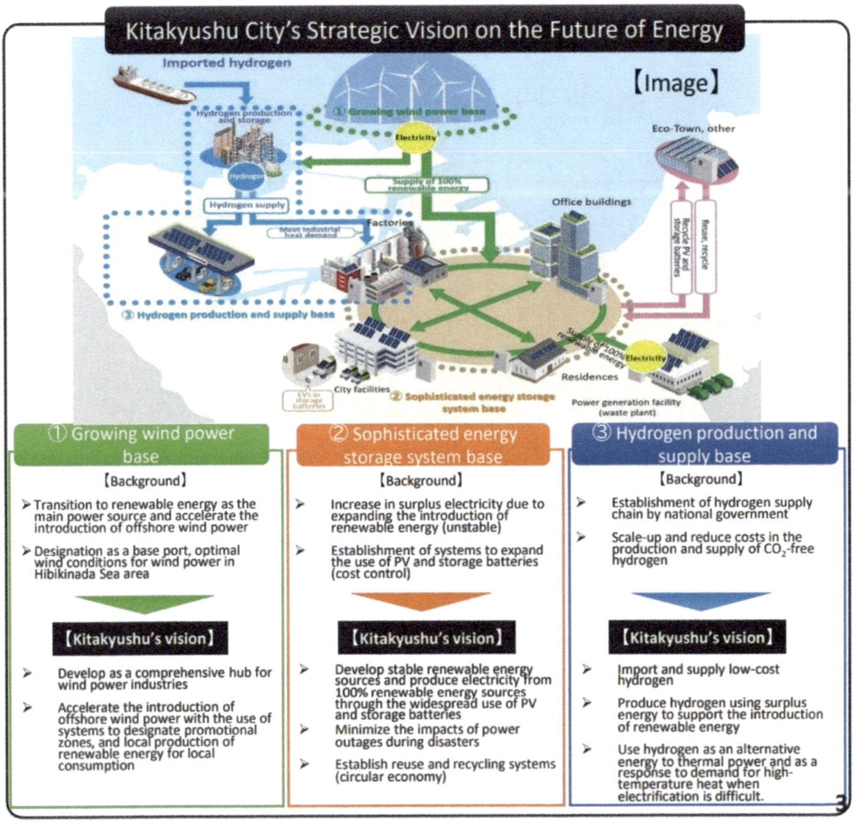


**Fig. 9** Kitakyushu City's strategic vision for the future of energy beyond the decarbonization advanced area. From the internet of "Kitakyushu City Action Plan for Global Warming Countermeasures. Green Growth Promotion Division, Environment Bureau." (Kitakyushu-city 2021)

Education, Energy (HOPE 2023a), solar power and storage batteries will be installed on the roofs of houses (on-site PPA), and solar power will be installed on the disaster site (off-site PPA). Meanwhile, all consumers in the area will be decarbonized. In addition, various support projects related to the purchase of storage batteries, EV mobility, energy-efficient home appliances, pellet stoves, etc. will be conducted, and comprehensive support for energy conservation, energy creation, and energy storage will be provided. Moreover, in some areas, private-line microgrids, solar power, and storage batteries have been introduced to strengthen resilience. In 2030, the electricity demand from the residential sector is expected to reach 9932 GWh, while the renewable energy supply is expected to reach 14,179 GWh.

Higashimatsushima Smart Disaster Prevention Eco-Town (HOPE 2023b) was built in the city as part of the "FutureCity" initiative. The target included 70 detached houses, 15 apartments, two large-scale hospitals, two small-scale hospitals, one

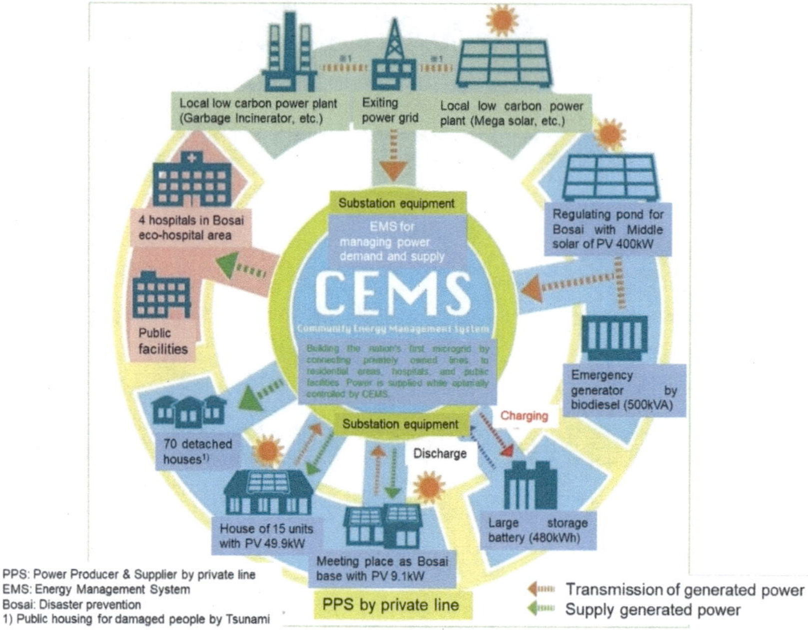


**Fig. 10** Energy management system of the town (HOPE 2023b). (Author translated the original chart written in Japanese)

public facility, and one assembly hall. They were connected by private lines to form a microgrid, which is Japan's first. In 2016, solar power generation reduced 307 tons of $CO_2$ annually, while realizing the local production and consumption of electricity within the area. Figure 10 shows the energy management system of the city (HOPE 2023a). Figure 11 shows the residential area in Eco-Town.

## 7 CASBEE-UD and CASBEE for Cities as Evaluation Methods

Finally, the author introduces a comprehensive assessment system for built environment efficiency (CASBEE)-urban development (UD) (IBECs 2023a) and CASBEE cities (IBECs 2023b) as methods for evaluating the environment of cities and communities.

The CASBEE (IBECs 2023c) developed by the Japan Sustainable Building Association is a method for evaluating and rating the environmental performance of buildings and the built environment. It is a comprehensive assessment of the quality of buildings and evaluates features such as interior comfort and scenic aesthetics, while considering environmental practices including the use of materials and equipment

**Fig. 11** One part of the site of Eco-Town. (This photo is provided by Higashimatsushima Organization for Progress and Economy, Education, Energy)

that conserve energy or generate lower environmental loads. The principles of the evaluation method are as follows: (1) a comprehensive assessment throughout the life cycle of a building; (2) an assessment of the built environment quality and load; (3) an assessment based on the newly developed built environment efficiency indicator. The CASBEE assessment is ranked based on five grades: superior (S), very good (A), good (B+ ), slightly poor (B−), and poor (C).

The CASBEE comprises assessment tools customized to different scales: construction (houses and buildings), urban (town development) and city management. These tools are collectively known as CASBEE Family. CASBEE-UD (Urban Development) is used to evaluate environmental performance of urban blocks and town development. CASBEE-City evaluates environmental performance on a local government scale and is widely used by domestic construction companies, design offices, building owners, and real estate investment institutions; currently, it is being used as a notification system by some local governments.

## 8 Summary

To realize a zero-carbon city/community, one must first conserve energy in a single building. This implies improving shelter performance by insulating and air-tightening a building, improving facility efficiency, and introducing renewable energy such as solar energy. Hence, environmental design must be pursued, such as planning the layout of buildings and greenery while considering local climatic conditions such as temperature, humidity, sunshine/insolation, wind direction/speed, and amount of rainfall.

At the regional level, one should consolidate urban functions, promote the use of public transportation, and promote area management as well as the use of greenery and energy. In addition, the awareness and behavior of citizens should be reformed toward decarbonization. In Japan, the planning, construction, and implementation of cities and communities are progressing rapidly in various locations owing to cooperation among the industry, academia, government, and the private sector toward carbon neutrality by 2050.

Herein, the author reviewed the relevant policies and proposals. Furthermore, he presented an overview of 26 regional proposals selected as "decarbonization leading areas" and four specific examples pertaining to plans for decarbonizing cities and regions. Finally, an overview of CASBEE UD and CASBEE cities as evaluation methods was presented.

## References


AIJ (Architecture Institute of Japan) (1997) Response to climate warming in the field of architectural research (Chairman's Statement). https://www.aij.or.jp/jpn/archives/971202.htm. (In Japanese) Accessed 4 Jan 2023

AIJ (Architecture Institute of Japan) (2000) Architecture charter for a global environment. http://news-sv.aij.or.jp/kensho/panfu.pdf. Accessed 4 Jan 2023

AIJ (Architecture Institute of Japan) (2009) Proposal 2050 vision: building-related measures to counteract global warming towards carbon-neutralization. https://www.aij.or.jp/scripts/request/document/20091222-1e.pdf. Accessed 4 Jan 2023

AIJ (Architecture Institute of Japan) (2015) Proposal of AIJ action plan toward a low carbon society in the year 2050-road map for the carbon neutralization of the built environment of Japan. https://www.aij.or.jp/scripts/request/document/20150413.pdf. Accessed 4 Jan 2023

CAO (Cabinet Office) (2023a) "FutureCity" initiative. https://future-city.go.jp/en/about/. Accessed 4 Jan 2023

CAO (Cabinet Office) (2023b) SDGs future city. https://www.chisou.go.jp/tiiki/kankyo/index.html). (In Japanese) Accessed 3 Jan 2023

CPLCS (Council for Promoting Low Carbon Society) (2020) Three proposals for carbon neutrality by 2050 -how to implement decarbonized and resilient human settlements by forming a stock-based society. http://news-sv.aij.or.jp/y900/link/suisin_teigen_20200702_%E7%A2%BA%E5%AE%9A%E7%89%88.pdf. Accessed 4 Jan 2023

CPLCS (Council for Promoting Low Carbon Society) (2023) Website of council for promoting low carbon society. http://news-sv.aij.or.jp/y900/ (In Japanese). Accessed 4 Jan 2023

HOPE (Higashimatsushima Organization for Progress and Economy, Education, Energy) (2023a) Website of HOPE. http://hm-hope.org/ (In Japanese) Accessed 5 Jan 2023

HOPE (Higashimatsushima Organization for Progress and Economy, Education, Energy) (2023b) Smart disaster prevention eco-town. http://hm-hope.org/?page_id=286 (In Japanese). Accessed 5 Jan 2023

IBECs (Institute for Built Environment and Carbon Neutral for SDGs) (2023a) CASBEE family and tools. CASBEE for urban development (CASBEE-UD). https://www.ibec.or.jp/CASBEE/english/toolsE_urban.htm. Accessed 6 Jan 2023

IBECs (Institute for Built Environment and Carbon Neutral for SDGs) (2023b) CASBEE family and tools. CASBEE for cities. https://www.ibec.or.jp/CASBEE/english/toolsE_city.htm. Accessed 6 Jan 2023

IBECs (Institute for Built Environment and Carbon Neutral for SDGs) (2023c) CASBEE basic concept. https://www.ibec.or.jp/CASBEE/english/basicconceptE.htm. Accessed 6 Jan 2023
IPCC (Intergovernmental Panel on Climate Change) (2018) Special report on global warming of 1.5 °C. https://www.ipcc.ch/sr15/. Accessed 3 Jan 2023
Kitakyushu-city (2021) Kitakyushu city action plan for global warming countermeasures. green growth promotion division. Environment Bureau. https://www.city.kitakyushu.lg.jp/files/000964812.pdf. Accessed 5 Jan 2023
Kitakyushu-city (2022) Selected by the national government as a "Decarbonization Leading Area". Green Growth Promot Div Environ Bur. https://www.city.kitakyushu.lg.jp/files/000978025.pdf (Japanese). Accessed 5 Jan 2023
METI (Ministry of Economy, Trade and Industry, Agency for Natural Resources and Energy) (1979) Act on Rationalizing Energy Use. https://www.japaneselawtranslation.go.jp/ja/laws/view/2857#je_ch5. Accessed 3 Jan 2023
METI (Ministry of Economy, Trade and Industry) (2020) Green growth strategy through achieving carbon neutrality in 2050. https://www.meti.go.jp/english/press/2020/1225_001.html. Accessed 3 Jan 2023
METI (Ministry of Economy, Trade and Industry, Agency for Natural Resources and Energy) (2021) Strategic energy plan. https://www.enecho.meti.go.jp/en/category/others/basic_plan/. Accessed 3 Jan 2023
MLIT (Ministry of Land, Infrastructure, Transport and Tourism) (2011) Low carbon city development guidance. https://www.mlit.go.jp/common/000136941.pdf. Accessed 3 Jan 2023
MLIT (Ministry of Land, Infrastructure, Transport and Tourism) (2012) Low carbon city promotion act. https://www.mlit.go.jp/common/001062852.pdf. Accessed 3 Jan 2023
MLIT (Ministry of Land, Infrastructure, Transport and Tourism) (2015) Act on improvement of energy consumption performance of buildings. https://elaws-e--gov-go-jp.translate.goog/document?lawid=427AC0000000053_20210401_501AC0000000004&_x_tr_sl=ja&_x_tr_tl=en&_x_tr_hl=en&_x_tr_pto=sc. Accessed 6 Jan 2023
MLIT (Ministry of Land, Infrastructure, Transport and Tourism) (2021) Is your town ecological? https://www.mlit.go.jp/toshi/city_plan/content/001343154.pdf (In Japanese). Accessed 3 Jan 2023
MLIT (Ministry of Land, Infrastructure, Transport and Tourism) (2022) Outline of certification system of low-carbon buildings based on the Eco-Town Law. https://www.mlit.go.jp/jutakukentiku/house/content/001514370.pdf. (In Japanese) Accessed 3 Jan 2023
MOE (Ministry of the Environment) (1998) Act on promotion of global warming countermeasures. https://www.japaneselawtranslation.go.jp/ja/laws/view/3796. Accessed 3 Jan 2023
MOE (Ministry of the Environment) (2021a) Plan for global warming countermeasures. https://www-env-go-jp.translate.goog/earth/ondanka/keikaku/211022.html?_x_tr_sl=ja&_x_tr_tl=en&_x_tr_hl=en&_x_tr_pto=sc. Accessed 3 Jan 2023
MOE (Ministry of the Environment) (2021b) Decarbonization leading area application guideline. https://www.env.go.jp/content/900512579.pdf (In Japanese). Accessed 4 Jan 2023
MOE (Ministry of the Environment) (2022a) 1st overview of decarbonization leading areas. https://www.env.go.jp/content/000039032.pdf (In Japanese). Accessed 4 Jan 2023
MOE (Ministry of the Environment) (2022b) State of declaration of net zero carbon dioxide emissions in 2050 by local governments. https://www.env.go.jp/policy/zerocarbon.html (In Japanese). Accessed 3 Jan 2023
NIES (National Institute for Environmental Studies) (2022) Japan's GHG emissions data (FY1990—2020), Greenhouse Gas Inventory Office of Japan. https://www.nies.go.jp/gio/archive/index.html. Accessed 3 Jan 2023
SCJ (Science Council of Japan) (2004) Policy proposals for reducing energy consumption in residential and commercial buildings. https://www.scj.go.jp/ja/info/kohyo/pdf/kohyo-20-t38-3.pdf (In Japanese). Accessed 4 Jan 2023

SCJ (Science Council of Japan) (2017) Pathway to low carbon and healthy lifestyle and cities/ buildings. https://www.scj.go.jp/ja/info/kohyo/pdf/kohyo-23-t249-2.pdf (In Japanese) Accessed 4 Jan 2023
SCJ (Science Council of Japan) (2023) Website of science council of Japan. https://www.scj.go.jp/en/index.html. Accessed 4 Jan 2023

# Urban Maintenance and Robustness

**Jian Zang, Manish Kumar, and Rakesh Kumar**

**Abstract** This chapter presents Indian case studies carried out (i) to identify socio-demographic determinants of water end-use consumption in households and (ii) to critically appraise an innovative decentralized water system, which combined rainwater harvesting (RWH) with wastewater reclamation to provide design alternatives and water management optimization opportunities. Studies present real-world sustainability analyses where the water infrastructure costs, blue and recycled water demands, electricity demands, and operational costs over a calendar year were collated to find the sustainability of RWH systems. This chapter also introduces policy and management related to city resilience and robustness. Scenario analysis recommended a water system design alternative with wastewater reclamation for ground maintenance only, which was predicted to yield similar environmental benefits. Rainwater harvesting had high infrastructure costs, and water policy in low- and middle-income countries should instead support wastewater reclamation as best practice in water management. In this study, we also enumerated the various ongoing uses of biowaste-derived char and proposed real-world practical IoT-based applications as an additive to control diffuse micro-pollutants and to increase city safety ecologically.

**Keywords** Maintenance · Rainwater harvesting system · End-use · Urban robustness · Policies and management

J. Zang
School of Civil Engineering, Chongqing University, Chongqing, China

M. Kumar (✉)
Sustainability Cluster, University of Petroleum and Energy Studies, Dehradun, Uttarakhand 248007, India
e-mail: manish.env@gmail.com

Escuela de Ingeniería y Ciencias, Tecnologico de Monterrey, Campus Monterey, 64849 Monterrey Nuevo Leon, Mexico

R. Kumar
Department of Biosystems Engineering, Auburn University, Auburn, AL 36849, United States

 
R. Yao (ed.), *Resilient Urban Environments*, Cities and Nature,
https://doi.org/10.1007/978-3-031-55482-7_17

## 1 Introduction

The scarcity of freshwater across the world has become one of the major crises exacerbated by a growing population, changing consumption patterns, and the rise in irrigated agriculture (Kumar et al. 2021). In light of this, the United Nations has dedicated two of its 17 "sustainable development goals" (SDGs) exclusively to the conservation of water, in particular, SDG 6, which is "clean water and sanitation" as the other SDG on water (SDG 14) focuses on aquatic life (Huang et al. 2021a). SDGs were adopted in 2015 to improve the standards of human well-being by 2030 (UN 2015) and call for the following goals related to water sustenance: (i) avoidance of water pollution, (ii) improvement and/or restoration of water quality and the protection and/or restoration of aquatic ecosystems across the world (Huang et al. 2021a). Yet, conventional water treatment methods and many of the green technologies in cities either lack precise management (Zhang et al. 2022; Kumar et al. 2021; Fang et al. 2023) or do not provide a feasible solution for "water infrastructure" expansion in low- and middle-income countries (Mohapatra et al. 2014). Thus, in this regard, the focus should be shifted towards decentralized methods involving the direct participation of the community.

It has been stated that "at the global level and on an annual basis", there is enough fresh water to meet demand. RWH has been in use in Japan prominently since the early 1980s; here the local governments in several regions have promoted the rainwater harvesting (RWH) method as a water harvesting and recycling method for countering water shortage in large cities and the problem of urban flooding (Chen et al. 2021). Thus, in many parts of the world with a well-defined rainy season, e.g., the monsoons in India, RWH can be considered one of the available decentralized water conservation mechanisms. Survey projects applying GIS tools have shown the potential for RWH in Botswana, Ethiopia, Kenya, Malawi, Mozambique, Rwanda, Tanzania, Uganda, Zambia, and Zimbabwe (Campisano et al. 2017). In recent years, RWH in African countries has been pioneered by mainly non-governmental organizations (UNESCO, SIDA, and UNEP) and other research organizations (Campisano et al. 2017). Interest in RWH for commercial and industrial usage is rising, but the approach is still not viewed as viable or sustainable for end uses like agriculture or cooling (Campisano et al. 2017).

In fact, studies have already cited that the application of existing indigenous knowledge like RWH, as well as the scholarly knowledge obtained through modern research, are equally important (Van Meter et al. 2014; Tolossa et al. 2020). Another advantage is that rainwater by nature is far cleaner than surface water; hence, integrating locally available-cum-affordable materials and methods for treatment, like adsorption with biochar, sand filters, etc., can genuinely fortify RWH as a viable method of water conservation (Nandi and Gonela 2022). This approach could prove beneficial in several notable cases; for example, in regions with a strong rainy season, like Bangladesh or West Bengal in India which are affected by geogenic groundwater contamination, RWH could be a potential solution.

This chapter introduces a comprehensive assessment of the current trends and applications of RWH, along with the challenges it faces and the future potential that it holds. Despite the potential and advantages of RWH, the technology has not been widely adopted; it has been underestimated and under-exploited (Bailey et al. 2018; Andualem et al. 2020; Huang et al. 2021b). In developing and under-developed countries, there may be a lack of awareness or a shift toward more "modern" centralized methodologies, while in developed nations, there is already an overemphasis on centralized options (Huang et al. 2021a).

## 2 People's Perception and End Users for Sustainability

To promote, correctly install, utilize, and maintain rainwater harvesting (RWH) systems, it is essential to comprehend the receptivity. Domínguez et al. (2017) evaluated the RWH system in low-income, low-consumption homes in a social housing complex in Colombia. Potential end users were involved in a subsequent consultation about the cost–benefit analysis of the suggested alternatives. For example, in Jaipur, the majority of participants (67%) reported just taking conventional bucket baths, whereas only 3% reported exclusively utilizing showers; 30% revealed using both techniques (Ramsey et al. 2017). Still, 5% of participating houses reported installing rainwater-harvesting devices, the majority during the last five years (Ramsey et al. 2017).

Despite their limited financial resources, potential end users prioritize investment requirements less than water conservation criteria. Water-saving habits that demand some work but no financial cost, such as reusing wastewater from filtration systems, were substantially less popular. A total of 73.7% of participants reported having some water treatment equipment at home for purifying drinking water (Ramsey et al. 2017). Further investigations in the real world are necessary to ascertain whether end users will act on their plans and make the investments they had intended (Domínguez et al. 2017). Thus, Ward and Butler (2016) have proposed a two-part analytical framework based on niche management and the safe, resilient, and sustainable (or "Safe and SuRe") paradigm was used to analyze social network analysis further. Ward and Butler (2016) depict RWH and social network analysis by integrating the infrastructure-organization model (Fig. 17.1).

As shown in Fig. 17.1, the social network related to the harvesting system is complex; it consists of several areas, including research, action, knowledge exchange, advocacy, technology list, and central government. Improving the sustainability or reliability of the system required systemic thinking as well as full-cover consideration for the above areas; any point missing could affect the robustness of the RWH green-infrastructure system. Figure 17.1 clearly shows the connections and relationship between all sectors, for instance, RWH supplier and installer (HydroINT) and well-established RWH supplier and installer (GrafUK) have a green circle, which indicates the unit developing the novel RWH product, processes, or service. Blue diamonds represent actively engaged universities, institutions, or project collaborations in areas

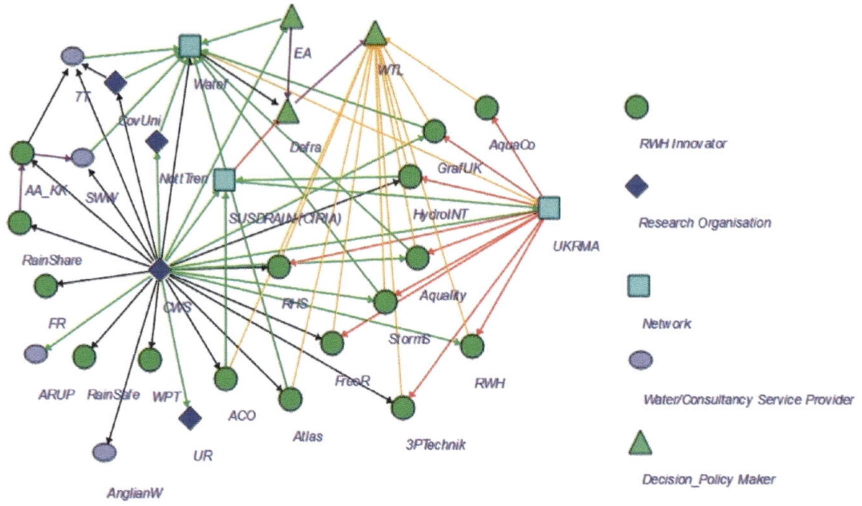


**Fig. 17.1** Interactions between infrastructure innovators and organizations in the UK through the RWH network model. Green = knowledge exchange (infrastructure providers); black = research; purple = action; red = advocacy/lobbying; orange = water technology list (WTL); green triangle = central government. Abbreviations: EA = Environment Agency; 7 T = Severn Trent Water; Cov Uni = Coventry University; Defra = Department for Environment Food & Rural Affairs; AA_ KK = Aqua Analytics and Cloud Keeper; UKRMA = UK Rainwater Management Association; StormS = Stormsaver; FR = FlushRain; UR = UrbanRain; FreeR = Freerain; SWW = South West Water; RWS = rain harvesting system; ARUP = ARUP/PIPEX—Flowstow; SUSDRAIN (CIRIA) = Construction Industry Research and Information Association; AnglianW = Anglian Water; WaterF = water efficiency network; NottTrent = Nottingham Trent University. Adopted from Ward and Butler (2016)

of RWH research. Networks (aquamarine blue square) indicate small-scale organizations relating to RWH. Large organizations (light blue oval) are also highlighted due to their importance in city resilience. The last marker (green triangle) represents the central government support in terms of decision or policy. All these networks and units make up a resilient infrastructure system.

### *2.1 Policies and Management for Increasing City Resilience in India*

India has developed legislation and policies that favour building rainwater collecting systems, but their implementation depends on public approval and willingness (Barthwal et al. 2014). In comparison, in India, Kusre et al. (2017) calculated the water consumption of the end users, evaluated the water availability during the dry season, and planned the volume of the storage structure to construct a viable water harvesting system. Related water stress issues in Sikkim (India) were discussed according to data collected from December to March, and the extreme weather exacerbates the impact on city robustness.

Rural or urban residential water management is crucial for lowering the health hazards associated with exposure to water. Thus, Sarkar (2021) have suggested RWH as the feasible solution for the remediation of arsenic in the region of potential high rainfall, as well as a technical disadvantage in Bengal regions due to rooftop materials, like kaccha materials straw or steel tiles. Therefore, Biswas et al. (2020) have analyzed primary data collected from 923 rural families in four blocks of North 24 Parganas district, West Bengal, India, and reported that 8.2% and 7.7% of participants who had faced serious health hazards were aware of the water quality, notably iron and arsenic pollution. However, it was shown that only 27.3% and 25.2% of individuals were aware of arsenic and iron exposure.

Holland-Stergar (2018) has examined how policy interventions lead to increased water conservation in India. It has been reported that adopting an RWH system was discouraged due to subsidized municipal water supply and no incentive for RWH in Banglore, India; this was also due to the high installation cost of RWH systems.

In addition, for almost 40 years, residents of the North 24 Parganas district in West Bengal, India, have been consuming water from shallow tube wells polluted with iron and arsenic. Thus, Biswas et al. (2021) established a link between rooftop rainwater gathering potential and socioeconomic level through the rigorous sampling of 923 rural households to address both qualitative and quantitative water stress, which confirmed statistically significant for the feasibility of rooftop rainwater harvesting systems. The government has yet to develop an arsenic-free, affordable, and user-friendly system. Barthwal et al. (2014) investigated how people in Dehradun, India, feel about rooftop rainwater collecting systems, their attitudes, and their acceptance. Recently, it has been reported that integrating RWH systems with solar purification is considered a low-cost drinking water system in India (Patange et al. 2020). This will also be helpful in the removal of contaminants from wastewater, and the younger generation will get exposed to RWH systems in rural and hilly areas.

## 2.2 *Energy Management Related Robustness Municipal Case Study in India*

It is often claimed that decentralized systems (DS) have environmental, financial, and social benefits compared to centralized systems (CS). As Table 17.1 shows, the performance of the investigated decentralized drinking water treatment plant (0.4 kWh/m$^3$) is better than the average of the centralized drinking water treatment plants but is within the range of the standard deviation of reported values (0.52 $\pm$ 0.36 kWh/m$^3$). For a decentralized system, the distance between the drinking water treatment plant and the point of use will be much shorter than in centralized systems. The performance of the investigated decentralized wastewater treatment (0.30 kWh/m$^3$) is also better than the average of the reviewed papers but within the range of the standard deviation (0.43 ± 0.32 kWh/m$^3$). The treatment train of the wastewater treatment plant at IITGN does not include an activated sludge process that requires air pumping into the wastewater and would have a high electricity demand. However, the wastewater reclamation process includes energy-intensive treatment steps, such as UV disinfection and pressurized sand filtration. For the later process, effluent from the planted wetland is lifted about eight meters high from below ground to the pressurized sand filtration units.

Table 17.1 indicates that the overall energy demand of the DS at IITGN 1.36 kWh/m$^3$ is substantially above the average overall energy demand of CS reported in the literature (1.12 ± 0.89 kWh/m$^3$). Possible reasons could be detailed pumping

**Table 17.1** Summary of energy consumption result of the decentralized water supply and treatment system at IITGN in comparison with published results for both centralized and decentralized systems

| DWTP (kWh/m$^3$) | | | WWTP (kWh/m$^3$) | | | System type, decentralized (DS) or centralized (CS) |
|---|---|---|---|---|---|---|
| Plant | Distribution | Total | Conveyance | Plant usage | Total | |
| – | – | – | 0.094 | 0.19 | 0.284 | CS |
| 0.06 | 0.31 | 0.37 | – | – | – | CS |
| 0.37 | 0.48 | 0.85 | – | 1.09 | – | CS (modelled) |
| 0.04 | 0.27 | 0.31 | 0.04 | 0.17 | 0.21 | CS |
| – | 0.17 | – | – | 0.14 | | CS |
| 0.64 | 0.15 | 0.79 | 0.21 | 0.87 | 1.08 | CS |
| 0.23 | 0.18 | 0.41 | 0.06 | 0.75 | 0.81 | CS |
| 0.55 | 0.50 | 1.05 | – | 0.30 | | CS (estimated) |
| 0.09 | 0.10 | 0.19 | 0.14 | 0.44 | 0.58 | CS |
| 0.21 | 0.18 | 0.39 | – | 0.31 | – | CS (modelled) |
| 0.34 | 0.11 | 0.45 | – | 0.21[a] | – | CS (modelled) |
| 0.21 | 0.18 | 0.39 | – | 3.87 | 3.87 | CS (calculation) |
| – | 0.60 | 0.60 | – | 0.47 | – | CS (estimated) |
| 0.08 | 0.24 | 0.32 | 0.06 | 0.41 | 0.47 | CS (modelled) |
| – | – | – | – | 0.67 | 1.09 | CS (estimated) |
| – | – | – | – | 0.24 | 0.24 | CS |
| 0.21 | 0.18 | 0.39 | 0.08 | 0.30 | 0.38 | Median |
| 0.26 | 0.27 | 0.52 | 0.17 | 0.43 | 0.60 | Average |
| 0.20 | 0.16 | 0.36 | 0.21 | 0.32 | 0.53 | Standard deviation |
| – | – | – | 0.24 | 0.18 | 0.42 | DS |
| 0.8[a] | 0.615 | 1.65 | 0.36 | 1.48 | 1.84 | DS (Birla campus, India) |
| 0.35 | 0.41 | 0.76 | 0.30 | 0.30 | 0.60 | DS (raw data) |

[a] All data collected in India (Zang et al. 2021)
Adopted from Zang et al. (2021)

data to the header tank (0.36 kWh/m$^3$) compared to other studies. The wastewater treatment plant at IITGN reclaims water for irrigation and other non-potable water usage. Similar to rainwater harvesting, the payback period is key to judging the sustainability of the system. Financially, if using PW for irrigation, only costs for chemicals, electricity in treating, raw water price, and pumping energy (0.14 kWh/m$^3$, assuming the same as RW distribution) were required. Reclaimed water system investment consists of UV lamb construction/operational cost, RW pipe price, and pumping energy cost, the total water demand on the campus is 1037 m$^3$/d (364 irrigation water demand, 604 river water incomings, and 69 rainwater collection).

As shown in Table 17.2, in the conventional scenario, the costs are an average of 194.3 £/d to first treat river water to reach a potable standard and then pump water through a WSC (DW-RW) pipe to satisfy daily, toilet and irrigation usage on the IITGN campus with final necessary wastewater treatment. In scenario A, investing in RWS on the campus, the result shows 0.127 £/m$^3$ is required because 364 m$^3$/d reclaimed water replaces conventional potable standard water (chemical, electricity, and raw water expenditure) was saved regarding irrigation. The discounted payback year is only 43 years. The reclaimed system seems beneficial and sustainable. Additionally, the traditional payback period for the investment in adding RWHS is more than 300 years in scenario B. The RWHS here seems unsustainable due to enormous capital investment and less rainfall. In IITGN (Scenario C), both RWHS and RWS were invested, the average operational cost for water, excluding staff salary, is 0.119 £/m$^3$, and the discounted payback period for the investment is about 135 years.

At IITGN, the whole university has access to affordable drinking water. Per unit electricity consumption for the DWTP (0.35 kWh/m$^3$) and WWTP (0.30 kWh/m$^3$) and its distribution data is higher than normal municipal usage, therefore wastes the city budget resulting in a negative effect on city robustness. Re-use applications and infrastructures were installed and conducted in Indian cities, but the efficiency is still low because of poor management and low robustness.

## 2.3 *Importance of Maintenance Work for Resilient Cities*

Potential risks to adequate water availability include rising populations, increased per capita water demand, climate change, and budgets for water conservation (Ray and Shaw 2016; Deshkar 2019; Ahmadi et al. 2020; Kumar et al. 2020; Talat 2021). Cities are particularly susceptible to water shortages due to the high population density and the need for a dependable water supply to support the rapid pace of human and economic activities (Ray and Shaw 2016). Intermittent supply, inappropriate pressure, substantial leakage loss, high energy usage, and reliance on groundwater are major shortcomings in the distribution and water supply systems (Ray and Shaw 2016; Ramsey et al. 2017). Furthermore, Govindarajulu (2020) reported that various urban cities, such as Patna, Kochi, Chennai, Guwahati, among others, face poor water supply, stormwater drainage, and sewerage systems, which require significant investment for infrastructure and resilience. Underneath the new paradigm of climate

**Table 17.2** Cost analysis for different scenarios for supporting IITGN water demand and their payback year by different methods

| | Extra investment required (£) | Total water operational cost except for salary (£/d) | Annuity compared to (£/d)[a] | Operational cost intensity excludes salary (£/m$^3$) | Period (method) | Years |
|---|---|---|---|---|---|---|
| Scenario Conventional | 0 | 194.3 | 0 | 0.187 | | |
| Scenario A (add reclaimed water system) | 514 K | 131.7 | 62.7 | 0.127 | Payback period (traditional) | 22 |
| | | | | | Discounted payback period (interpolation method) | 43 |
| Scenario B (add RWHS) | 1041 K | 186.0 | 8.42 | 0.179 | Payback period (traditional) | 339 |
| | | | | | Discounted payback period (interpolation method) | > 800 |
| Scenario C (add both RWHS and RWS) | 1555 K | 123.3 | 71.1 | 0.119 | Payback period (traditional) | 60 |
| | | | | | Discounted payback period (interpolation method) | > 134 |

*Note* NPV (net present value) and NFV (net future value) will not be more than 0 (start benefit) within 600 years, no point to search the real payback period for NFV; interest was 6.5% in 2018. Assuming salary remains in a different scenario

[a] Annuity water bill for RWHS is the water bill price if you buy the same amount of water as rainwater collected from an outside company (10.7 Rs for 25,202 m$^3$/year), a total of £3064.3/year. The annuity for RWTP is the water bill saved to buy river water directly in irrigation

Adopted from Supporting Information Zang et al. (2021)

change and urbanization, it is necessary to assess the vulnerability and resilience of urban water from both quantitative and qualitative angles (Kumar et al. 2019). In Jaipur, households with inadequate water sources were less likely to agree to a flat percentage reduction in water use (Ramsey et al. 2017). Maintenance of existing water supply systems is crucial for city resilience, as water supply mechanisms depend on potential stress and interventions to increase urban resilience, in long-term accountability influenced by service providers, stakeholders, and durability of water supply systems (Ray and Shaw 2016). In addition to being essential for providing communities with services and fostering economic development, Deshkar (2019) summarized that basic infrastructure and services, like water supply and sanitation, are also crucial for defining urban resilience in the face of stressful circumstances due to uncertain climatic variability or lacking adequate maintenance. Thus, managing urban water infrastructure is of utmost importance for building urban resilience in India via provisioning, developing economies, and financing eco-friendly services, like convenient and appropriate drainage service for stormwater and regular maintenance of leakages in distribution systems. In another survey, respondents witnessed water leaks in their surroundings to gauge overall water awareness; 42% of respondents reported experiencing water leaks in their communities. This indicates that these individuals had a baseline level of water awareness, which supports the notion that poorly maintained water infrastructure causes considerable water loss in Jaipur (USAID 2013). The findings showed that shortening the length of water delivery may promote a scarcity mindset, in which individuals are less inclined to engage in water-conservation actions (Ramsey et al. 2017). Ramsey et al. (2017) reported that only 27% of households have agreed to cut water use and reported reusing wastewater, compared to 44% on average observed in Jaipur for reusing wastewater, according to USAID (2013).

## 2.4 RWH Systems for Urban Water Management

The robustness of building with green infrastructure is decreasing due to poorly managed infrastructure. In India, according to Rainwater (Harvesting and Storage Bill) 2016, it is mandatory to install RWH system and storage tanks in all buildings, such as residential and commercial complexes. In addition, the Tamil Nadu district administration has significantly introduced active and passive RWH systems. Whereas in Chennai, the government bodies were not providing subsidies on harvested water but benefitted in terms of home loans by banks to those unable to afford installation costs. In this study, compared with India, Tucson (US), have alternative financial assistance to low-income families for RWH installation in terms of loans and grants. To encourage residents, the municipal corporation of Indore has offered 6% rebate on property tax after implementing RWH systems on rooftops/buildings. In a case study, CSE, India has reported that 50% of harvested water can fulfil the water demand-supply gap in the national capital city, New Delhi. Recently, Sarkar (2021) reported that RWH technology had improved the water supply in the Purulia District of West Bengal, India, and results were supported with a logistic regression model to provide accurate information on WTP for better water services. It is essential to consider factors like income, RWH water as the primary source of household usage, distance from the RWH source, the time required to collect water from it, and general knowledge of RWH.

RWH systems are considered an efficient and sustainable approach regarding water quality, as harvested water can be easily used as potable and non-potable utilities, significantly reducing pressure on the processed water supply. Mehrabadi et al. (2013) analyzed the applicability and performance of RWH systems to meet water demand for non-potable water utilities in Iran. This study revealed that non-potable water supply was fulfilled via harvest rainwater, i.e., 75% of non-potable water demand in humidity through 70% of maximum duration, 75% in a Mediterranean climate for 45% of maximum duration, and 40% in arid climates for 23% of maximum duration. Therein, in Bangladesh, Rahman et al. (2014) analyzed various water quality parameters of harvested water through RWH systems under permissible limits, which offer an efficient reduction in power consumption and contribute to energy savings. However, increased faecal coliform was observed in stored unused water due to inadequate water harvesting during June 2011. In addition to water pollution, in Brazil, across different cities, Silva and Ghisi (2016) have reported several uncertainties due to varying rainfall patterns and water storage tank capacity in daily water demand on the performance of RWH systems. To provide a reliable supply of clean drinking water while addressing the increasing burden that climate change is placing on water resources, climate-resilient water safety plans have been established. Páez-Curtidor et al. (2021) summarized that planning for climate change adaptation strategies may result in increased vulnerabilities or decreased overall resilience of urban systems.

## 2.5 Green and Sustainable Technologies for RWH Systems

Rainwater harvesting is affected by yearly rainfall patterns, water-retaining capacity, and roof runoff coefficient. The extensive green roof potentially holds up to 55% of the yearly rainfall, resulting in the capacity to regulate urban stormwater runoff, thereby decreasing the probability of urban flooding (GhaffarianHoseini et al. 2016).

Despite the many advantages of using rainwater harvesting to assist sustainable development and green built environments, there is still a need to integrate these design elements to effectively produce consistent and sustainable results (GhaffarianHoseini et al. 2016). In South Africa, Ndeketeya and Dundu (2019) recently suggested three policy reforms: standard, incentive, and result-based policies for sustainable RWH systems in urban areas. Besides, as Germany and Australia have shown, the policy is crucial to enable the success and sustainability of RWH systems (Chubaka et al. 2018). To stimulate the adoption of RWH, there are still significant policy gaps at the local and national levels that must be addressed. Implementation challenges may result from a lack of enabling institutional structures and regulations.

Conventional RWH systems have limitations like pollution levels in urban areas significantly affecting water quality. Interestingly, Verma (2022) has introduced the aid of sensors and the Internet of Things (IoT), which can ensure quality checking, monitoring, and water segregation. To urban water management at a low-cost but effective, several cheap and locally available materials, like sand filters, charcoal, PVC, and sponge filters, could be implemented in filtration units either individually or mixed in different proportions (Pradhan and Sahoo 2019). In Rajgir, South Bihar

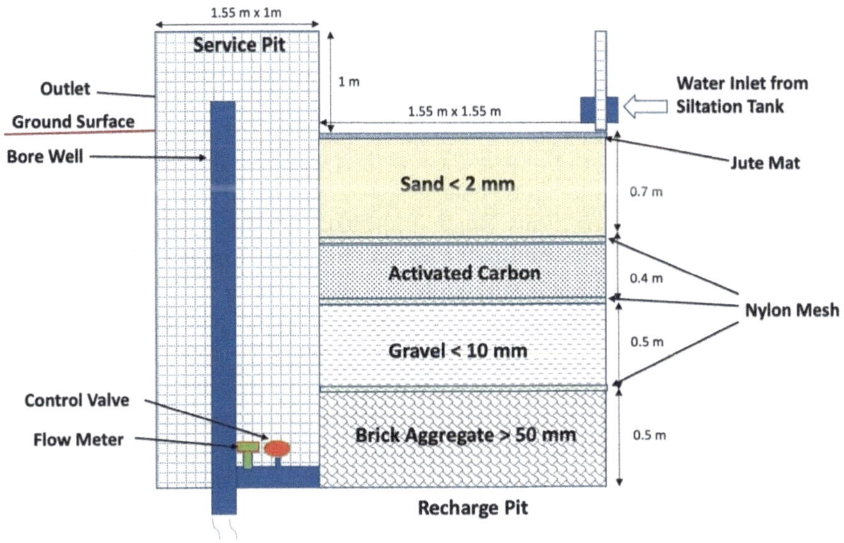


**Fig. 17.2** Representing different layers of sand, activated carbon, gravel, and brick aggregate in the filtration unit of the recharge pit. Adopted from Bandyopadhyay et al. (2021)

of India, Bandyopadhyay et al. (2021) have used harvested rainwater in artificial groundwater recharging, but before injection, rainwater was treated in a filtration unit consisting of sand, gravel, bricks, and activated charcoals in different proportions (Fig. 17.2). Thus, to test water quality, biochar could be potential and most efficient novel materials for harvested water treatment, before potable or non-potable purpose or groundwater recharging. Several studies have reported that biochar has been utilized as a biosorbent in stormwater treatment (Boehm et al. 2020; Cuong et al. 2022).

Biochar amendment in green roofs has been reported effective in exploring the stormwater management of dual-layer substrates with biochar (Huang et al. 2020). Gan et al. (2021) and Gan et al. (2022) have reported that biochar is amended with green roofs through soil columns to treat water quality for stormwater and rainwater management. Also, Bentley and Summers (2020) have utilized biochar synthesized from pine and biosolids to remove micropollutants from contaminated surface water, stormwater, and wastewater, as low-cost treatment methods in low-income countries. Importantly, Ouedraogo et al. (2023) have summarized the application of biochar in the treatment of stormwater from highways, in which various heavy metals are removed significantly, including nutrients, suspended solid particles, and organic contaminants.

## 3 Challenges and Opportunities for Robustness in Urban Water Supply and Management

Intermittent piped water supply, an inevitable concern of water scarcity, can cause waterborne diseases due to contamination in the pipes or during residential storage, dangerous water sources during outages, and inadequate water availability for hygiene (Bivins et al. 2020). Thus, Bivins et al. (2017) have summarized that low- and middle-income nations frequently have intermittent water supplies, which increases microbial contamination and may raise the risk of contracting a waterborne disease.

In Hubli-Dharwad, India, Ercumen et al. (2015) reported that waterborne diseases have been observed due to continuous and intermittent water supply systems. A total of 3922 households in 16 neighbourhoods with children < 5 years old were analyzed and reported bloody diarrhoea, cholera, gastrointestinal illness, typhoid fever, hepatitis, and deaths of children < 2 years old. Through socio-economic status, in lower-income households, children < 5 years old with continuous supply had a 37% lower prevalence of bloody diarrhoea compared to intermittent supply households. However, in higher-income households, there was no significant association between continuous versus intermittent supply and child diarrheal illnesses. With only 1% of samples testing positive for *E. coli*, thereby failing to satisfy WHO drinking water quality recommendations, compared to 32% under intermittent service, tap water quality was much higher under continuous supply. Further, transitioning from intermittent to continuous water supplies impacts sustainability, equality, and net benefits (Burt et al. 2018). Similar to this study, in Nagpur, India, Bivins et al. (2021) also reported *E. Coli* and various waterborne pathogens in household taps due to intermittent water supply systems. As a result, an intermittent water supply is less likely to meet the need for drinking purposes that can be sustainably managed. Burt et al. (2018) examined the net advantages at the household level, the equality of their distribution, and the accessibility to water under continuous supply. In this study, Burt et al. (2018) concluded that the positive net advantages of the present trial cannot support the city-wide project, and the upgrade may be perceived as highly inequitable when CWS is not expanded up to the rest of Hubli-Dharwad either due to capacity limits or capital restrictions.

Water demand increases from continuous supply may be considered via upgrading intermittent supplies to continuous supplies, and improved leak control and supply system governance can make urban water supply and management more resilient and robust (Ercumen et al. 2015). Overall, robustness in urban water supply management has been observed due to shifting from intermittent to continuous water supply in India, resulting in considerable improvements in microbiological water quality, turbidity, and pathogens (Bivins et al. 2021). These findings imply that transitioning from intermittent to continuous water supply in Nagpur significantly decreased faecal pollution in drinking water and genetic components linked to waterborne diseases. Regarding policy interventions, Bivins et al. (2021) demonstrated that enhanced billing and metering practices in Nagpur, India raised income for the utility, enabling it to make infrastructure investments that have improved residential drinking water quality.

Interestingly, Vaidya et al. (2022) reported chlorine decay in intermittent and continuous water supply networks in India. Besides, overall degradation rates for ductile and cast-iron pipes were 1.5 times greater in intermittent networks than in continuous networks. However, very few chlorine decays have been investigated in the drinking water supply. Chlorination has been implied in the disinfection process, though treated water supplied to pipes which can produce toxic disinfected by-products, could create human health risks (Kumar et al. 2023).

## 4 Summary

In summary, the city is facing challenges and opportunities because of the conflict between continuous daily living supply demand increases and because insufficient attention is paid to city robustness in a developing country (especially in policy/maintenance/management work). City resilience improvement work could increase the well-being and robust living quality of citizens. However, the developing country (e.g., India) case studies here demonstrate that those policy, maintenance and management improvement works are more effective in increasing city resilience than some green technologies themselves, only huge capital cost green infrastructures and technologies could not help India in improving city resilience. This review have summarized a widely accepted renewable resources reusing infrastructure failure case in India because of lacking proper management. Additionally, robustness in urban water supply management has been observed due to the shifting from intermittent to continuous water supply in India which has resulted in considerable improvements in city resilience and well-being. So, to conclude this chapter, this work offers and summarizes five recommendations to increase city resilience effectively and ecologically: (a) education for endpoint users should be improved, (b) people's perspectives should never be ignored, (c) the policymaker should track the effectiveness of their work after implementation, (d) maintenance work plays a big role in city robustness, and (e) proper management should be considered as the primary effective solution before investing huge capital cost green infrastructures.

## References

Ahmadi MS, Sušnik J, Veerbeek W, Zevenbergen C (2020) Towards a global day zero? Assessment of current and future water supply and demand in 12 rapidly developing megacities. Sustain Cities Soc 61:102295. https://doi.org/10.1016/j.scs.2020.102295

Andualem TG, Hagos YG, Teka AH (2020) Rainwater harvesting potential assessment for non-potable use in urban areas. Sustain Water Resour Manage 6(6):1–8. https://doi.org/10.1007/s40899-020-00464-x

Bailey RT, Beikmann A, Kottermair M, Taboroši D, Jenson JW (2018) Sustainability of rainwater catchment systems for small island communities. J Hydrol 557:137–146. https://doi.org/10.1016/j.jhydrol.2017.12.016

Bandyopadhyay S, Sharma A, Sahoo S, Dhavala K, Sharma P (2021) Potential for aquifer storage and recovery (ASR) in south Bihar, India. Sustainability 13(6):3502. https://doi.org/10.3390/su13063502

Barthwal S, Chandola-Barthwal S, Goyal H, Nirmani B, Awasthi B (2014) Socio-economic acceptance of rooftop rainwater harvesting—a case study. Urban Water J 11(3):231–239. https://doi.org/10.1080/1573062X.2013.765489

Bentley MJ, Summers RS (2020) Ash pretreatment of pine and biosolids produces biochars with enhanced capacity for organic micropollutant removal from surface water, wastewater, and stormwater. Environ Sci: Water Res Technol 6(3):635–644. https://doi.org/10.1039/C9EW00862D

Biswas S, Debsarkar A, Pal M (2020) Water insufficiency, health hazards and rainwater harvesting in North 24 Parganas, West Bengal, India: results of a socio-economic survey. Arthaniti: J Econ Theory Practice 21(2):236–254. https://doi.org/10.1177/0976747920963399

Biswas S, Sahoo S, Debsarkar A, Pal M (2021) Assessment of adoption potential of rooftop rainwater harvesting to combat water scarcity: a case study of North 24 Parganas district of West Bengal, India. Arab J Geosci 14:1–16. https://doi.org/10.1007/s12517-021-07989-1

Bivins AW, Sumner T, Kumpel E, Howard G, Cumming O, Ross I, Nelson K, Brown J (2017) Estimating infection risks and the global burden of diarrheal disease attributable to intermittent water supply using QMRA. Environ Sci Technol 51(13):7542–7551. https://doi.org/10.1021/acs.est.7b01014

Bivins A, Lowry S, Murphy HM, Borchardt M, Coyte R, Labhasetwar P, Brown J (2020) Waterborne pathogen monitoring in Jaipur, India reveals potential microbial risks of urban groundwater supply. NPJ Clean Water 3(1):35. https://doi.org/10.1038/s41545-020-00081-3

Bivins A, Lowry S, Wankhede S, Hajare R, Murphy HM, Borchardt M, Labhasetwar P, Brown J (2021) Microbial water quality improvement associated with transitioning from intermittent to continuous water supply in Nagpur, India. Water Res 201:117301. https://doi.org/10.1016/j.watres.2021.117301

Boehm AB, Bell CD, Fitzgerald NJ, Gallo E, Higgins CP, Hogue TS, Luthy RG, Portmann AC, Ulrich BA, Wolfand JM (2020) Biochar-augmented biofilters to improve pollutant removal from stormwater–can they improve receiving water quality? Environ Sci: Water Res & Technol 6(6):1520–1537. https://doi.org/10.1039/D0EW00027B

Burt Z, Ercümen A, Billava N, Ray I (2018) From intermittent to continuous service: costs, benefits, equity and sustainability of water system reforms in Hubli-Dharwad, India. World Dev 109:121–133. https://doi.org/10.1016/j.worlddev.2018.04.011

Campisano A, Butler D, Ward S, Burns MJ, Friedler E, DeBusk K, Fisher-Jeffes LN, Ghisi E, Rahman A, Furumai H, Han M (2017) Urban rainwater harvesting systems: Research, implementation and future perspectives. Water Res 115:195–209. https://doi.org/10.1016/j.watres.2017.02.056

Chen W, Gao W, Jiang J, Wei X, Wang R (2021) Feasibility analysis of decentralized hybrid rainwater-graywater systems in a public building in Japan. Sustain Cities Soc 69:102870. https://doi.org/10.1016/j.scs.2021.102870

Chubaka CE, Whiley H, Edwards JW, Ross KE (2018) A review of roof harvested rainwater in Australia. J Environ Public Health 6471324. https://doi.org/10.1155/2018/6471324

Cuong DV, Wu PC, Liou SYH, Hou CH (2022) An integrated active biochar filter and capacitive deionization system for high-performance removal of arsenic from groundwater. J Hazard Mater 423:127084. https://doi.org/10.1016/j.jhazmat.2021.127084

Deshkar S (2019) Resilience perspective for planning urban water infrastructures: a case of Nagpur City. In: Ray B, Shaw R (eds) Urban drought. Disaster risk reduction. Springer, Singapore, pp 131–154. https://doi.org/10.1007/978-981-10-8947-3_9

Domínguez I, Ward S, Mendoza JG, Rincón CI, Oviedo-Ocaña ER (2017) End-user cost-benefit prioritization for selecting rainwater harvesting and greywater reuse in social housing. Water 9(7):516. https://doi.org/10.3390/w9070516

Ercumen A, Arnold BF, Kumpel E, Burt Z, Ray I, Nelson K, Colford JM Jr (2015) Upgrading a piped water supply from intermittent to continuous delivery and association with waterborne illness: a matched cohort study in urban India. PLoS Med 12(10):e1001892. https://doi.org/10.1371/journal.pmed.1001892

Fang B, Yu J, Chen Z, Osman AI, Farghali M, Ihara I, Hamza EH, Rooney DW, Yap PS (2023) Artificial intelligence for waste management in smart cities: a review. Environ Chem Lett 21(4):1959–1989. https://doi.org/10.1007/s10311-023-01604-3

Gan L, Garg A, Wang H, Mei G, Liu J (2021) Influence of biochar amendment on stormwater management in green roofs: experiment with numerical investigation. Acta Geophys 69(6):2417–2426. https://doi.org/10.1007/s11600-021-00685-4

Gan L, Garg A, Huang S, Wang J, Mei G, Zhang K (2022) Experimental and numerical investigation on rainwater management of dual substrate layer green roofs using biochar-amended soil. Biomass Convers Biorefinery 1–10. https://doi.org/10.1007/s13399-022-02754-0

GhaffarianHoseini A, Tookey J, Yusoff SM, Hassan NB (2016) State of the art of rainwater harvesting systems towards promoting green built environments: a review. Desalin Water Treat 57(1):95–104. https://doi.org/10.1080/19443994.2015.1021097

Govindarajulu D (2020) Strengthening institutional and financial mechanisms for building urban resilience in India. Int J Disaster Risk Reduc 47:101549. https://doi.org/10.1016/j.ijdrr.2020.101549

Holland-Stergar B (2018) The law and policy of rainwater harvesting: a comparative analysis of Australia, India, and the United States. UCLA J Environ Law Policy 36:127. https://doi.org/10.5070/L5361039903

Huang S, Huang D, Garg A, Jiang M, Mei G, Pekkat S (2020) Stormwater management of biochar-amended green roofs: Peak flow and hydraulic parameters using combined experimental and numerical investigation. Biomass Convers Biorefinery 1–12. https://doi.org/10.1007/s13399-020-01109-x

Huang J, Zhang Y, Bing H, Peng J, Dong F, Gao J, Arhonditsis GB (2021a) Characterizing the river water quality in China: recent progress and on-going challenges. Water Res 201:117309. https://doi.org/10.1016/j.watres.2021.117309

Huang Z, Nya EL, Rahman MA, Mwamila TB, Cao V, Gwenzi W, Noubactep C (2021b) Integrated water resource management: rethinking the contribution of rainwater harvesting. Sustainability 13(15):8338. https://doi.org/10.3390/su13158338

Kumar M, Chaminda T, Honda R, Furumai H (2019) Vulnerability of urban waters to emerging contaminants in India and Sri Lanka: resilience framework and strategy. APN Sci Bull https://doi.org/10.30852/sb.2019.799

Kumar M, Deka JP, Kumari O (2020) Development of water resilience strategies in the context of climate change, and rapid urbanization: a discussion on vulnerability mitigation. Groundw Sustain Dev 10:100308. https://doi.org/10.1016/j.gsd.2019.100308

Kumar M, Sharma A, Tabhani N, Otaki Y (2021) Indoor water end-use pattern and its prospective determinants in the twin cities of Gujarat, India: enabling targeted urban water management strategies. J Environ Manage 288:112403. https://doi.org/10.1016/j.jenvman.2021.112403

Kumar M, Shekhar S, Kumar R, Kumar P, Govarthanan M, Chaminda T (2023) Drinking water treatment and associated toxic byproducts: Concurrence and urgence. Environ Pollut 121009. https://doi.org/10.1016/j.envpol.2023.121009

Kusre BC, Bora PK, Rai D, Suranjoy SS, Tana A, Subba S (2017) Combating water scarcity through roof water harvesting: planning and design with stakeholders' perception in Sikkim (India). Water Sci Technol Water Supply 17(3):799–810. https://doi.org/10.2166/ws.2016.180

Mehrabadi MHR, Saghafian B, Fashi FH (2013) Assessment of residential rainwater harvesting efficiency for meeting non-potable water demands in three climate conditions. Resour Conserv Recycl 73:86–93. https://doi.org/10.1016/j.resconrec.2013.01.015

Mohapatra S, Sargaonkar A, Labhasetwar PK (2014) Distribution network assessment using EPANET for intermittent and continuous water supply. Water Resour Manage 28(11):3745–3759. https://doi.org/10.1007/s11269-014-0707-y

Nandi S, Gonela V (2022) Rainwater harvesting for domestic use: a systematic review and outlook from the utility policy and management perspectives. Utilities Policy 77:101383. https://doi.org/10.1016/j.jup.2022.101383

Ndeketeya A, Dundu M (2019) Maximising the benefits of rainwater harvesting technology towards sustainability in urban areas of South Africa: a case study. Urban Water Journal 16(2):163–169. https://doi.org/10.1080/1573062X.2019.1637907

Ouedraogo AS, Yuzhu FuG, Yunus AI (2023) Treatment of highway stormwater runoff using sustainable biochar: a review. J Environ Eng 149(2):03122005. https://doi.org/10.1061/JOEEDU.EEENG-7049

Páez-Curtidor N, Keilmann-Gondhalekar D, Drewes JE (2021) Application of the water–energy–food nexus approach to the climate-resilient water safety plan of Leh Town, India. Sustainability 13(19):10550. https://doi.org/10.3390/su131910550

Patange SD, Nimbkar N, Rajvanshi AK (2020) Low cost drinking water technology–rainwater harvesting with solar purification. Curr Sci 118(6):872

Pradhan R, Sahoo J (2019) Smart rainwater management: new technologies and innovation. In: Bobek V (ed) Smart urban development, IntechOpen, p 83. https://doi.org/10.5772/intechopen.86336

Rahman S, Khan MTR, Akib S, Din NBC, Biswas SK, Shirazi SM (2014) Sustainability of rainwater harvesting system in terms of water quality. Sci World J 721357. https://doi.org/10.1155/2014/721357

Ramsey E, Berglund EZ, Goyal R (2017) The impact of demographic factors, beliefs, and social influences on residential water consumption and implications for non-price policies in Urban India. Water 9(11):844. https://doi.org/10.3390/w9110844

Ray B, Shaw R (2016) Water stress in the megacity of Kolkata, India, and its implications for urban resilience. In: Shaw R, Atta-ur-Rahman, Surjan A, Parvin GA (eds) Urban disasters and resilience in Asia. Butterworth-Heinemann, pp 317–336

Sarkar SM (2021) Improved water supply through rainwater harvesting technology in the Purulia District of West Bengal, India. In: Prasad R, Jhariya MK, Banerjee A (eds) Advances in sustainable development and management of environmental and natural resources. Apple Academic Press, Palm Bay, pp Vol2–355

Silva AS, Ghisi E (2016) Uncertainty analysis of daily potable water demand on the performance evaluation of rainwater harvesting systems in residential buildings. J Environ Manage 180:82–93. https://doi.org/10.1016/j.jenvman.2016.05.028

Talat N (2021) Urban water-supply management: indirect issues of climate change leading to water scarcity scenarios in developing and underdeveloped nations. In: Thokchom B, Qiu P, Singh P, Iyer PK (eds) Water conservation in the era of global climate change. Elsevier, Amsterdam, pp 47–71. https://doi.org/10.1016/B978-0-12-820200-5.00009-9

Tolossa TT, Abebe FB, Girma AA (2020) Rainwater harvesting technology practices and implication of climate change characteristics in Eastern Ethiopia. Cogent Food Agric 6(1):1724354. https://doi.org/10.1080/23311932.2020.1724354

UN (2015) Goal 6: ensure access to water and sanitation for all

USAID (2013) Municipal water use efficiency guideline: Jaipur. United States Agency for International Development, Washington, DC, USA

Vaidya R, Kumar KS, Mohan Kumar MS, Rao L (2022) Chlorine degradation in continuous and intermittent drinking water supply networks. Urban Water J 1–14. https://doi.org/10.1080/1573062X.2022.2139275

Van Meter KJ, Basu NB, Tate E, Wyckoff J (2014) Monsoon harvests: the living legacies of rainwater harvesting systems in South India. Environ Sci Technol 48(8):4217–4225. https://doi.org/10.1021/es4040182

Verma G (2022) Low cost smart ground system for rainwater harvesting for Indian houses using IoT technology. Wireless Pers Commun 126(4):3317–3332. https://doi.org/10.1007/s11277-022-09866-w

Ward S, Butler D (2016) Rainwater harvesting and social networks: visualising interactions for niche governance, resilience and sustainability. Water 8(11):526. https://doi.org/10.3390/w8110526

Zang J, Kumar M, Werner D (2021) Real-world sustainability analysis of an innovative decentralized water system with rainwater harvesting and wastewater reclamation. J Environ Manage 280:111639. https://doi.org/10.1016/j.jenvman.2020.111639

Zhang M, Hong Y, Zhu B (2022) Does national innovative city pilot policy promote green technology progress? Evidence from China. J Cleaner Prod 363:132461. https://doi.org/10.1016/j.jclepro.2022.132461

# A Long-Term Integrated Approach to Plan and Implement Green and Liveable Cities: Case Studies in the UK

**Wei Yang**

**Abstract** Urgent action is needed to tackle climate change, biodiversity decline, pollution, and waste as immediate threats to humanity and the planet. This chapter emphasises the importance of Spatial Planning as a critical mechanism for achieving net zero, environmental net gain, and a green industrial revolution. However, successful implementation of Spatial Planning requires systemic changes to effectively engage communities, reconcile conflicting interests, promote sustainable land use, and develop long-term mechanisms. The chapter presents case studies on Cornwall Council's inclusive and integrated policymaking process and Milton Keynes' long-term strategy for city-wide green infrastructure, offering practical insights into the application of a long-term integrated approach to plan and implement green and liveable cities.

**Keywords** Long-term integrated approach · Grand challenges · Spatial planning · Policymaking · Milton keynes · Green infrastructure

## 1 Introduction: The Quest of a Long-Term Integrated Approach to Tackle the Grand Challenges

According to the UN's Intergovernmental Panel on Climate Change (IPCC) Sixth Assessment Report *Climate Change 2022* (IPCC 2022a), currently approximately 3.3 to 3.6 billion people live in places that are highly vulnerable to climate change. More than 45% of the world population live in areas of high conservation importance, which are priority areas for nature conservation. Despite this, the world population is growing fast with a projected increase to approximately 8.6 billion by 2030 (United Nations 2017). Based on the medium-variant projection, the world population is expected to increase by 2.2 billion people between 2017 and 2050, reaching 9.8 billion people in 2050 (United Nations 2017). Scientists from the IPCC and the

W. Yang (✉)
Wei Yang & Partners, London, UK
e-mail: info@weiyangandpartners.co.uk



R. Yao (ed.), *Resilient Urban Environments*, Cities and Nature,
https://doi.org/10.1007/978-3-031-55482-7_18

Intergovernmental Science-Policy Platform on Biodiversity and Ecosystem Services (IPBES) emphasise that climate change and biodiversity decline must be urgently addressed together, along with pollution and waste, as they pose immediate threats to humanity and the planet (IPBES and IPCC 2021). Treating climate, biodiversity, and human society as coupled systems is key to achieving successful outcomes from policy interventions (IPBES and IPCC 2021).

The IPCC *Climate Change 2022* report assessed the impacts, risks and adaptation to climate change worldwide, with over 34,000 cited references (IPCC 2022b). It is evident that 'people's health, lives and livelihoods, as well as property and critical infrastructure, including energy and transportation systems, are being increasingly adversely affected by hazards from heatwaves, storms, drought and flooding as well as slow-onset changes, including sea level rise'. The report emphasised the urgent need for systemic transitions, enabled by transformations in energy, urban and infrastructure, industrial, land, and ocean systems, as well as in society through Climate Resilient Development Pathways (IPCC 2022e). It also highlighted the absolute necessity of managing climate risks in an integrated manner. IPCC urged to reconcile and integrate climate action, resilience and sustainable development activities through a better-connected policy and practice framework.

As a key mechanism for delivering net zero and environmental net gain, spurring the development of a circular economy, and achieving a green industrial revolution for a fairer society, Spatial Planning was identified as one of the adaptation options requiring the largest systemic changes (IPCC 2022d). Additionally, with a place-based whole systems approach, Spatial Planning has powerful leverage to engage communities to create beautiful and liveable places. With the perspective for understanding and addressing issues in an integrated manner, Spatial Planning enables us to reconcile conflicting priorities and interests among different stakeholders and government departments. It achieves these objectives by defining land use, protecting the environment, engaging in design, and stimulating social and economic activities (Batty and Yang 2022).

A long-term integrated approach is crucial for planning and implementing green and liveable cities. On the one hand, this requires cross-departmental and multidisciplinary collaboration, as well as wider citizen engagement. On the other hand, it necessitates a long-term strategy that considers the entire lifecycle of creating sustainable and resilient places, starting from visioning, planning, implementation, enforcement, management, and maintenance, as well as monitoring and feedback.

This chapter includes two case studies to discuss how the long-term integrated approach can be utilised. The first case study focuses on Cornwall Council and its inclusive and integrated policymaking process. The second case study examines Milton Keynes and its long-term strategy for planning and implementing city-wide green infrastructure.

## 2 Case Study One: Integrated Policymaking Process in Developing Carbon–Neutral Development Strategy in Cornwall

> Robust decision-making, using pathway analyses to determine 'no regrets' options among trade-offs, has been shown to be a useful starting point under deep uncertainty (IPCC 2022d).

Cornwall is a rural and coastal county in South West England with a population of half a million people. As the unitary local authority for Cornwall, Cornwall Council declared a climate emergency in January 2019 and committed to achieving carbon neutrality by 2030. After consulting with thousands of residents across Cornwall, the Council's *Climate Change Action Plan* (Cornwall Council 2019) was unanimously approved by its Cabinet in July 2019. The plan provides a roadmap for achieving carbon neutrality in Cornwall ahead of the UK Government's target of achieve Net Zero by 2050.

To create a shared vision and a collective response to tackle the complex interconnected challenges, the council has developed a new tool called the Cornwall Development and Decision Wheel. This in-house tool aims to reshape project and policy development and decision-making by formally incorporating a comprehensive assessment of environmental, social, and economic factors that were previously not thoroughly considered. The tool assists officers and decision-makers in understanding the complexity of these multifaceted challenges, ensuring that future decisions do not have negative impacts on residents. Since its formulation, the wheel has been utilised in all Council decision-making processes to assess the environmental and social implications of decisions (Cornwall Council 2020) (Fig. 1).

The essence of the Wheel was derived from the concept of Doughnut Economics, a theory developed by economist Kate Raworth (2018), which assesses various social and environmental boundaries within the framework of sustainable development. The Carbon Neutral Cornwall Team identified the relevant areas pertinent to its main policy drivers and powers as a local authority and customised different segments of the doughnut model to align with local needs (Rainbow 2022). Experts within the council were engaged to assist in tailoring questions for each segment, prompting users to consider potential impacts in these priority areas as a means to evaluate decision-making. The Wheel was designed to ensure that the outer environmental sections and inner social sections are considered in conjunction, providing a comprehensive perspective on the decision-making process (Cornwall Council 2020).

The Wheel has proven to be highly successful in assisting officers and politicians in early identification of significant negative impacts and their integrated mitigation. It fosters a new mindset that encourages individuals to comprehend the interconnectedness of various issues. One of the key advantages of the Wheel is its ability to highlight evidence and knowledge gaps. By obtaining necessary evidence from other teams within the council, it helps break down departmental silos within the organisation. The visually represented coloured wheel graphic simplifies the

**Fig. 1** An example of the cornwall development and decision wheel showing a range of impacts (Image source: Rainbow, A. Cornwall Council 2022)

communication between decision-makers and council officers, overcoming professional barriers. Decision-makers can quickly grasp key impacts through visually summarised assessments (Rainbow 2022).

Following the approval of its *Climate Change Action Plan*, the Carbon Neutral Cornwall initiative identified the creation of a new integrated strategic planning document, the *Climate Emergency Development Plan Document* (DPD), as one of its priority projects, alongside Renewable Energy and Nature Recovery Networks (Cornwall Council 2022).

The *Climate Emergency DPD* is intended to assist the council in planning measures to address and mitigate climate change, while strengthening the policies of the *Cornwall Local Plan*. In February 2021, the Pre-Submission version of the document was published for consultation purposes. A *Sustainability Appraisal* and *Habitats Regulations Assessment* were conducted to inform the policies included in the DPD. The Development and Decision Wheel was used to evaluate the DPD (Cornwall Council 2021). Additionally, apart from planning policies, the council acknowledged the need for systemic changes in other key actions that fall outside of planning regulations, such as behaviour change, energy efficiency standards, agricultural practices, funding, and retrofitting existing buildings, among others.

Key planning policies in the *Climate Emergency DPD* (Cornwall Council 2021; RTPI 2021) cover the following areas:

**Natural Climate Solutions**: Proposed policies encompass Green Infrastructure Design and Maintenance, Biodiversity Net Gain, Tree Canopy, and Local Nature Recovery Network. The policies stipulate that Green Infrastructure should have a central role in scheme design, ensuring the integration of wildlife-friendly and people-oriented spaces, and establishing a multifunctional network of areas and activities. Additionally, all major development types must achieve a minimum Biodiversity Net Gain of 10%. It is also required that all major developments provide tree canopy coverage equivalent to at least 15% of the site area, achieved through the retention of existing trees and/or the establishment of new ones. Furthermore, development is expected to contribute to the Nature Recovery Network.

**Rural Development and Agriculture**: Proposed policies endorse proposals for diversification and/or enhancements to agricultural holdings that contribute to the management, reduction, or absorption of carbon or other emissions. These policies also aim to provide public benefits and sustain a viable and active countryside within farms and land holdings. Additionally, low impact residential development is encouraged as part of a regenerative approach to land use.

**Town Centres, Design and Density**: Proposed policies encourage the diversification of uses and increase in the number of residential dwellings in town centres, aiming to foster their long-term sustainable, social, and economic stability. This can be accomplished through measures such as change of use, redevelopment, improvement of the public realm and historic built environment, and the promotion of sustainable lifestyles. These policies also encourage the development of public transport and cycling infrastructure, as well as the provision of green spaces.

**Community Resilience (Coastal Change and Flood Management)**: Proposed policies aim to spatially identify Coastal Vulnerability Zones and Candidate Coastal Change Management Areas, ensuring that developments are not proposed in areas that will be exposed to coastal change in the future. Development proposals are expected to be designed in a way that reduces flood risk to the application site and its surroundings. Sustainable Drainage Systems (SuDS) proposals are required to prioritise the use of above non-buried SuDS, including retrofit SuDS and where feasible within existing developments.

**Sustainable Transport**: Proposed policies advocate for the design and location of new developments that minimise the necessity for travel, while promoting a modal hierarchy that prioritises walking, cycling, public transport, car clubs, electric vehicles, and, finally, private fossil-fuelled vehicles. Additionally, former railway track beds and other railway land are required to be safeguarded from development that could hinder their reuse for railway purposes, the establishment of new travel or distribution networks, or the creation of sustainable transport connections and facilities.

**Renewable Energy and Strategic Sites**: Proposed policies support proposals for renewable and low carbon energy generation and distribution networks within the framework of sustainable development and climate change objectives, with the aim of contributing to Cornwall's target of achieving a 100% renewable electricity supply by 2030. The policies also prioritise the protection of strategic sites for renewable energy development.

**Sustainable Construction and Energy Efficiency**: Proposed policies mandate that development proposals integrate the Energy Hierarchy into building design, prioritising building fabric, orientation, and landscaping to minimise energy demand for heating, lighting, and cooling. All proposals should explore opportunities for incorporating solar PV and energy storage. New developments of all types are required to meet high environmental standards. Considerable emphasis will be placed on the benefits of development that significantly improves energy efficiency and reduces carbon emissions in existing buildings. The council will support proposals that enhance resilience to climate change and ensure a sustainable future for historic buildings and other designated and non-designated heritage assets. The council will also support the use of domestic and non-residential renewables, such as solar panels, while considering measures to minimise visual impact. All dwellings, including conversions, reversions, and changes of use, should aim for a water consumption of no more than 110 L per person per day by incorporating water-saving measures where feasible. Development proposals should minimise material usage, waste creation, and promote opportunities for a circular economy.

Cornwall Council submitted the *Climate Emergency DPD* for independent examination in November 2021, as required by planning legislation in England. Upon approval, these planning policies will play a crucial role in facilitating sustainable and resilient developments that will benefit the community in Cornwall in the long term.

As demonstrated in this case study, Cornwall Council adopted an innovative approach to foster a shared vision and enhance collaboration through inclusive governance and integrated planning policies. The creation of the Cornwall Development and Decision Wheel and the *Climate Emergency DPD* facilitated the necessary systemic changes within the council, aligning climate actions with the support of the local community to achieve net zero emissions. The lessons learned from this experience, particularly the strategies employed to drive systemic change, can be applicable in various locations worldwide. Further follow-up studies to analyse the policy's impact post-approval and implementation would be useful.

## 3 Case Study 2: City-Wide Green Infrastructure Framework in Milton Keynes

Longer term urban planning and design, including Nature Based Solutions (NBS) to reduce urban heat effects has been identified as a global level transformational adaption mechanism (IPCC 2022d).

Milton Keynes is a city located in the South East of England with a population of around 287,000 people (ONS 2022). It was formally designated as a New Town in 1967 with a target population of approximately 250,000. Now, 55 years after its designation, Milton Keynes has become the fastest growing city and is among the highest productivity cities in the UK.

The Milton Keynes Development Corporation was responsible for the planning and development of the city from 1967 to 1992. The city has benefited from a consistent green infrastructure-led Master Plan, which broke new ground in establishing a landscape framework for development, as well as a long-term commitment to green infrastructure management and maintenance through land value capture mechanisms.

The city area of Milton Keynes, designated by the UK Government in 1967, was approximately 9 $Km^2$ in size, which included four existing towns and 13 villages. Apart from the existing settlements, the new city boundaries enclosed a predominantly agricultural area. Major man-made and natural features within the city boundary were the A5, a major national road, the main London Euston to Birmingham railway line, the Grand Union Canal, and the valleys of the Great Ouse River and its tributaries (Milton Keynes Development Corporation 1992). The topography of the city is relatively gentle, with the valleys of Ouzel and Loughton Brook running from north to south in parallel (Fig. 2).

The Master Plan of the city was developed in 1970 and formally updated in 1989. It functioned as a strategic framework with considerable flexibility, capable of responding to changing needs. The Master Plan had many distinctive characteristics, such as its 'grid' pattern of main roads serving dispersed land uses and an extensive network of pedestrian and cycle routes, which would require an entire separate discussion on their design intentions and real outcomes. In this case study, particular attention is given to Milton Keynes' city-wide green infrastructure, which is one of the major outstanding characteristics of the city. Using the principles of Nature-Based Solutions, the Master Plan created a structural landscape framework for the new city and provided the 'predominant continuity of green space' (Milton Keynes Development Corporation 1992). The landscape framework was designed to be multifunctional, providing co-benefits for flood defence, wildlife conservation, visual amenity, as well as offering active and passive leisure and recreation facilities.

From a city-wide strategic scale, the landscape framework was designed to emphasise and enhance the natural character of the area, using mainly native species and incorporating existing features where possible (Milton Keynes Development Corporation 1992). Parkland and woodland were planned as linear parks along the valleys of the three major water courses that ran through the city. Major flood defence works were incorporated into the form of connected balancing lakes and ponds. They simultaneously provided additional nature reservation areas, as well as leisure and recreation facilities. The landscape framework was also intended to provide shelter, soften the impact of extensive new development, and create landmarks for the new city. During the design stage, consideration was given to the future growth, management, and maintenance of the landscape so that it would remain effective and continue to evolve as the city matures (Milton Keynes Development Corporation 1992). Planting was carefully balanced between fast-growing species for early

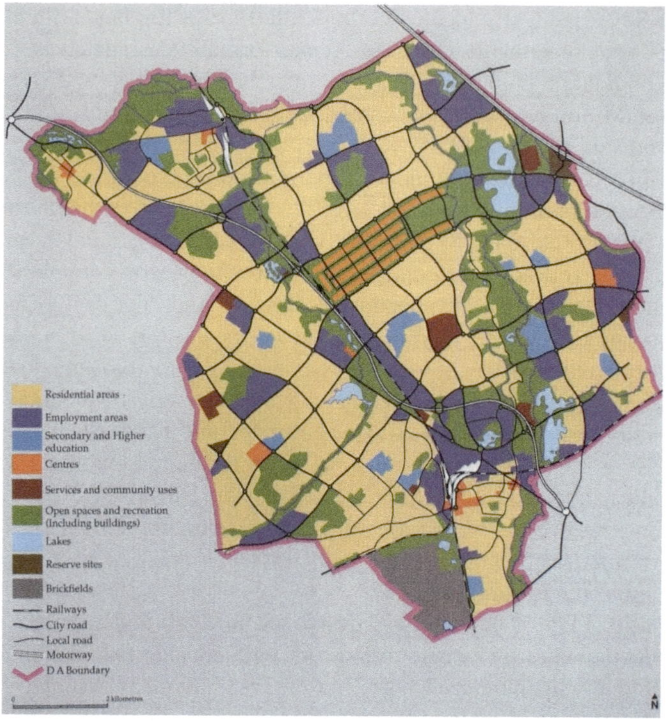


**Fig. 2** Plan for Milton Keynes 1970 (Image source: Milton Keynes Development Corporation 1992)

maturity and longer-lived forest trees that may reach their peak in up to 100 years' time. For established woods, the design reconciled public access with the conservation of wildlife. A typical wood contains open rides and glades, coppiced areas, mature woodland, and a few impenetrable areas where wildlife is undisturbed (Milton Keynes Development Corporation 1992).

When the city was designated, the availability of drainage, water supply, and other services was a major concern. The Development Corporation played a key role in coordinating and ensuring the establishment of a comprehensive, modern system. Surface water drainage and foul water drainage were designed as two separate systems (Milton Keynes Development Corporation 1992). Surface water is discharged into watercourses in the linear parks as part of the land drainage system, while foul water is collected and treated at sewage treatment works before being discharged into the River Ouzel.

The linear parks within the city were planned as primary opportunities for outdoor leisure activities, including formal pitch sports, water sports, and informal recreational activities in open spaces such as camping, picnicking, sledging, dog walking, or kite flying (Milton Keynes Development Corporation 1992). These activities were supported by a network of footpaths, bridgeways, and cycle lanes. The landscape along the city's main grid roads also formed an integral part of the extensive green infrastructure. The road reservations were designed to provide landscape features, noise attenuation, and accommodate additional main services and other forms of transport if required (Milton Keynes Development Corporation 1992).

At the neighbourhood level, local parks were designed to provide a diverse range of open spaces within predominantly residential areas. Distinctive landscape design elements were used to create both visual focal points and functional open spaces within each residential area. Children's play areas were strategically located based on catchment areas, with neighbourhood play areas designed for children aged eight to fourteen, and local play areas for children under eight years old. Allotments, local sports fields, and pavilions were integrated into the open space network (Milton Keynes Development Corporation 1992). Space-defining boundary planting utilised local hedgerow species to provide shelter and food for wildlife (Fig. 3).

In line with the concept of a 'forest city,' Milton Keynes Development Corporation planted over 20 million trees from its own nursery between 1967 and 1992. Overall, Milton Keynes' parks and green spaces account for approximately 25% of the new city area, including river valleys, woodlands, lakesides, parks, and landscaped areas (Milton Keynes Development Corporation 1992). This extensive and multifunctional

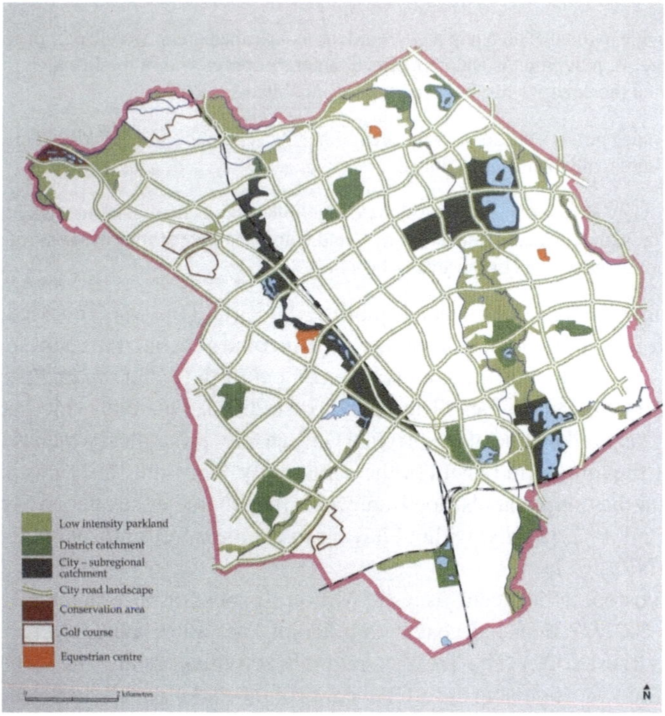


**Fig. 3** Landscape structure of Milton Keynes (Image source: Milton Keynes Development Corporation 1992)

green infrastructure has become an invaluable asset for Milton Keynes. However, the planners and designers of Milton Keynes also recognised that the long-term success of the city-wide green infrastructure would depend on proper maintenance regimes supported by adequate funding.

In 1992, the Development Corporation was directed by the UK Government to wind up its operations. To ensure that the city's vision could be implemented and achieved in the future, the Development Corporation published *The Milton Keynes Planning Manual* (1992). This manual recorded the original design intentions and specified the expected requirements for different areas in the city.

Another significant achievement of the Development Corporation in 1992 was the creation of the Parks Trust, an independent charity entrusted with the expert care of over 2.4 $Km^2$ of parks and green spaces in Milton Keynes. Following the winding up of the Development Corporation, ownership of the landscapes along the grid roads and major parks was transferred to the Parks Trust, which operates separately from the municipal authority and is intended to resist pressures to develop these parks over time. The Parks Trust inherited most of the Development Corporation's responsibilities in the parks but not its planning powers (NIC 2017).

The creation of the Parks Trust was forward-thinking and ahead of its time. As recorded in *The Milton Keynes Planning Manual*, the objectives of the Parks Trust were (Milton Keynes Development Corporation 1992):

> a. To provide maintain and equip parks, gardens, landscaped areas, woodlands, open spaces, playing fields, playgrounds and recreational amenity spaces within the Borough of Milton Keynes and the environs for the benefit of the inhabitants.
>
> b. To advance public education in the area of benefit with particular regard to horticulture, arboriculture, wild plants and wildlife.
>
> c. To provide or assist in the provisions of facilities for recreation or other leisure time occupation in the interests of social welfare and with the object of improving the conditions of life for the inhabitants of the area of benefit.

To ensure its long-term financial stability, the Parks Trust was endowed with cash and a portfolio of commercial properties. The income generated from these assets has been used to fund the nurturing and enhancement of the extensive green spaces and landscapes in Milton Keynes (The Parks Trust 2022). This city-wide maintenance model was inspired by one of the original Garden City principles, which is land value capture for the long-term benefit of the community. This innovative mechanism has ensured that the unique landscape framework that characterises the city is managed and protected in perpetuity, without having to compete for funds with other council priorities (NIC 2017).

As the custodian of green spaces in Milton Keynes for 30 years, the Parks Trust ended the 2021/22 financial year with a healthy net asset value of £150.3 million (The Parks Trust 2022). The Trust's charitable activities resulted in a small surplus of £0.7 million. Recognising the effectiveness of the Parks Trust's model, new growth areas in Milton Keynes have begun allocating additional green spaces to the Trust, along with an additional Endowment Fund of £20.3 million in 2021/22. The Parks Trust's *Annual Report and Financial Statement for 2021–22* highlighted some of their key activities and achievements over the past 12 months. These tangible outcomes and intangible benefits perfectly summarise the success of the long-term vision and innovative mechanism created by the Milton Keynes Development Corporation. Some of these achievements are listed below:

**Climate and Biodiversity Resilient Actions** (The Parks Trust 2022):

**Five-year Biodiversity Action Plan**: The Trust has conducted ongoing surveys of the local landscape supporting the diversity of flora and fauna and guiding its future projects. Significant progress has been made in fulfilling the commitments outlined in the Trust's five-year biodiversity action plan, which aims to conserve local species and habitats.

**New Tree Planting**: During the 2021 tree planting season, the Trust planted over 35,000 new trees and shrubs. The Trust's Outdoor Learning team organised sessions for local schools, interest groups, and residents, providing opportunities to participate in tree planting and learn about nurturing young trees as they grow. In response to the impact of Ash dieback, a fungal disease that led to the loss of numerous valuable trees in the parks, the Trust planted 400 new oak saplings generously donated by local individuals who responded to their call to plant acorns.

**Floodplain Forest Nature Reserve**: In 2022, the Floodplain Forest Nature Reserve in Milton Keynes welcomed the nesting of the Marsh Warbler, one of the UK's rarest breeding birds. This marked the first recorded instance of this bird species nesting in Buckinghamshire, highlighting the significance of the reserve in providing an enriched habitat for wildlife. The opening of the Floodplain Forest by the Trust in 2016 has evidently contributed to achieving its objective of supporting and nurturing diverse wildlife populations.

**Outdoor Learning for All**: The Trust's Outdoor Learning team dedicated their efforts to promoting environmental education and wildlife awareness, in line with one of the charity's core objectives. In 2021, they successfully delivered more than 560 educational sessions and activities catering to individuals of all ages. Additionally, they developed an extensive array of resources such as activity sheets and informative boards placed at various points of interest across the sites.

**Health and Wellbeing** (The Parks Trust 2022):

**Events for All**: The parks hosted a diverse array of events and activities aimed at serving the entire community. Over the course of 12 months, the Trust successfully organised and supported more than 1,700 events, drawing in nearly 100,000 attendees. These events ranged from guided walks and play days to heritage experiences and festivals celebrating the city's rich diversity. By fostering a sense of ownership among local residents and fostering community cohesion, these events played a crucial role in bringing people together within the parks.

**Water Recreation**: Willen Lake, the largest attenuation lake in Milton Keynes, stands as one of the city's most beloved leisure destinations, attracting approximately 1 million visitors annually, both on and off the water. Over the past several years, the Parks Trust has made significant improvements to the facilities, such as upgrading the city's largest outdoor play area and establishing a fully accessible Water Sports Centre. Additionally, a new electric community boat has been introduced on the Grand Union Canal, offering an exciting opportunity to explore the parks from a unique water perspective.

**Outdoor Exercises**: The Trust has not only expanded the variety of gym and play equipment available throughout the city but also ensured their regular maintenance, enabling people to engage in free outdoor exercise at any time. Additionally, the Trust remains committed to

> promoting the health and wellbeing of residents through organised events such as 'Walking for Health'.

The model of Milton Keynes' Parks Trust has been recognised as an innovative mechanism to ensure the long-term fulfilment and maintenance of the initial design intentions. In 2021, for the fifth consecutive year, the Parks Trust received a Green Flag Award for the high standards they maintain in parks throughout the city. This prestigious award scheme, managed by environmental charity Keep Britain Tidy under licence from the UK Government, acknowledges and rewards well-managed parks and green spaces, setting the benchmark standard for their management across the UK and around the world (The Parks Trust 2022).

## 4 Summaries

> Settlements bring together people, buildings, economic activities and infrastructure services, and thus integrated, cross-sector, adaptation actions offer the best way to build resilience to climate change impacts (IPCC 2022c).

This chapter features two case studies highlighting the creative mechanisms that can be employed in the process of planning and implementing sustainable and resilient places. The first case study in Cornwall focuses on the early strategic policy-making stage, while the second case study in Milton Keynes showcases a successful implementation that has spanned decades. Both cases serve as exemplary examples to demonstrate the breadth and depth of spatial planning, emphasising the crucial need to adopt an integrated and long-term approach right from the outset of the process.

However, the full potentials of spatial planning and the necessary systemic changes urged by numerous reports and experts are not being realised adequately or quickly enough, mainly due to a lack of understanding and inadequate best practice sharing. The *Climate Change 2022 report* (IPCC 2022e) highlights that there is limited evidence of proposed transformational adaptation across all regions, with very little evidence of implementation. To address this significant issue, it is crucial to view spatial planning as a cyclic sequence that includes not only policymaking and high-level land use plans but also the entire loop of engagement, design, implementation, maintenance, monitoring, and feedback. Gathering evidence through monitoring and feedback stages can facilitate effective knowledge sharing among different projects, cities, and countries. As a result, the planning process transforms from a linear procedure with segregated actions carried out by different actors into an integrated process driven by long-term visions shared by all, aiming to plan and implement green and liveable cities.

## References


Batty M, Yang W (2022) A digital future for planning—Spatial planning reimagined. Digital Task Force for Planning, London

Cornwall Council (2019) Climate Change Plan: creating the conditions for change through direct action and a new form of place-based leadership for Cornwall to become net carbon neutral

Cornwall Council (2020) Climate emergency Cornwall's call to action—My world, Our planet

Cornwall Council (2021) Climate emergency development plan document: pre-submission consultation

Cornwall Council (2022) Our ACTION PLAN (webpage) https://www.cornwall.gov.uk/environment/climate-emergency/our-action-plan/

IPBES and IPCC (2021) Pörtner HO, Scholes RJ, Agard J, Archer E, Arneth A, Bai X, Barnes D, Burrows M, Chan L, Cheung WL, Diamond S, Donatti C, Duarte C, Eisenhauer N, Foden W, Gasalla MA, Handa C, Hickler T, Hoegh-Guldberg O, Ichii K, Jacob U, Insarov G, Kiessling W, Leadley P, Leemans R, Levin L, Lim M, Maharaj S, Managi S, Marquet PA, McElwee P, Midgley G, Oberdorff T, Obura D, Osman E, Pandit R, Pascual U, Pires APF, Popp A, ReyesGarcía V, Sankaran M, Settele J, Shin YJ, Sintayehu DW, Smith P, Steiner N, Strassburg B, Sukumar R, Trisos C, Val AL, Wu J, Aldrian E, Parmesan C, Pichs-Madruga R, Roberts DC, Rogers AD, Díaz S, Fischer M, Hashimoto S, Lavorel S, Wu N, Ngo HT IPBES-IPCC Co-sponsored workshop report on biodiversity and climate change. IPBES and IPCC

IPCC (2022a) Climate change 2022: impacts, adaptation, and vulnerability. Contribution of working group II to the sixth assessment report of the intergovernmental panel on climate change [In Pörtner H-O, Roberts DC, Tignor M, Poloczanska ES, Mintenbeck K, Alegría A, Craig M, Langsdorf S, Löschke S, Möller V, Okem A, Rama B (eds)]. Cambridge University Press

IPCC (2022c) Dodman D, Hayward B, Pelling M, Castan Broto V, Chow W, Chu E, Dawson R, Khirfan L, McPhearson T, Prakash A, Zheng Y, Ziervogel G (2022) Chapter 6 cities, settlements and key infrastructure. In: Climate change 2022: impacts, adaptation, and vulnerability. Contribution of working group II to the sixth assessment report of the intergovernmental panel on climate change [Pörtner H-O, Roberts DC, Tignor M, Poloczanska ES, Mintenbeck K, Alegría A, Craig M, Langsdorf S, Löschke S, Möller V, Okem A, Rama B (eds)]. Cambridge University Press

IPCC (2022d) New M, Reckien D, Viner D, Adler C, Cheong S-M, Conde C, Constable A, Coughlan de Perez E, Lammel A, Mechler R, Orlove B, Solecki W (2022) Chapter 17 decision making options for managing risk. In: Climate change 2022: impacts, adaptation, and vulnerability. Contribution of working group II to the sixth assessment report of the intergovernmental panel on climate change e [Pörtner H-O, Roberts DC, Tignor M, Poloczanska ES, Mintenbeck K, Alegría A, Craig M, Langsdorf S, Löschke S, Möller V, Okem A, Rama B (eds)]. Cambridge University Press

IPCC (2022e) Schipper ELF, Revi A, Preston BL, Carr ER, Eriksen SH, Fernandez-Carril LR, Glavovic B, Hilmi NJM, Ley D, Mukerji R, Muylaert de Araujo MS, Perez R, Rose SK, Singh PK, 2022: Chapter 18 *climate resilient development pathways*. In: Climate change 2022: impacts, adaptation, and vulnerability. Contribution of working group II to the sixth assessment report of the intergovernmental panel on climate change e [Pörtner H-O, Roberts DC, Tignor M, Poloczanska ES, Mintenbeck K, Alegría A, Craig M, Langsdorf S, Löschke S, Möller V, Okem A, Rama B (eds)]. Cambridge University Press

IPCC (2022b) IPCC press release: climate change: a threat to human wellbeing and health of the planet—Taking action now can secure our future. (28 Feb 2022, Berlin)

Milton Keynes Development Corporation (1992) The milton keynes planning manual. Grillford Limited, Milton Keynes

National Infrastructure Commission (2017) Partnering for prosperity: a new deal for the Cambridge-Milton Keynes-Oxford Arc

Office for National Statistics (ONS) (2022) How the population changed in Milton Keynes: Census 2021

Rainbow A (2022) Carbon-neutral development: cornwall development and decision wheel (CDDW). In: Batty M, Yang W (eds) A digital future for planning—Spatial planning reimagined. Digital Task Force for Planning
Raworth K (2018) Doughnut economics: seven ways to think like a 21st-century economist. Random House, London
Royal Academy of Engineering (2019) Sustainable living places—a systems perspective on planning, housing and infrastructure
Royal Academy of Engineering (2020) Net zero: a systems perspective on the climate challenge
Royal Town Planning Institute (2021) Place-based approaches to climate change
The Parks Trust (2022) Annual report and financial statement: 2021–22
UN-Habitat (2016) World cities report 2016: urbanization and development—emerging futures. UN-Habitat, Nairobi
United Nations Department of Economic and Social Affairs, Population Division (2019) World urbanization prospects: the 2018 revision (ST/ESA/SER.A/420). United Nations, New York
United Nations Department of Economic and Social Affairs, Population Division (2017) World population prospects 2017—Data Booklet (ST/ESA/SER.A/401)

**Wei Yang** is an internationally renowned town planner and urban designer, who champions a place-based whole systems approach to tackle the grand challenges. She is Chairman of Wei Yang & Partners, Co-Founder and CEO of the Digital Task Force for Planning. She is a Fellow of Academy of Social Sciences, Chair of the UK Construction Industry Council, and an Honorary Professor at University College London. She was President for 2021 of the Royal Town Planning Institute.

# Wellbeing and Social Health in the Built Environment

**Ziona Strelitz**

**Abstract** The physical environment is critical to human experience and sustainable habitation, with multiplex impacts on people's experience. Physical settings influence health, safety, economic and social life, sense of community, and aesthetic nourishment. For cities to involve sustainable resource use, they should be efficiently used, and to promote efficient use, they should be convenient and appealing. Drawing on a rich repertoire of research in people's use of buildings and urban spaces, this chapter looks beyond well-recognised factors like thermal quality, addressing life style and balance, and social influences on 'wellbeing in use', including: location, spatial scale, biophilia, 'active by design', community and social memory.

**Keywords** End-users · Master-plan · Post occupancy evaluation · Social health · Wellbeing · Resilience

## 1 Physical and Mental Health: Widespread Concerns

The net of factors that shape human experience in the built environment is myriad and complex. The contextual physical influences—climate, terrain, water, mineral resources—are well-recognised determinants, especially in extreme environmental conditions where the severity and risk of their impacts is hard to ignore. At the scale of buildings and spaces, the very many component elements involved in design and construction have differential respective impacts on health and wellbeing—from benign to negative. And their combination in buildings has separate effects, relating to design decisions on aspects like siting, orientation, sectional height, plan depth, façade treatment, light and shade, thermal flow, and sound attenuation. This is the locus of opportunities to create spatial and environmental quality to generate positive outcomes for occupants' experience, contingent on choices relating to dimensions and materials—elementally and in inter-relationship.

Z. Strelitz (✉)
ZZA Responsive User Environments, London, UK
e-mail: ziona.strelitz@zza.co.uk



R. Yao (ed.), *Resilient Urban Environments*, Cities and Nature,
https://doi.org/10.1007/978-3-031-55482-7_19

There is now wide knowledge regarding technical outcomes—like emissions, pollutants, solar gain, and physical health effects, and a constantly expanding toolkit of approaches to avoid and/or mitigate harmful results. The other chapters in this book are comprehensive in proposing approaches to eliminate or limit this raft of untoward impacts.

But built environment conditions also affect social and mental wellbeing, and despite an increased cultural focus on 'happiness' (Layard 2005) as both an existential goal and index of social good, as well as converse concerns about the risks and incidence of poor wellbeing, the built influences on these conditions—especially the positive—are less recognised, researched and addressed. This gap includes lack of attention to 'sense of community' as a widely shared objective.

Discussion of wellbeing requires some definition. Is Maslow's hierarchy of needs an instructive baseline to frame it? Starting with physiological needs, it rises through safety, love and belonging, then esteem, to self-actualisation (Maslow 1954). Covid advanced awareness of environmental conditions for physical health, spotlighting the large disparities in urban living conditions between wealthy and poor city-dwellers. The pandemic's urban lockdowns also sharpened recognition of mental wellbeing, of people's needs to feel comfortable in their environment, to be productive, to participate, and have social connection. These are experiential aspects that relate to Maslow's framing, and the contemporary light that global communication shines on the diversity of human environments also highlights the universal validity of Maslow's hierarchy. The challenge is on ways to realise it.

The topic of city resilience must necessarily embrace the contrasting realities of rich and poor regions, with their wider and narrower respective options relating to urban conditions. Can one talk about a contemporary opera house or landmark broadcasting building in Beijing, an advanced sports stadium in the United Kingdom, a thoughtful social housing estate in Hong Kong, and a life-changing school built on low-cost, vernacular principles in northern Malawi in the same meaningful discourse? These assorted examples reflect not just a notable mix of building typologies, but significant divergence of economic capacity across localities. Does design for wellbeing represent and mean the same across such a heterogeneous range of settings?

This question matters as economies with more complex and deep-pocketed design norms promote development internationally, and as they educate future professionals—some of whom will later work in less well-resourced environments, often disseminating a hegemonic concept of 'good design' that is effectively unaffordable in terms of building life cycle performance in those settings. More significantly for wellbeing, this approach is often irrelevant and superfluous to local requirements. A re-direction of perspective is powerfully advocated by the architect Francis Kéré from Burkina Faso (Lepik 2010). His honouring with the 2022 Pritzker Prize is timely recognition of the core design approach offered by perceptive engagement and alignment with local activities and structures.

## 2 Wellbeing: Learning from Research on Building Use

Sense of wellbeing as it relates to built conditions is a consistent thread in ZZA's three decades of research on people's experience in using buildings and external spaces. Deriving from a distinctive disciplinary mix of social anthropology, urban planning, and interior architecture, the long-term, twin focus of these studies has been social and spatial—centred on individuals, organisations and culture on one hand, and the buildings, spaces and settings in which people live, work, learn and play on the other. This research therefore encompasses social and physical phenomena, and since the rise of ubiquitous technologies with their impactful reach, that nexus is now triangulated to incorporate virtual experience and potential as well.

This body of inquiry involves many buildings and spaces, from the perspectives of the cross-sections of individuals who use them in varied ways (Strelitz 2019). This includes what are called 'end-users'—people in their capacities as residents, workers, educators, students, and visitors, as well as those using buildings in specialist roles—like retailers, space managers, porters, security personnel, and cleaners, and also less direct 'stakeholders' like neighbours and passers-by on whose experience of their towns and cities, buildings also impinge. In all cases, our investigations are tailored to the given building, space or campus, encompassing sets of systematic questions specific to both the index setting and the site context. The responses for every study are coded for evaluation, aggregated for quantitative analysis, and illuminated with respondents' qualitative commentaries. From the outset, this research has engaged with issues that are core to sustainable living—in essence, to wellbeing.

In 2008, this cumulative learning on positive effects from building design was distilled in a suite of principles for 'Buildings that Feel Good', the title of a book analysing 20 case-study projects that were all exemplary in demonstrating aspects that promote wellbeing (Strelitz 2008). That knowledge base has now been enriched and expanded by extensive learning from many studies undertaken since. And given the speed with which technology catalyses change in behavioural modalities, cultures, and people's expectations across all facets and activities, as well as in approaches to designing, constructing and managing buildings and cities, the continued research contributes to necessary evolution of the evidence base on user experience.

The principles to shape design for wellbeing are inter-related; they can be 'mapped' as a connecting network of briefing or programme requirements to optimise any project's provision for people's wellbeing. The now intensified urgent need to address climate change as the primary built environment driver for human health gives the principles an order of precedence that is reflected in the following clusters, illustrated by case studies that span the spatial scale from building design to master-planning.

# 3 Inside-Outside

## *3.1 Passive Design, Simple Systems, User Control*

Thermal comfort is a fundamental condition for feeling good in buildings and urban spaces. Grounding architecture in response to the sun's path is key to sustainable design, and by extension, design for wellbeing. Too often other criteria are prioritised—orienting a building frontage for a prestigious 'address' or premium view are examples, or master-planning an ensemble of buildings as a visual 'pattern' or composition, whereas the buildings' dispositions read that way only on plan or from an aerial view that is typically unseen in most people's actual use. Examples of sophisticated buildings that are thoroughly predicated on passive design principles are therefore important as evidence that this principle can be effectively incorporated with other development ambitions, even in premium locations. The Heron Tower (now named Salesforce Tower) by KPF in a high-value London location exemplifies a successful project aimed at a discerning commercial market, with a design that derives directly from a simple diagram of the building's orientation relative to the sun (Strelitz 2005).

Despite widespread awareness that building occupants value control of their indoor comfort, natural ventilation—as the fulcrum of passive design—has taken time to gain recognition in amenable moderate environments where reliance on mechanically conditioned, sealed buildings had become normative. In 2010, the London School of Economics (LSE) purchased a nearby building that had been built in 1903–1913 for the UK government's Land Registry, which then occupied it for the next 100 years. The building's magnificent façade and interiors were recognised in its 'listing' as Grade II on the National Heritage List for England. The effect of such a designation is that the exterior and many interior features were not permitted to be altered. Despite the School's £56 million investment to transform the building for twenty-first century academic use—introducing classrooms, two Harvard style lecture theatres, a cafeteria, breakout spaces, offices for faculty and professional services staff, and a new accessible entrance, many of its original features were necessarily retained to comply with heritage requirements.

Following the LSE's adaptation and use, a Post Occupancy Evaluation (POE) researched the building's effectiveness for users. A key finding was the newly found pleasure that occupants who had previously occupied sealed buildings described in the age-old simple amenity of being able to open their windows. Illustrative user perspectives include the following assessment by an administrator: *"I think it's better for a lot more people here. Not everyone has great views, but they all have windows that they can open. That's a big improvement—open plan sealed office spaces are environments from hell."* A faculty member commented: *"The window's great—it's the best thing about the whole office, having the window options."*

The building faces the central London Park of Lincoln's Inn Fields, originally laid out in the 1630s, and currently operated by a municipal authority, the London Borough of Camden. Covering 7.25 acres/29,340 sq. m, it is the biggest public square in London. The framing of the administrator's statement quoted above that all building occupants can open a window is relevant. While everyone with offices that face either the park or the neighbouring Inns of Court enjoys high quality views to a mature treescape, for numerous offices the outlook is not so prepossessing, yet the operable windows rather than the views were presented as the key benefit.

In POE research at another London site, where the buildings address a highly developed external landscape, some users expressed frustration that the extensively glazed—but sealed—façade provided an appealing green vista, but one that they could only engage with in a more sensory way by leaving the building. 'Feeling the outside'—rather than just seeing it—required occupants first to negotiate the building's large floor plate, then its elevators, then the security cordon in the entrance lobby, before they could reach the outdoors, with the same sequence in reverse when they re-entered the building.

Current industry practice shows progression in building design for wellbeing. For example, since the LSE's adaptation of the Land Registry, the School has continued to enact an ambitious programme of campus development, curating and enhancing its collection of buildings, many of which derive from a history of ad hoc acquisition. Reflecting the School's adoption of zero carbon as a goal, a subsequent new building in the heart of the campus is heavily predicated on passive design. Designed by RSHP, the £125 million Centre Building incorporates offices for academics and administrators, lecture halls and other teaching spaces, a large student 'learning commons', and external terraces. It has almost full reliance on natural ventilation with operable windows, except in the teaching spaces where room temperature and air quality are centrally maintained and linked to occupancy levels. A wall-mounted control panel in each office and open plan area enables occupants to adjust the temperature and fresh air intake via the windows. To maximise efficiency and prevent loss or damage to rooms or equipment in the event of inclement weather, occupants are asked to close their windows when they leave their offices (LSE Estates).

Following the building's opening for use in the summer of 2019, a POE by ZZA covered the occupants' experience in the initial summer and winter conditions. In the core research undertaken in winter, a high majority of the interview sample evaluated the temperature in their work or study area as 'feeling mostly right now', with some reports of the space being somewhat cold. This contrasted with notable reporting of the space having been hot in summer. In fact, the summer research was during a heatwave, with external temperatures reaching the atypical high for London of 38°, during which the building interior with its fresh air performed favourably relative to outside conditions. The period since the POE has allowed for the building's controls—including window operability—to be optimised, and for users to become accustomed to its servicing mode.

The relevance of this exemplar is the School's deployment of learning from structured POE, and their commitment to natural ventilation. The Centre Building's architecture and servicing strategy demonstrate adherence to this principle; the design incorporated no fall-back provision—no 'Plan B'. Occupants are expected to relate their expectations of thermal comfort to climate imperatives, and to enhance their comfort by appropriate dress—essentially with 'layers' of garments that they can add or remove according to variations in temperature. This approach reflects the view that

**Fig. 1** London School of Economics, Centre Building, designed with low energy servicing based on operable windows. Exterior image (left:) © Mark Gorton/RSHP; Interior image: Daylight-filled student learning commons, with active Ground Floor frontage and external view (right) © Joas Souza (joas.co.uk)

environmental comfort involves adaptive occupant behaviours as well as effective building design, giving users agency as partners both in their own wellbeing and the wellbeing of future users (Fig. 1).

## *3.2 Natural Light, Views, and Sightlines*

There are many respects in which an active interface between inside and out affects wellbeing. ZZA's research abounds with the significance that building occupants place on natural light and external aspect. In our studies across varied building types, the narrative quotations on people's appreciation of natural light are extensive. The POE of the LSE Centre Building evidenced natural light as occupants' most cited favourite thing about the building, instanced by the range of capacities in which they use the building—students in the learning commons, and academics, administrators and PhDs with workspace on the building's upper floors.

Arising from a series of POEs on new buildings that were developed for police forces, a notable evolution involved the incorporation of a skylight above a custody desk (Strelitz 2013). 'Custody'—where people who have been arrested for alleged wrongdoing are detained pending further inquiry—is a stressful environment both for detainees and police officers. It therefore serves everyone's interests to lessen the tension. But custody environments have traditionally been sealed 'boxes' without windows, on the basis that glazing in the building envelope was presumed to make these facilities less secure. In the case of the study building, the official who had commissioned its procurement explained that the report on the first POE in a series for his authority had registered that lack of visual connection to the outside made the policing job harder, whereas a view to external aspect—enabling occupants to see outdoor weather conditions and perceive the time of day—could reduce stress, promote officers' wellbeing, and facilitate better work. This was the impetus to revise the brief for police station design to incorporate a skylight—a seemingly modest change to the building, but one with high positive impact on occupants' experience. Seeing cloud movement in the sky is a meaningful connection to the external environment, touching lives and enhancing wellbeing.

Daylight, clear views, and sightlines are notable influences on people's comfort in buildings, as well as attributes that boost confidence in movement round neighbourhoods, strengthening social as well as individual wellbeing. To promote a sense of assurance, the rail station designed by Wilkinson Eyre Architects at Stratford in East London was expressly built with transparency and unobstructed lines of sight. Its upward sweeping profile creates a positive and inviting external presence that defines a sense of place and projects a tone of optimism. Internally, the station's spaces are flooded with natural light by day, and radiate as an illuminated beacon after dark. These design attributes helped foster positive local identity prior to the subsequent siting of the 2012 Olympic Games in Stratford.

Similarly, the City of Manchester Stadium was predicated on design principles for safety and security. This too was translated in terms of sightlines, with clear viewing of the game for all spectators, averting crowd frustration. The circulation ramps were designed to avoid the hazard of people crushing on arrival and leaving the stadium. Wellbeing is further promoted by the roof that gives weather protection to all seats, while incorporating daylight and fresh air from openable windows around the upper perimeter of the stadium 'bowl'.

Clear sightlines also contribute to wellbeing by promoting orientation within buildings, with visible markers that signal where people are. Designing interiors with a sequential narrative not only promotes intuitive way-finding, but enriches people's experience in using the building. The design of the school for the Royal Society of Dramatic Art achieved all these purposes with an articulated 'user journey' through its varied new and legacy spaces. This leveraged the architecture for narrative and animation, alternating expansiveness with spatial compression, and providing a legible 'reading' of the building's sequence of social, performance and teaching functions.

Another significant new LSE development, the Marshall Building, provides clear views through the span of its large Ground Floor footprint, aiding orientation and navigation across the irregular street pattern of the campus. This also offers sightlines to the green landscape of Lincoln's Inn Fields, bringing a softening glimpse of the trees at the north of the building to the more hard-edged streetscape to its south.

## 4 Recycling Buildings

Research on perceptions of greener living identifies multiple measures that people consider relevant. These include greater care to minimise resource-intensity in using electronics, paper and packaging, increased use of renewable energy, more recycling, and less travel. What is generally considered in reference to recycling are goods and consumables, but recycling buildings themselves—illustrated by the LSE's effective re-use of the former Land Registry—has moved up the agenda to strengthen future resilience.

The case for adaptive re-use was highlighted in 'Buildings that Feel Good', with a third of the case-study analyses showcasing recycled buildings. These cover varied typologies—including a meat warehouse, water filtration plant, and post office sorting station as former uses, and education, offices, leisure and retail as uses that followed these buildings' adaptations. A notable benefit is the capacity of legacy buildings to nourish social memory. In the current era of intensive transformation of culture and the urban fabric, incorporating and perpetuating familiar references is grounding; it promotes wellbeing.

With the growing impetus to address climate change, adaptive re-use has gained significant traction for the range of aims it can serve for both environmental sustainability and social wellbeing (Strelitz 2021). ZZA's research on 'recycled environments' includes a recent POE of an architectural studio that successfully accommodates 130 people below street level in the former car-park of a Central London office building (Strelitz 2024). Not only has this workplace harnessed significant embodied carbon, but the people who work there are overwhelmingly pleased with their current work environment. This satisfaction includes their specific pleasure in the retained physical features of the space that express its previous car-park function—the entrance ramp, and the locational identifiers for parking spaces that are painted on the columns. The relevance of this example is in evidencing the wide scope for buildings that may be feasible to recycle, reducing waste, generating the physical wellbeing benefits of climate mitigation, and promoting social wellbeing with legible connection to symbols of the changing cultural landscape (Fig. 2).

## 5 Biophilia

People's draw to the natural world calls for nature's presence in the built environment to enhance wellbeing. Ken Yeang has been a distinctive exponent of planting integrated with buildings, including vertical greening in high-rise development (Yeang 1995). While such greening is typically associated with iconic tall structures, widespread benefit can be attained by incorporating planting in lower rise buildings.

An example of such impact is evidenced by the POE of Make Architects' Teaching and Learning Building at the University of Nottingham. Five storeys high, the

**Fig. 2** Make Architects' studio, London. Adaptive re-use of a former car-park: The collaborative studio environment (top); Processional multi-purpose entrance providing space for displays and informal seating along the original vehicle ramp (lower left); Marking on the column showing the parking location in the previous use of the space, now providing cultural context in the current studio setting (lower right). All three images © Make Ltd

building is situated in a part of the campus with mature trees. The building's architecture uses this, providing views to the exterior landscape at all interior levels. The following student quotations, from a larger pool of similar observations, are explicit in highlighting how this outlook to nature supports their psychological wellbeing (Fig. 3):

> "The natural light with the trees and the natural scenery – you don't feel enclosed or trapped."

> "Of course – the view outside with so much green. It's really nice, it helps me connect with nature. There's green all over the place – it's wonderful."

> "I like seeing nature. If you're writing an essay, you need to think, and I prefer seeing trees to a wall."

**Fig. 3** Teaching and Learning Building, University of Nottingham. Exterior showing the architectural disposition relative to the mature trees on site (top); Beneficial outlook to this landscape, enhancing student experience inside the building (below). Both images © Make Ltd.

"Because it seems more natural, it's nice. You feel you're in nature – it's calm and peaceful when I'm studying."

"Having a nice view is pleasant, it has a feel of working outside, and having lots of greenery is good for your psyche."

"I feel relief from depression when I come upstairs and look out over trees."

"It helps a lot. It's good for the lecturer and the students psychologically. It provides a break if you're doing an exercise, especially if you see trees and green."

The wellbeing value of connection to nature is fundamental. Market realism on people's attraction to nature is frequently reflected in high-value commercial schemes where providing ready visual and physical access to nature from apartments, offices, and hotel rooms can command a premium. In successive POEs at a highly successful 'mid-urban' business park, the research evidenced the large lake at the centre of the development as the pre-eminent feature that users value. Surrounded by verdant planting and populated by wildlife, this central body of water offers a unique environment for quiet reflection or social gathering. It affords distinctive biophilic benefits for both the occupants who work in the office blocks that are arranged in a ring around this oasis, as well as local residents who visit the park. The key to achieving this result is the master-plan, which placed the buildings side by side at relatively close proximity, allowing for a critical mass of open space in the middle to be developed for high landscape impact. A second vital strategy to achieve this soothing environment was making the central zone vehicle-free (Fig. 4).

A similar result has been reached in a social housing development in Hong Kong. On the Upper Ngau Tak estate in Kowloon, the residential blocks are arranged in a close linear configuration, side by side, making most of the site available for hard and soft landscaped amenity use by residents and visitors. The design incorporates a rich palette of planting, play settings for children, facilities for senior residents to exercise and socialise—promoting both residents' fitness and social wellbeing. Purposeful deployment of the site's gradient provides a choice of pedestrian routes based on strategies for 'active by design'. There are stairs for more energetic circulation and ramps for gentler ambulation. The lower part of the development is designed for community events, with an informal arena. The flourishing planting is in view from all parts of the estate and its surroundings.

Being outdoors affords access to sun, shade, thermal movement, colour, reflection and precipitation—all of which are sensory and aesthetic biophilic resources. The ground plane is typically the most accessible for direct engagement by the greatest

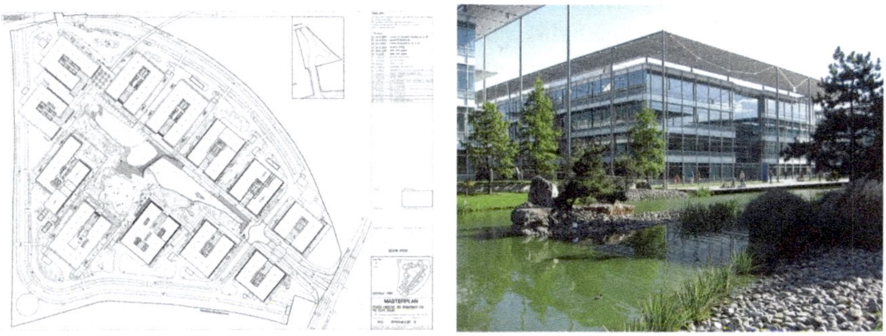

**Fig. 4** Chiswick Park, London, master-planned with the buildings arranged efficiently side by side, and vehicle access and parking confined to the perimeter, allowing the vehicle-free central zone to be developed as a biophilic oasis. Master-plan (left): Image © RSHP. The park's central zone of verdant planting and extensive, calming water feature, both greatly valued by people who work on the park and others who live nearby (right): Image © ZZA Responsive User Environments

number of people, as well as being visible in views from above. But it is not the only relevant level for contact with nature. Outdoor amenity areas in the form of upper-level terraces and rooftop amenity areas are valuable zones on which to provide wellbeing benefit across residential, office, education, cultural and retail buildings, particularly in densely developed urban areas, and in large buildings—not just high-rise, where the cumbersome circulation required to experience direct contact with the external environment at ground level can be a deterrent.

Several specific considerations of feasibility and potential are relevant in considering the provision of 'nature spaces'. One is life cycle resilience. Installations like green walls that are challenging and costly to maintain cause disappointment if they fail. In contrast, planting that occupants can maintain themselves is a more resilient strategy. Effective design can achieve strong green massing, based on planted areas that individual households or tenants can nurture. A striking example is Bosco Verticale in Milan, comprising two adjacent residential towers, with lush planting on every balcony (Bosco Verticale, Wikipedia). In combination, the cascading green effect is like the Hanging Gardens of Babylon, extending the biophilic experience to people well beyond the building and its immediate precinct (Lubell 2020).

A second point relates to the scale of planted areas. Not every building can have planting incorporated within it, or around it, and in urban conditions many of the opportunities available for plants involve small spaces. This does not detract from their value, nor their aggregate impact in greening a city, as is evident in the dense urban streets of Tel Aviv. But bigger spaces are more amenable for cultivation. The UK's system of allotments makes sizeable areas of land that is owned by public, private or ecclesiastical entities available to individual citizens on a low-cost leased basis, for cultivation of vegetables, fruit, and flowers. Located close to residential areas, allotments are typically situated in groups of plots, so that allotment-holders obtain not just the wellbeing benefits related to gardening—physical activity, time spent outdoors in green surroundings, the satisfaction of growing, and contributions to food security, but they are also advantaged by company and advice from their allotment neighbours. The demonstrated benefits of gardening, camaraderie and social bonds for physical and mental health are considerable (Dobson et al. 2021), and the allotment model lends itself to wide application where public authorities want to generate these benefits.

A third point on space for nature concerns large-scale sites. While planted landscape is evidentially beneficial, more does not necessarily mean better. The grain of development affects its impact, and too big an expanse of open space can be counter-productive. A case in point was a business park with extensive green landscaped zones between the buildings. While this extent of greening originated in the groundwater strategy for the site, the development team envisaged it as an appealing amenity for building occupants. However, some users' POE feedback reported too much open space for convenience and comfort: too far to walk from the car-park or transport stop, too much distance to reach the shop or restaurant, too exposed in wind or rain. An excess of outdoor space also risks being thinly populated, and too diffuse to promote a sense of community, with people feeling insecure in the evening an added deterrent to use. When master-planning new developments, the density of buildings relative to the scale and disposition of the spaces between and around them is key to consider in terms of massing, micro-climate, and perception.

## 6 Spatial Scale and Community

Spatial scale is relevant to community. The significance of community for wellbeing has largely been recognised in its absence, when redevelopment has painfully separated and dispersed neighbours who had been living on the same estate or in shared neighbourhoods for many years—in some cases for generations (Young and Willmott (2011). Community has value for people's sense of place and belonging, for affection, support, and mutual aid. While cities have long been characterised as places of anonymity that people may be drawn to because they want freedom from oversight and constraining norms, the benefits of community for wellbeing are notable (Coulombe and Krezni 2019).

A study on the reaction of people whose workplace was relocated from a big, new, award-winning office building to residual space in an old, partially redundant technical equipment centre illuminated key factors concerning scale, including its link to community. The new building was well specified, suffused with daylight, and provided excellent amenities for employees' use—a gym, a shop, a café, elegantly furnished informal areas, even a music room and hairdresser. In the company's technology sector in the UK, this was notably generous provision. The reasons for vacating this building that had only recently been procured were financial. With such prime space, the organisation was confident of finding a tenant to take over its lease commitments, and it also owned buildings where technological evolution had condensed the requirement for floor space, generating spare capacity. While these equipment centres had never been intended for office use nor to have a brand image, the available space there could be re-purposed for office work, albeit at a simple standard. When operations were moved from the prime building to these legacy buildings, this afforded a test-bed of people's preference that identified influential factors relating to building size and location.

The office building had essentially been a headquarters, with its large scale underpinning the fine facilities it offered. However, this also involved over 600 workers travelling in from a wide catchment area. In contrast, the older technical buildings were smaller structures, with a number distributed across the region, and on moving from the big building, employees were given the choice to work in the centre closest to their residence. In essence, therefore, the move from new to old meant that the workers traded the quality and amenities of a headquarters for a much more basic, refurbished workplace, but one that was also smaller, more compact and closer to where they lived.

Asked to compare their experience of the two buildings after the move, most interviewees in the study building rated the space they had moved to as meeting their needs—"improved or makes no difference". An aspect particularly linked to wellbeing was: "Feel comfortable and 'at home' when you come to this workspace", to which 95% of the sample responded affirmatively. The reasons why people endorsed the simpler workplace were its provision for community and convenience. Unlike the prime building with its big floor plates, the smaller building with just one floor enabled occupants to know the other 90 workers, affording face-to-face community and social cohesion. The smaller building was also preferred for ease of entry and exit—people could step outside for a break or fresh air more quickly. The case study highlights the influence of building scale both on sense of community and fluid movement between the interior and outdoors.

## 7 Location: Sustainability and Work–Life Balance

This research also evidenced the benefits of shorter journeys when workplaces are located closer to home. Exploring this further an international study in 2011 sought to learn where individuals choose to work when they have the choice of work location, and where they consider themselves to work most productively. Based on an online sample of 17,800 business managers from over 60 countries, and 87 structured, qualitative interviews in Greater New York, Greater Paris, Greater London, Hong Kong and Mumbai, "Why Place Matters in the Digital Age" examined contemporary issues at the nexus of technology, people, and place (Strelitz 2011). The core questions were if, and how, a specific physical location may feature for people who can select their work setting, in parallel with use of technology.

For over 70% of the interviewees, the most frequent place of work was neither their home nor a company workplace, but a 'third place' of their choice—a business centre, library or café. Their choices were shaped by three high level factors: facilitating productive work, safeguarding work-life balance, and easing the interface between their work and family life. People reported it being hard to get started when they tried to work at home, and difficult to avoid distractions. In contrast, working away from home freed them from disruption by family members, provided a barrier that stopped them from 'always working', and provided motivation from other people who were also working. Importantly, the interviewees also wanted to work *near* to home, in order to avoid the negative effects of commuting—the time and expense involved, and the stress related to delays and other contingencies, and to be close by for child-care, their children's schools, and / or elder-care. All these are ingredients of wellbeing.

In the decade since this research was published, the enormous growth of co-working spaces internationally has evidenced the impetus for people to obtain the benefits described. This has been compounded since the Covid pandemic, with many more people exposed to the wellbeing advantages of working locally to where they live. With both governments and employers looking to facilitate work–life balance (Strelitz 2017), a spatial strategy that distributes workplaces can better address people's need to live manageable lives, rather than employment that is concentrated in single large-scale buildings, no matter how highly these may be specified (Strelitz 2010).

## 8 Urban Villages for Today: The Fifteen Minute City

In evidencing the factors that promote wellbeing and social health at the scale of buildings and cities, community is an important contributor (Wiseman and Brasher 2008). How can community be facilitated in urban environments, and conversely, how can the negative conditions of isolation, loneliness and anomie be mitigated? The concept of 'villages in the city' is well established and critiqued (Biddulp et al. 2003), and continually referenced for its desirable attributes. Areas like London's urban villages are recognised for their small scale of buildings, mix of uses, convivial streetscape, and typically also for historic charm (How to Find the Village Feel in London, 2021).

More recently, Professor Carlos Moreno's concept of the 'Fifteen Minute City' offers a model that affords individual, social and environmental benefits, based on convenient physical access to the core life aspects of residence, work, shopping, education, and play, plus face-to-face engagement (Moreno 2024; Whittle 2021). This is about heterogeneity of function at a walkable urban grain, as opposed to extensive tracts of city that are defined by mono-use—whether focused on industry, offices, or housing. In addition to its benefits of environmental efficiency, the Fifteen Minute City also facilitates community, practical caring and social cohesion. For its relevance as a model to promote wellbeing and social health, it has wide potential for application.

## 9 Wellbeing in Cities: Summarising the Case for an Holistic Approach

In focusing on social and mental wellbeing, this chapter references key aspects of human experience—work–life balance, community, and social memory—that have lain peripheral to traditional perspectives of architecture and engineering, but are increasingly appreciated as integral to the contemporary design agenda. In discussing these experiential dimensions as they are affected by factors like location, spatial scale, planning and zoning strategies, the chapter implicitly proposes a breadth of professional expertise to shape healthy environments.

The scope to create environments that promote wellbeing is amplified by the applicability of relevant learning to legacy as well as new environments. As the case study of the former car-park now adapted to an architecture studio shows, recycling buildings can dramatically transform prior uses, introducing contemporary functionality, often with the additional benefit of retained social memory. The demonstrated potential to retrofit existing buildings can be extended to urban districts, transforming areas that have been characterised by a single or predominant typology to environments based on fifteen-minute principles, thereby scaling positive impact for wellbeing.

Adaptive re-use is not always feasible, viable or relevant. When life cycle analysis shows that recycling a building is unjustified in terms of carbon, likely usage or cost-effectiveness, new development will follow. This affords the opportunity to design and construct on principles for social wellbeing like those described here, with buildings and exterior spaces based on architecture that is disposed to occupants' health, spaces—inside and out— where people can experience nature, and urban planning that addresses the physical scale, grain, density and functional mix to afford individual balance and resilient, healthy communities.

## References

Biddulph M, Franklin B, Tait M (2003) From concept to completion: a critical analysis of the Urban Village, the town planning review 74, 2:165–193. Liverpool University Press, https://doi.org/10.3828/tpr.74.2.2

Bosco Verticale, https://en.wikipedia.org/wiki/Bosco_Verticale

Coulombe S, Krezni A (2019) Associations between sense of community and wellbeing: a comprehensive variable and person-centered exploration. J Community Psychol 47(5):1246–1268. https://doi.org/10.1002/jcop.22186,WileyOnlineLibrary,OnlineISSN:1520-6629

Dobson MC, Reynolds C, Warren PH, Edmondson JL (2021) My little piece of the planet: the multiplicity of well-being benefits from allotment Gardening. Br Food Jo 123(3):1012–1023, Emerald Publishing Limited, Bingley

How to Find the Village Feel in London (2021) https://bridgesandballoons.com/villages-london/

Layard R (2005) Happiness: lessons from a new science, 1st edn. Penguin, New York

Lepik A (2010) Small scale, big change: new architectures of social engagement. Museum of Modern Art, New York

LSE Estates, Centre building: building user guide, https://info.lse.ac.uk/staff/divisions/estates-division/Assets/Documents/CBR/19-0191-Centre-Building-Brochure-FINAL.pdf

Lubell S (2020) How Milan's bosco verticale has changed the way designers think about sustainable design, architectural digest, 3 November, https://www.architecturaldigest.com/story/how-milans-bosco-verticale-has-changed-the-way-designers-think-about-sustainable-design

Maslow A (1954) Motivation and personality, 1st edn. Harper & Row, New York

Moreno C (2024) The 15-Minute city: a solution to saving our time and our planet. Wiley

Strelitz Z (ed) (2005) Tall buildings, a strategic design guide, the british council for offices and RIBA Publishing, London

Strelitz Z (2008) Buildings that feel good. RIBA Publishing, London

Strelitz Z (2011) Why place still matters in the digital age: third place working in easy reach of home. Regus, London

Strelitz Z (2019) How is it for you? building design as experienced by users and makers, architectural design, 03, 89, 259: 60–67. Wiley, Oxford

Strelitz Z (2021) Roll on retro first, new london quarterly, 48: 44–45. New London Architecture, London

Strelitz Z (2010) Liveable lives: office push and pull, common employee predicaments, Regus, London

Strelitz Z (2013) Guiding building innovation in a distinct functional sector: POE of new police accommodation. Intel Build Int 5(3). https://doi.org/10.1080/17508975.2012.737702, Taylor & Francis

Strelitz (2017) The physical workplace and work-life balance: perspectives from practice, work-life balance in times of recession, austerity and beyond, Routledge, New York and Oxford

Strelitz Z (2024) Making it work, new london quarterly. Vol 57. New London Architecture, London, pp 50–53

Urban Village. https://en.wikipedia.org/wiki/Urban_village

Whittle N (2021) The 15 minute city: global change through local living. Luath Press, Edinburgh
Wiseman J, Brasher K (2008) Community wellbeing in an unwell world: trends, challenges, and possibilities. J Publ Health Policy 29, Palgrave Macmillan Ltd.
Yeang K (1995) Designing with nature: the ecological basis for architectural design. McGraw-Hill, New York
Young M, Willmott P (2011) Family and kinship in east london, 1st edn. Routledge Revivals, London

# Reading, UK: A Case Study in Urban Futures

**Timothy J. Dixon**

**Abstract** Decision-makers in cities and towns around the world are facing a 'perfect storm' of climate change impacts and the continued socio-economic consequences of the COVID-19 pandemic. In these challenging circumstances, it is even more important to take a long-term perspective of how our towns and cities can become smarter, more sustainable, and resilient to future shocks and impacts. 'Urban futures' thinking, which is at the heart of 'city visioning', can help urban planners and other decision-makers understand what cities and urban areas will be like in the long-term (beyond 20 years), how they will operate, what infrastructure and governance systems will underpin and co-ordinate them and how they are best shaped and influenced by their primary stakeholders (civil society, governments, businesses and investors, academia, and others). This chapter takes the example of a major town, Reading, in the UK, and using a quadruple helix (QH) conceptual framework, examines how stakeholders in the town have come together to produce (i) the Reading 2050 vision; (ii) the Climate Emergency Strategy (2020–2025), focusing on a net zero resilient Reading by 2030; and (iii) an economic post-COVID recovery strategy. Drawing on previous research, and using a case study-based narrative, the chapter provides an overview of current thinking and research on urban futures and city visioning, including city foresight techniques, before describing the QH framework in detail, and examining the three recent integrated 'urban futures' elements in more detail. The chapter concludes by providing a summary of the lessons learned and the implications for other towns and cities.

**Keywords** Quadruple helix approach · Integrated approach · Stakeholders · All-level strategy · Urban futures

---

T. J. Dixon (✉)
Sustainable Futures in the Built Environment, School of the Built Environment, University of Reading, Reading, UK
e-mail: t.j.dixon@reading.ac.uk

 
R. Yao (ed.), *Resilient Urban Environments*, Cities and Nature,
https://doi.org/10.1007/978-3-031-55482-7_20

## 1 Introduction

Decision-makers in cities and towns around the world are facing a 'perfect storm' of climate change impacts and the continued socio-economic consequences of the COVID-19 pandemic. In these challenging circumstances, it is even more important to take a long-term perspective of how our towns and cities can become smarter, more sustainable, and resilient to future shocks and impacts (Dixon and Tewdwr-Jones 2021). After all, the majority of the world's population lives in cities, and this is set to grow to more than 70% by 2050 (UN 2018a, 2018b). Because of this, while cities create a majority of the world's emissions, they can also act as hubs for innovation and transformation in the energy, water and waste transmissions needed to transition to a sustainable future.

However, if we are to make the changes needed to the built environment in our cities, not only do we need people to change their current ways of living, and this applies especially to those living in developed countries on relatively higher incomes and with more consumption-based lifestyles (Wiedmann et al. 2020; Ivannova et al. 2020), but we also need to develop clear 'visions' of how we will achieve our goals; for example, net zero emissions in a particular city by 2030 or 2050. In this sense a vision, or a shared and desirable view of the future, is an important element in 'urban futures' thinking. 'Urban futures' thinking, which is at the heart of 'city visioning', can help urban planners and other decision-makers understand what cities and urban areas will be like in the long-term (beyond 20 years) (Dixon and Tewdwr-Jones 2021).

Since the 1980s, there has been an increase in the interest in developing city visions in many cities around the world. This has increased even more recently as 'climate emergencies' have been declared in many municipalities around the world: for example, in 2022 there were declarations in 2037 jurisdictions and local governments covering 1 billion people (Climate Emergency Declaration 2022). This chapter therefore begins by examining the background and context to the emergence of urban futures thinking, before focusing on a specific town in the UK: Reading. Using a quadruple helix (QH) conceptual framework, the chapter examines how stakeholders in the town have come together to produce (i) the Reading 2050 vision; (ii) the Climate Emergency Strategy (2020–2025), focusing on a net zero resilient Reading by 2030; and (iii) an economic post-COVID recovery strategy. Drawing on previous research, and using a case study-based narrative, the chapter provides an overview of current thinking and research on urban futures and city visioning, including city foresight techniques, before describing the QH framework in detail, and examining the three recent integrated 'urban futures' elements in more detail. The chapter concludes by providing a summary of the lessons learned and the implications for other towns and cities.

## 2 Background and Context

Some previous authors have argued that that the inherent complexity and unpredictability of cities means that although we can develop models of cities as complex systems (which can help us understand how cities have evolved and how they behave in what is termed a 'science of cities') we cannot predict their future with any degree of certainty because we, as inhabitants of a city, are all an inherent part of that same future (see, for example, Batty 2018). On the other hand, it can be argued that although the future may itself not be 'predictable', it is crucial to find other ways of developing desirable and shared visions for our future cities in the light of the many complex and 'wicked' problems (or problems that may seem impossible to solve because of complex, incomplete or contradictory factors) that we face (Dixon 2021; Dixon and Tewdwr-Jones 2021).

Therefore, to overcome the frequent disconnection between relatively short-term planning horizons of 5–10 years and long-term environmental changes (20 years or more), city decision-makers and planners need to develop specific longer-term 'visions' that open up a possibility space to explore multiple futures, and also provide a detailed' roadmap' of how to achieve a shared and desirable future. This does not dismiss recognising the inherent complexity of cities, the continued need for immediate and short-term political decision-making, or even the vital role that the 'science of cities' plays in our understanding of cities. But it does require us to develop new ways of seeing and planning for a transition to a sustainable urban future.

This is what can be termed 'urban futures', which is a term used to 'imagine what cities and urban areas will be like in the long-term, how they will operate, what infrastructure and governance systems will underpin and co-ordinate them, and how they are best shaped and influenced by their primary stakeholders (civil society, governments, businesses and investors, academia and others)' (Dixon and Tewdwr-Jones 2021: p14).

Urban futures thinking requires city stakeholders to work together in terms of co-creating a city vision in a highly participatory way. This means that four main groups need to work together to build and develop city visions: namely civil society, local government, academia, and business in what is known a 'quadruple helix' partnership (Goddard and Tewdwr-Jones 2016; Dixon et al. 2018). As part of 'urban futures' thinking, city visioning is the formal process of creating a 'city vision', or a shared and desirable future for a particular city or urban area. In practical terms the city vision might relate to a single preferred urban future, or it could explore a variety of different and alternative urban futures. City foresight, which includes city, is therefore the 'science of thinking about the future of cities' (GOfS 2016) and includes a range of futures-based methods and tools to help build and develop a city vision: for example, 'backcasting', which starts with defining a desirable future and then works backwards to identify policies and programmes and pathways that will connect the present with the specified future, and 'three horizons' (3H) thinking, which is designed to help visioning participants think about three overlapping waves (for example, short (now), medium (near future) and long term (far future)) into the future (GofS 2017).

Visions are essential in developing our thinking about the future and often related to preferred or desirable futures and to a shared sense of change and transformation. Early examples of what might be termed humanistic visionary thinking emerge in the writings of Plato (fourth century BC) and later on, Thomas More's city-based Utopia (sixteenth century), and this sense of 'futurism' is also seen in the writings of Patrick Geddes and Ebenezer Howard, two of the early visionary planners in the late 19th/early twentieth centuries, who developed particular generic visions of what an ideal city should be (Dixon 2021).

In the context of urban planning, the idea of 'city visioning' (or having a clear and formal sense of where a particular city wants to be in the long term future) emerged during the 1980s and 1990s, particularly in the USA, not only as a way of understanding the future, but also to plan for a desirable, or preferred, set of sustainable outcomes (see for example, Atlanta, and Portland) (Dixon et al. 2018; Dixon and Tewdwr-Jones 2021). Newman and Jennings (2008) also highlight 'successful' examples of city visions in Perth, Vancouver and Chicago during this period. This emergence of thinking about the future of cities also reflected a growing body of literature focusing on 'visioning sustainability' in a range of other contexts, such as energy futures (Wiek and Iwaniec 2014). Since the early 2000s, we have also seen the development of more 'formal' visioning processes (or what might be termed 'city foresight' methods) in many cities and urban areas which have been used to develop city visions (see, for example, Phoenix, Johannesburg and Vancouver or, in the UK, Reading (Dixon et al. 2018) and Newcastle (Tewdwr-Jones et al. 2015; Dixon and Tewdrw-Jones 2021). This has also been a feature in other parts of the world, including Norway and Africa, and impacts on city regions as well as cities (Dixon et al. 2022).

The UK Government Office for Science (GOfS) Future of Cities Programme (2013–16) highlighted the importance of 'city foresight', which was founded on the science of thinking about the future of cities, and which can be used to enable city stakeholders to explore urban futures not only in a local and regional context, but as part of a wider connected network of cities (GOfS 2016). Several UK city visions were created as part of this programme, resulting from partnerships based on the 'quadruple helix' model of innovation (Arnkil et al. 2011; Goddard and Tewdwr-Jones 2016). Some of these visions have also linked with, and underpinned, the existing statutory local plans in cities (see for example, Dixon et al. 2018). A particular example of urban futures thinking is Reading, located in Berkshire, about 40 miles west of London in the UK. The chapter will now examine this case study in more detail.

# 3 Reading, UK: Urban Futures Case Study

## 3.1 *Overview*

Although Reading is not yet officially a 'city',[1] it forms part of one of the most economically vibrant and connected urban areas in the UK: Reading (population 174,200 in 2021 (ONS 2022)), as part of a wider Reading/Wokingham urban area (including Arborfield, Woodley, Theale (West Berkshire), Crowthorne, Earley), has a population of 318,000 (based on 2011 ONS data), and this is set to grow to 362,000 by 2037 (Dixon and Cohen 2015; Dixon and Farrelly 2020) (Fig. 20.1). Following its early historical development from a small Saxon town in the sixth century, Reading's economic success and prosperity is based on its physical as well as its virtual connectivity nationally and internationally, but this also presents natural resource and pollution challenges, as Reading seeks to maintain and enhance its position in an increasingly globalised and competitive marketplace. This, however, also presents important challenges in maintaining a competitive edge for Reading and dealing with the important environmental and socio-economic issues arising from a focus on continued economic growth.

In this next part of the chapter, we will examine three main elements of 'urban futures' thinking which have been developed in Reading (Fig. 20.2) to help address some of these issues. These comprise:

- Smart and Sustainable Reading 2050 Vision.
- Reading Climate Emergency Strategy 2020–2025.
- Reading Economic Recovery Strategy 2021–2024.

## 3.2 *Smart and Sustainable Reading 2050 City Vision*

Developing a Reading 2050 vision which was both 'smart' (making the best use of technology) and 'sustainable' (creating a truly sustainable city) was seen an important step in supporting longer-term planning and development in Reading (Fig. 20.3). The starting point for this vision was provided through a formal definition of a smart and sustainable city as one (ITU 2014: 12–13):

'...that leverages the ICT infrastructure to:

- Improve the quality of life of its citizens.
- Ensure tangible economic growth for its citizens.
- Improve the well-being of its citizens.
- Establish an environmentally responsible and sustainable approach to development.
- Streamline and improve physical infrastructure.

[1] Reading made an unsuccessful bid for city status in 2022.

**Fig. 20.1** Reading, Berkshire, UK (https://commons.wikimedia.org/wiki/File:Reading,_Berkshire_montage.jpg)

- Reinforce resilience to natural and man-made disasters.
- Underpin effective and well-balanced regulatory, compliance and governance mechanisms.'

In 2013 the Reading 2050 project brought together The University of Reading (led by the School of the Built Environment), Barton Willmore (a major planning and design consultancy), and Reading UK (the economic development unit for Reading[2]) to lead the development of the vision. Drawing on previous research which had

[2] Now Reading's Economic and Destination Agency (REDA) (see https://www.visit-reading.com/business).

**Fig. 20.2** Three examples of 'Urban Futures' thinking in Reading, UK

**Fig. 20.3** Reading 2050 vision. *Source* Reading 2050 website (www.visit-reading.com/business/sustainable-future) Image courtesy of Reading 2050—a collaborative initiative, jointly led by Barton Willmore, Reading UK and the University of Reading

scoped out retrofit visions for Cardiff and Manchester (Eames et al. 2013; Dixon et al. 2014) the Reading 2050 project combined elements of a smart city with those of a sustainable city. This was because Reading already had a long-term aspiration to be 'low carbon' by 2050, but also has a strong technology and green technology focus in its existing economy. Moreover, a 2050 time-horizon provides space to think beyond today's immediate problems and facilitates a greater sense of strategic thinking by identifying desirable as well as undesirable outcomes.

The visioning process which ran from 2013 to 2017, adopted a 'quadruple helix' approach which brought together business, local government, civil society, and higher education (Arnkil et al. 2011), and was based on workshops and the adoption of a backcasting approach. This is where a desirable future is co-created with stakeholders through a participatory based foresight approach, and then look stakeholders work together to look backwards from that future to the present to strategise and to plan how it could be achieved. Over the course of its work the Reading 2050 programme has so far engaged with 21,000 people and more than 400 businesses with some 15 linked events (Dixon and Farrelly 2020).

As a result, three interrelated urban futures were developed for the Reading 2050 vision[3]:

- 'Green Tech City': A city that builds upon its established technology focus. It celebrates and encourages diversity through business incubation units, 'Ideas Factories' and a city centre University campus through which to exhibit and test cutting edge ideas and approaches, no matter what discipline they are emerging from.
- 'City of Diversity and Culture': A city that builds on the success of the iconic Reading Festival to deliver arts and culture to people of all ages and ethnicities. Reading would facilitate community interaction and opportunity. The city would integrate, enhance and celebrate our heritage, bringing it to life through modern interpretations and uses of space as well as preservation.
- 'City of Rivers and Parks': A city that recognises how water has shaped much of Reading would celebrate its waterways, opening them up to offer recreational spaces such as animated parks, a lido, food production opportunities and city centre waterside living.

The vision is strongly linked with the development of the new Reading Borough Council Local Plan (which looks ahead to 2036) and is directly referenced within it as an important longer-term framework for Reading (Reading Borough Council 2019). A similar synergy is highlighted in the Corporate Plan where the council describes its endorsement of the vision and its commitment to integrating the 2050 ambitions into its priorities. Finally, the vision also links with the Reading Climate Emergency Strategy (2025–2030) which targets net zero emissions by 2030 and the Economic Recovery Strategy (2020) which are now discussed in more detail.

[3] See https://www.visit-reading.com/business/sustainable-future.

### 3.3 *Reading Climate Emergency Strategy 2020–2025*

Recent research in the UK has shown how important it is for UK cities and towns to develop 'climate commissions' (or partnerships) as part of a place-based approach to climate change strategy (Creasy et al. 2020). As with many other cities in the south of the UK, Reading faces an increasing impact from climate change—for example, as the climate warms, projections for the UK show that we are likely to experience milder and wetter winters, and hotter, drier summers, with an increasing frequency and intensity of extreme events (e.g. storms and heatwaves). Reading is also located in one of the most water stressed parts of the country, with a similar water availability per head to some communities in the Middle East. These impacts mean that Reading needs to become more resilient to a wider range of conditions than in the past. This is a particular concern for vulnerable people in Reading because of their relatively higher exposure to the impacts of a changing climate.

Following the Paris COP Conference in 2018, and increasing lobbying from climate activists and local people, Reading Borough Council (along with many other municipalities and local authorities globally) declared a climate emergency in February 2019 and made a pledge in to "… playing as full a role as possible – leading by example as well as by exhortation – in achieving a carbon neutral Reading by 2030". Following this, a consultation draft Reading Climate Emergency Strategy for 2020–25 was published in March 2020, and the strategy was launched in November 2020 (RCCP 2020).

The Reading Climate Emergency Strategy, 2020–25 (RCCP 2020) sets out the action required during this five-year period to meet the goal of a 'net zero'[4] and 'resilient'[5] Reading by 2030 and considers how Reading can adapt to the impacts of a changing climate. The strategy was produced through engagement and discussion with stakeholders, and a series of events and initiatives to engage the wider public, which led to a formal public consultation with more than 200 responses. Production of the strategy was co-ordinated by the Reading Climate Change Partnership (a community partnership established in 2009 to lead the response to climate change in the Reading area with community and voluntary sector, statutory sector, private sector, health sector and academic representatives), and with input from over 100 stakeholders across Reading. The Partnership Board sits at the head of a wider Reading Climate Action Network which includes an extensive range of voluntary sector bodies, community groups and individuals, between which there is a strong track record of effective collaboration. Although the Partnership has a role in co-ordinating strategy development, advocating for implementation and monitoring progress, it is not a delivery body—the responsibility for delivery sits with the partners who are identified in the strategy as owning individual actions.

[4] Defined as achieving an overall balance between CO2 emissions produced and CO2 emissions taken out of the atmosphere (RCCP, 2020).

[5] Defined as the ability to anticipate, prepare for, and respond to hazardous events, trends, or disturbances related to climate change (RCCP, 2020).

The strategy is based around the following six themes and linked action plans (underpinned by a communications and engagement plan):

- 'Energy and low carbon development', with a focus on retrofitting existing buildings[6] and building new net zero energy efficient homes and commercial property, and smart energy solutions.
- 'Transport', focusing on decarbonising transport systems by reducing use and switching more active mobility and EVs.
- 'Resources', with a particular emphasis on reducing consumption and waste and increasing recycling.
- 'Water', with a particular emphasis on reducing consumption and adapting to flood risk.
- 'Nature', using nature-based solutions (green infrastructure) to reduce emissions through natural 'carbon sinks' which will also help improve quality of life.
- 'Health and wellbeing', with the emphasis on living healthily in a changing climate and addressing health inequities by supporting the most vulnerable communities.

Clearly it is premature to assess the strategy's success, but since 2005, carbon emissions per person have fallen by 52% in the borough, the fourth largest reduction of any local authority in England (Fig. 20.4) and Reading Borough Council has reduced its own carbon footprint by 62.5% since 2008. The reduction represents the impacts of (i) national decarbonisation of the grid and the move towards renewables; (ii) Reading's relatively strong focus on the service sector rather than manufacturing; and (iii) local actions by key stakeholders. Reading was also one of just 95 towns or cities worldwide and 10 in the UK in 2021 to receive a top score on climate action in the Carbon Disclosure Project's 'gold standard' of environmental reporting. However, there is much more to do—the current rate of emissions reduction will need to approximately double to achieve net zero by 2030. Some of this will be helped by continuing national UK decarbonisation, but the remainder will need to be achieved through local action by all stakeholders in Reading, which will be challenging.

### *3.4 Reading Economic Recovery Strategy 2021–2024*

The final element in Reading's urban futures thinking is the economic recovery strategy (Reading UK 2020). Despite its strong economic structure, based on science,

[6] The council's 2019 Local Plan requires that all new residential developments of ten or more homes are built to zero carbon standards if possible. If reaching the zero-carbon standard is not possible (as determined by the developer), the development must deliver a 35% or greater reduction in carbon emissions compared to minimum UK standards, and the developer must pay £1,800 per tonne of carbon emissions to offsetting in Reading. Other residential development in Reading, those made up of nine or fewer homes, is required to deliver a 19% carbon reduction over national minimum standards. As set out in the Local Plan, if this standard cannot be reached, offsets may be possible through planning contributions to be ring fenced for carbon savings elsewhere (Ashden, 2022).

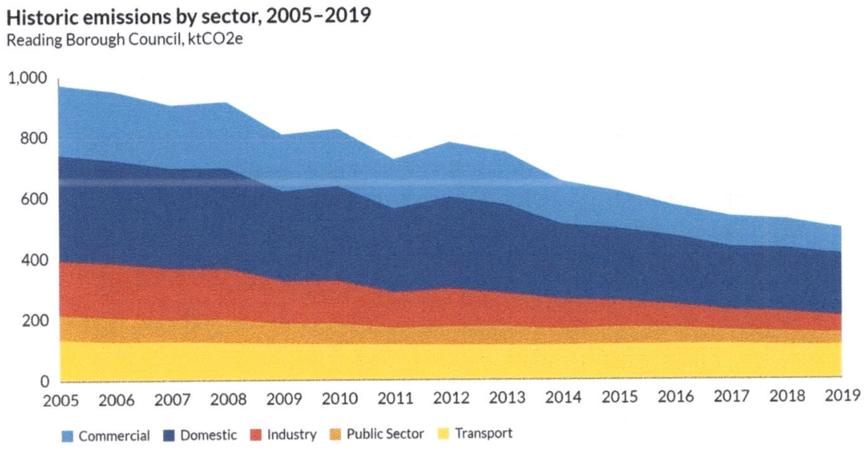


**Fig. 20.4** Carbon emissions (2005–2019): Reading. *Source* Data from BEIS: https://www.gov.uk/government/statistics/uk-local-authority-and-regional-carbon-dioxide-emissions-national-statistics-2005-to-2019

knowledge and professional services, and with a skilled workforce, global connectivity, and trade links, like many towns and cities in the UK, and elsewhere, Reading was impacted by the COVID crisis; for example, in 2020 footfall in the town was 50% of pre-COVID levels, and the17–24 and 50+ age groups were disproportionately affected by COVID-19 job losses with public transport also suffering revenue loss.

Although Reading has started to recover from the pandemic some sectors remain below pre-COVID levels. For example, reflecting pre-pandemic patterns, the food and drinks sector has been the most exposed to shifts to homeworking. The sector has taken a direct hit in places with many office workers, although these might be classified as 'stronger centres': in the city centres of London, Reading and Milton Keynes for instance, weekday sales in the food and drinks sector are still 15–20% down compared to their 2019 levels, although in contrast, at weekends, trade is back to, and above, pre-pandemic levels (Quinio 2022).

The new normal of 'hybrid' working after the pandemic therefore poses problems as well as opportunities for many town and city centres. While emissions might have fallen during the pandemic, and some people were able to enjoy greenspaces and more active lifestyles, the economic resilience of towns and cities has been severely tested. In response to the impact of COVID, Reading UK (a private sector led partnership promoting economic development in Reading) worked with Reading Borough Council and other stakeholders to develop the Economic Recovery Strategy ('Powered by the People') (Reading UK 2020). Linking closely to the Reading 2050 Vision and the Climate Emergency Strategy, the Economic Recovery Strategy aims to (Reading UK 2020:2):

"…to build back an economy that is more inclusive, smarter and more sustainable, and to continue to raise the profile of Reading as a great place to live, work, enjoy, visit and do business". The strategy has three main elements:

- 'Inclusive Reading'—Employment, skills, and training for local people.
- 'Smart Reading'—A successful, resilient economy, businesses and city centre fit for the future focusing on digital and green technology and infrastructure projects.
- 'Destination Reading'—Promoting Reading as a place to work, live, visit and do business focused on a net zero and inclusive growth future.

## 4 Discussion and Conclusions

The path to urban futures thinking in Reading has evolved over the last ten years, involving key partners and stakeholders in what might be termed a quadruple helix approach. The Reading case study highlights several important transferable lessons from which other towns and cities can learn and which are now discussed.

- Integrated approach: It is essential to ensure all high-level strategies and plans link with each other and provide an integrated approach to climate change and environmental impacts in urban areas. A city vision should be linked to climate change, energy, the economy, and to people, and the vision should be participatory and inclusive, analytically sound, and politically viable.
- Role of stakeholders: As trusted and independent institutions, Universities have a key role to play in helping formulate urban futures thinking, but it is also crucial to involve all stakeholders who are based in the town and city in the development of the strategy or the vision. It can sometimes be difficult to engage with people and to consult inclusively but new technologies and digital platforms have a role to play here in reaching wider audiences, although digital inclusion is also important to maintain.
- Power relations: Those developing and leading such visions must always reflect on the starting point and the inclusivity of their vision, and what this means for all groups of stakeholders in a city. For example, although visions tend to reflect dominant power relations (and the vision described in this paper is primarily academic and business-led), the Reading 2050 partners have sought to engage inclusively with a variety of groups to develop the vision (adopting a co-created approach which has tried to incorporate reflexivity in research and practice), and the support of Reading Borough Council has been crucial in this respect.
- R&D opportunities: Having an integrated urban futures framework for a place helps provide prospective R&D funders and those seeking funding with a sense of purpose and strategic thinking. The urban futures thinking described in this chapter have already led to several funded projects involving partners in the Reading quadruple helix—recent examples include the Thames Valley Berkshire Live Lab project (£4.8 m, 2019–21) (ADEPT 2022) and the Thames Valley Smart City Cluster Project (£2.1 m, 2018–20) (Thames Valley Berkshire LEP 2022), both of

which brought the six local authorities across Berkshire together. This was also helped by the fact that Reading was ranked among the UK's top 20 leading "smart cities" according to the second UK Smart Cities Index, commissioned by Huawei UK (Huawei 2017).

- COVID and resilience: Although it is clear towns and cities will continue to play an important role in national economies, the impact of COVID is still influencing the performance of urban economies and town centre 'ecosystems', and we have also seen the recent war in Ukraine and price inflation affecting the same economies. Developing future economic and climate change resilience is vital therefore to guard against known risks and 'black swan' events, and this must be embedded within coherent and integrated long-term city strategies and plans.
- Net Zero: defining what we mean by net zero and having a consensus in the relevant plans and strategies to achieve this ambitious goal are crucial. It is noticeable for example, that within Berkshire both the target dates and the ambitions on net zero vary markedly across the six local authorities (Bioregional 2021)

As to the future, this case study has highlighted the role which smaller towns and cities can play in the battle against climate change and environmental impact. After all, today more than 50% of the world's population lives in cities of less than 500,000 (with a higher figure of two thirds in Europe), and smaller and medium cities will continue to be important for growth in the future (Moir et al. 2014; UN 2018a, 2018b). City governance, as we have seen, is important in shaping the future of a city, and this links closely to the overall concept of urban futures thinking, and the cities with the most robust and effective 'hard' and 'soft' climate change governance structures will be the cities that continue to attract investment. Ultimately people, businesses, local government, academia, and other key city stakeholders must all be part of helping shape that future. As Sir David Attenborough said: "The final chapter is ours to write. We know what we need to do. What happens next is up to us."

## References

ADEPT (2022) Thames Valley live lab: final review. ADEPT, London. Accessed July 2022: https://www.adeptnet.org.uk/documents/thames-valley-live-lab-final-review

Arnkil R, Jarvesivu A, Koski P, Piirainen T (2011) Exploring quadruple helix: outlining user-oriented innovation models. Working paper, University of Tampere, Tampere

Ashden (2022) Reading's local plan ensures big housing developments aim for zero carbon. Accessed July 2022: https://ashden.org/news/readings-local-plan-ensures-big-housing-developments-aim-for-zero-carbon/

Batty M (2018) Inventing future cities. MIT Press, Cambridge Mass/Cambridge, England

Bioregional (2021) Berkshire net zero carbon research, bioregional Thames Valley Berkshire LEP. Accessed July 2022: https://www.thamesvalleyberkshire.co.uk/getfile/Berkshire%20Net%20Zero%20Research%20Gap%20Analysis%20and%20Recommendations.pdf

Climate Emergency Declaration (2022) Climate emergency website. Accessed July 2022 at: https://climateemergencydeclaration.org/climate-emergency-declarations-cover-15-million-citizens/

Creasy A, Lane M, Harvey-Crawford R (2020) Climate commissions as a stimulus for place-based action: an evidence synthesis from existing UK case studies. Place-Based Climate

Action Network. Accessed March 2024: https://pcancities.org.uk/climate-commissions-stimulus-place-based-action
Dixon T (2021) City visions: toward smart and sustainable urban futures. In: Brears RC (ed) The Palgrave encyclopedia of urban and regional futures. Palgrave Macmillan, London
Dixon T, Cohen K (2015) Towards a smart and sustainable Reading 2050 vision. Town and Country Planning, 20–27 January
Dixon T, Farrelly L (eds) (2020) Reading 2050: a smart and sustainable city. School of the Built Environment University of Reading, Reading, UK. Accessed July 2022: https://centaur.reading.ac.uk/93297/1/B26463%20Reading%202050%20WEB2.pdf
Dixon T, Eames M, Hunt M, Lannon S (eds) (2014) Urban retrofitting for sustainability: mapping the transition to 2050. Routledge, London
Dixon T, Montgomery J, Horton-Baker N, Farrelly L (2018) Using urban foresight techniques in city visioning: lessons from the Reading 2050 vision. Local Econ 33(8):777–799
Dixon TJ, Tewdwr-Jones (2021) Urban futures: planning for city foresight and city visions. Policy Press/Bristol University Press, Bristol
Dixon TJ, Karuri-Serina G, Ravetz J, Tewdwr-Jones M (2022) Re-imagining the future: city region foresight and visioning in an era of fragmented governance. Reg Stud. https://doi.org/10.1080/00343404.2022.2076825
Eames M, Dixon T, May T, Hunt M (2013) City futures: exploring urban retrofit and sustainable transitions. Build Res Inform 41(5):504–516
Goddard M, Tewdwr-Jones M (2016) City futures and the Civic University. Newcastle University, Newcastle
Government Office for Science (GOfS) (2016) Future of cities: foresight for cities. GOfS, London. Accessed July 2022: https://www.gov.uk/government/publications/future-of-cities-foresight-for-cities
Government Office for Science (GOfS) (2017) The futures toolkit: tools for futures thinking and foresight across UK government. GOfS, London. Accessed July 2022: https://assets.publishing.service.gov.uk/government/uploads/system/uploads/attachment_data/file/674209/futures-toolkit-edition-1.pdf
Huawei (2017) UK smart cities index, 2017. Huawei, London. Accessed July 2022: https://www.itu.int/en/ITU-T/ssc/resources/Documents/Huawei_2nd_Smart_Cities_Index_2017_FINAL.pdf
ITU (2014) Smart sustainable cities—an analysis of definitions. International Telecommunication Union, Geneva, Switzerland
Ivanova D, Barrett J, Wiedenhofer D, Macura B, Callaghan M, Creutzig F (2020) Quantifying the potential for climate change mitigation of consumption options. Environ Res Lett. https://doi.org/10.1088/1748-9326/ab8589
Moir E, Moonen T, Clark G (2014) The future of cities: what is the global agenda? Government Office for Science, London. Accessed July 2022: https://assets.publishing.service.gov.uk/government/uploads/system/uploads/attachment_data/file/429125/future-cities-global-agenda.pdf
Newman P, Jennings I (2008) Cities as sustainable ecosystems: principles and practices. Island Press, Washington, DC
ONS (2022) Population and household estimates, England and Wales: Census 2021. Accessed July 2022: https://www.ons.gov.uk/peoplepopulationandcommunity/populationandmigration/populationestimates/datasets/populationandhouseholdestimatesenglandandwalescensus2021
Quinio V (2022) Homeworking and the high street How important is it for city centres that workers return to the office? Centre for Cities, London. Accessed July 2022: https://www.centreforcities.org/wp-content/uploads/2022/06/Homeworking-and-the-high-street.pdf
Reading Borough Council (2019) Local Plan. Reading Borough Council. Accessed July 2022: https://images.reading.gov.uk/2019/12/Local_Plan_Adopted_November_2019.pdf
Reading Climate Change Partnership (2020) Climate Emergency Strategy. RCCP. Accessed July 2022: https://readingcan.org.uk/about-us/

Reading UK (Now Reading's Economic and Destination Agency (REDA) (https://www.visit-reading.com/business)) (2020) Reading economic recovery strategy. Reading UK, London. Accessed July 2022: https://web.livingreading.co.uk/recovery

Tewdwr-Jones M, Goddard J, Cowie P (2015) Newcastle city futures 2065: anchoring universities in cities through urban foresight. Newcastle Institute for Social Renewal, Newcastle University: Newcastle

Thames Valley Berkshire LEP (2022) Smart city cluster website. Accessed July 2022: https://www.thamesvalleyberkshire.co.uk/tvbsmartcity

UN (2018a) The world's cities in 2018. UN, New York. Accessed July 2022 at: https://www.un-ilibrary.org/human-settlements-and-urban-issues/the-world-s-cities-in-2018_c93f4dc6-en

UN (2018b) World urbanisation prospects: the 2018 revision. Key Facts. UN, New York. Accessed July 2022 at: https://population.un.org/wup/Publications/Files/WUP2018-KeyFacts.pdf

Wiedmann T, Lenzen M, Keyßer LT, Steinberger J (2020) 'Scientists' warning on affluence'. Nat Commun 11:3107. https://doi.org/10.1038/s41467-020-16941-y

Wiek A, Iwaniec D (2014) Quality criteria for visions and visioning in sustainability science. Sustain Sci 9:497–512

**Timothy J Dixon** is Emeritus Professor at the University of Reading and Visiting Fellow (Kellogg College) and Research Associate (Global Centre on Healthcare and Urbanisation), University of Oxford. He is a qualified chartered surveyor (FRICS) and has nearly 40 years of experience in built environment research, teaching, and practice. His research encompasses city foresight, futures studies, sustainable futures in the built environment, urban regeneration, and social sustainability.

# Index




R. Yao (ed.), *Resilient Urban Environments*, Cities and Nature,
https://doi.org/10.1007/978-3-031-55482-7

**H**



**I**



**L**



**M**



**N**



**O**

GPSR Compliance

*The European Union's (EU) General Product Safety Regulation (GPSR) is a set of rules that requires consumer products to be safe and our obligations to ensure this.*

*If you have any concerns about our products, you can contact us on ProductSafety@springernature.com*

In case Publisher is established outside the EU, the EU authorized representative is:

Springer Nature Customer Service Center GmbH
Europaplatz 3
69115 Heidelberg, Germany

**Batch number: 10761599**

Printed by Printforce, the Netherlands